THE CRAFTING OF THE 10,000 THINGS

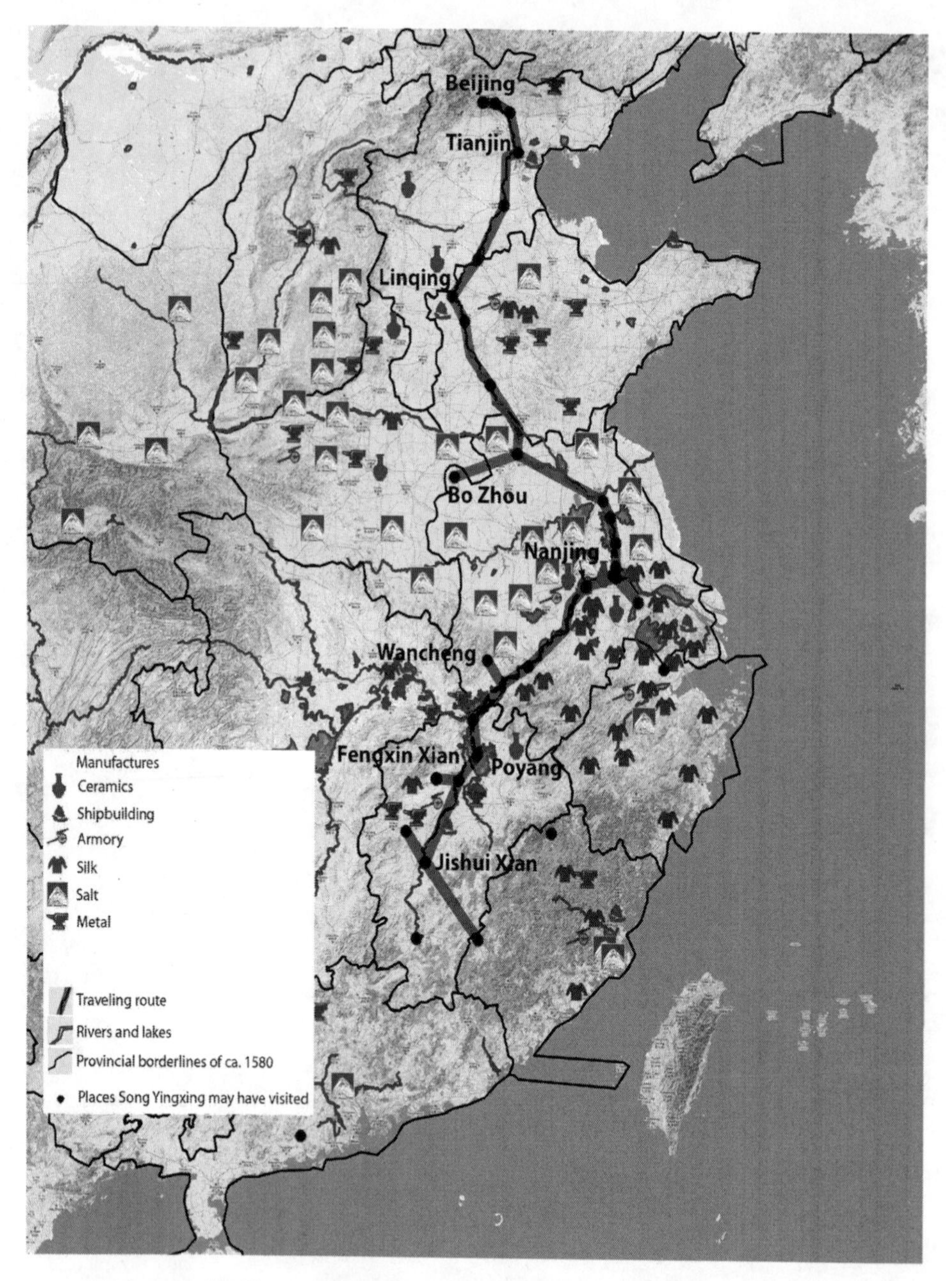

Song's geographical horizon of experience, as described in chapter 1. The map also shows the route candidates of this region traveled to participate in the highest examination that opened the door for a career in the central government in Beijing. It also points to the industries located along the way and the places mentioned by Song in *Work of Heaven*. A dynamic map can be found at http://chinagis.mpiwg-berlin.mpg.de/permanent/TGKW_Ming.html.

THE CRAFTING OF THE 10,000 THINGS

Knowledge and Technology in Seventeenth-Century China

DAGMAR SCHÄFER

THE UNIVERSITY OF CHICAGO PRESS
CHICAGO AND LONDON

The University of Chicago Press, Chicago 60637
The University of Chicago Press, Ltd., London

Paperback edition 2015
Printed in the United States of America

24 23 22 21 20 19 18 17 16 15   2 3 4 5 6

ISBN-13: 978-0-226-73584-9 (cloth)
ISBN-13: 978-0-226-27280-1 (paper)
ISBN-13: 978-0-226-73585-6 (e-book)
10.7208/chicago/9780226735856.001.0001

Library of Congress Cataloging-in-Publication Data
Schäfer, Dagmar.
The crafting of the 10,000 things : knowledge and technology in seventeenth-century China / Dagmar Schäfer.
p. cm.
Includes bibliographical references and index.
ISBN-13: 978-0-226-73584-9 (cloth : alk. paper)
ISBN-10: 0-226-73584-2 (cloth : alk. paper) 1. Song, Yingxing, b. 1587. Tian gong kai wu. 2. Science—China—History—17th century. 3. Technology—China—History—17th century. 4. Cosmology, Chinese. 5. Qi (Chinese philosophy) I. Title.
Q127.C5S825 2011
609.51909032—dc22

2010022421

♾ This paper meets the requirements of ANSI/NISO Z39.48–1992 (Permanence of Paper).

CONTENTS

INTRODUCTION

Picture a seventeenth-century Chinese shadow-puppet theatre. As the shadows deepen to darkness in the narrow streets of a provincial town, the backlit golden glow of the mulberry paper screen is a seductive magnet pulling in passersby already drawn by the blustering clash of cymbals and drums. This evening's performance tells the well-known story of the world-weary Emperor Shizong (1507–67, reign Jiajing 1521–67), who, searching for the elixir of immortality, neglects his sacrifices to heaven and leaves all state affairs to his grand secretary, Yan Song (1481–1565). Freed from restraints, the minister turns the court into a den of corruption. In the Chinese shadow theatre, the puppet depicting the minister is given the evil smirk of a demon when, in a bid for the throne, he orders a craftsman to make him the imperial regalia (a *ruyi* scepter). At this high point of dramatic tension, when the world trembles on the brink of utter chaos, loyal officials enter the scene. They put the maker of the regalia in custody, urging him to be their witness against the powerful prime minister. In the next two scenes the high officials run rings around the prime minister, lure him into a trap, and make him look like a fool. With the favorable applause of the audience, the play ends with a good fight in which the loyal officials thrash the villain vigorously for his sins.[1]

Relying on their extensive repertoire of historical dramas and traditional stories, shadow theatres spread news and communicated ideals and values to every level of premodern Chinese society.[2] The artisan made an essential contribution to the play; carving out the silhouettes of trees, temples, and houses and painting the faces and props for the puppets, his hand gave shape to the shadows and exemplified the material assets and cultural values of this era on the mulberry paper screen. And yet, as the play illustrates, the craftsman's contribution was rarely brought center stage. This era revered the scholarly ideal and relegated the craftsman to a minor role, not only in the oral culture of the shadow theatre, but even more so in the written account. The scholars were the protagonists whose cunning strategy finally brought order back to the world, to the play, to society, and to the state. Illuminating symbolic figures, the shadow theatre showed its varied audience how the world ought to be, which roles and themes were at stake. By displaying such matters, this performative arena corroborated the order of heaven, man, and nature in the grand play of the history of Chinese civilization. The narration demonstrated the ideals and concerns that governed this era and their cultural achievements.

One man who artfully rearranged these concerns in his literary legacy was Song Yingxing (1587–1666?), a minor local official living in southern China. It must have been early spring in the year 1637 when Song, in the white heat of indignation provoked by a political scandal, decided to have his thoughts published. In the following two years Song frequented the printer's workshop, busying the carvers who, squatting with stooped backs over low frames, engraved his words character by character, line by line, into pristine jujube woodblocks. Finally the craftsmen's hands had committed a set of six arresting texts to the memory of paper: one study on philology (*Huayin guizheng,* A Return to Orthodoxy); a collection of poetry (*Silian shi,* Yearnings); one sociopolitical critique (*Ye yi,* An Oppositionist's Deliberations); one on the heavens (*Tan tian,* Talks about Heaven); one delineating natural phenomena of all kinds, from sound to meteorology, as a reverberation of *qi* (*Lun qi,* On Qi); and, the largest work, a documentation of eighteen selected fields of practical knowledge (*Tiangong kaiwu,* The Works of Heaven and the Inception of Things).[3] The fields covered by the latter were (1) farming (*naili,* literally, the growing of grains), (2) cloth manufacturing (*naifu*), (3) dyes (*zhangshi,* literally colorings and adornments on cloth for the distinction of status), (4) the processing of grains (*suijing,* literally, the essence, i.e. the refinement of seeds), (5) salt (*zuoxian*) and (6) sweeteners (*ganshi*), (7) ceramics (*taoyan*), (8) smelting and casting (*yezhu*), (9) the construction of boats and carts (*zhouche*), (10) hammer forging (*chuiduan*), (11) the calcination of stones (*fanshi*), (12) fats and oils (*gaoye*), (13) papermaking (*shaqing*), (14) the five metal ores (*wujin*) and (15) weapon (*jiabing*) manufacture, the making of (16) vermilion and ink (*danqing*), (17) yeasts and wine starters (*qu'nie*) and (18) jewelry (*zhuyu*). Song's oeuvre pulled craftwork to the front of the stage. His life and work, the motifs and ideals that drove him, are the central issues of this book.

My study places Song's achievement squarely in its original context, utilizing the shadow theatre as a metaphor that both represents and embodies the seventeenth-century Chinese world. In this world issues that we would consider "scientific" or "technological" were approached in terms of *li, qi, yin-yang,* or the five phases (*wu xing*). Scholars pondered talents, skills, knowing, and the making of things in terms devoid of the nature/culture divide that informs modern minds. Making knowledge about natural phenomena and material inventiveness, Chinese culture was concerned with "knowledge and action" (*zhi xing*), the "investigation of things" (*gewu*) and "the inception of things and accomplishment of affairs" (*kaiwu chengwu*). Song was a man who, situated within the shadow theatre, reflected on the themes brought up in the play: state and society, rulership and administration, order and chaos, knowledge and truth. Defining the roles of emperors, scholars and craftsmen, heaven and men, artisanal work, and scholarly activity, Song laid out a radical view of cosmological processes and their relation to human action.

Song lived at the end of the Ming dynasty, a time when a sophisticated society felt haunted by specters of chaos and decay visible through a veil of material luxury and intellectual sophistication. These specters were the breakdown of social order and moral values, insecurity, and corruption. The shadow play indicates the first appearance of these specters in the Ming world almost seventy years before Song stepped on the historical stage: in 1567 the prime minister Yan Song, freed from all moral obligations by his infelicitous ruler, had almost led the country into ruin. Song's fate was intimately interwoven with this event. During this era his clan reached their social peak and began their slide into decay, together with their dynasty. Living on the eve of dynastic collapse, Song expounded in his writings a comprehensive program to restore the world to order, allocating a crucial role to issues we identify as "technological" and "scientific."

Song's legacy offers an exceptional opportunity to inquire into a seventeenth-century Chinese intellectual's approach to natural processes and material inventiveness—a chance to unveil what "scientific knowledge" and "technology" meant to this era. Examining how Song included these issues in written culture reveals practical knowledge and technology to be an "object of knowledge" for a scholarly elite who passionately discussed bookish learning and observational practice, a world that pondered the validity of studies of the mind and the role of bodily experience in man's quest for knowledge. Elucidating how Song shaped arguments, facts, and evidence and how he employed unity and singularity offers significant glimpses into this time's approaches to rationality, truth, and belief in the study of nature and culture alike. The purpose of writings on crafts and technologies and their effects on and role in premodern Chinese culture; the forms and fabrics of labor, the relationship of artisanal knowledge and scholarly knowing; nature, man's and heaven's place: the interweaving of all these makes up the complex knot of issues tackled in this book.

Focused on the life and work of an individual, this study utilizes a significant development in the historical analysis of science to expose some of the amorphous qualities of the "cultural" and "historical" that affected Chinese scientific and technological knowledge in the making. Recent research on early modern (mostly Western European) cultures addresses the combination of factors that contributed to the emergence of the various fields of science and technology. In a fresh new way, these studies concentrate on the diverse spectrum of men and women inquiring into nature and material objects, the rhetorical and epistemological embedding of knowledge, and how these factors fed into a "scientific revolution."[4] Particularly influential features are: what was considered practical or theoretical and why; the role in which people identified themselves; which issues were accepted as facts and which were seen as beliefs; and on what basis thoughts on nature or material production were authorized or dismissed. This new research employs a variety of methods, including some from the sociology of science, science and technology studies, philosophy, philology, art, history, and cultural studies. One recent trend has been to put the category of the historical actors into the foreground. The editors of the study *The Mindful Hand: Inquiry and Invention from the Late Renaissance to Early Industrialization,* Lissa Roberts, Peter Dear, and Simon Shaffer, for example, promote in the introduction the use of terms that were used in the periods themselves; "theory, practice, ingenuity, economy and inquiry." They reject a distinction "between the production of knowledge and some category such as science," emphasizing that concepts such as "science" and "technology" actually pose significant obstacles to our view on historical knowledge production.[5] Their shift from a historically exclusive to a comprehensive concept lifts previous restrictions that defined science and technology in the limited terms of the trajectory of the post–eighteenth-century European scientific revolution. The roots of their approach can be traced to the thoughts of historians researching premodern Chinese and other non–Western European cultures, a once alternative argument that has since demonstrated its validity throughout the field.[6] These studies create space for the wide spectrum of what "science" and "technology" actually are, a culturally and historically embedded making of knowledge about nature and matter.

A significant consequence of this new research is the transformation of the "narrative of the scientific revolution" in Western Europe from a consistent story about changing theories to a complex and highly dynamic texture of "science and technology as a cultural activity," as Pamela H. Smith has remarked.[7] Indeed great opportunities and huge challenges lie in these efforts with regard to cross-cultural perspectives on knowledge production. One opportunity is provided by an increase in our sensitivity to the diversity and change in the knowledge-making process, the great potential of such developments at different moments in time, in different places, and for each individual. The challenge is to implement this growing awareness to full effect, prizing open not only the obvious but also the elusive preconceptions that no historian is immune to.

Song's oeuvre is used in my study as a test case. The largest of his works, his document of craft know-how, the *Works of Heaven*, has long been manipulated to fit different conceptions of a Chinese trajectory of scientific and technological thinking. Those who employed the nineteenth-century standard of the European scientific revolution as a yardstick, attempted to find an equivalent, its mirror image, in Chinese history. Narrations thus assigned Song's effort an ambiguous role. Some scholars used the mere existence of this book as vital evidence of a Chinese interest in craft knowledge, technology, and practical endeavor. Others argued that, because not a single scholar followed Song's comprehensive approach to craft know-how (which is indeed true), this proved a lack of intellectual interest in practical things and technological considerations. Or, they took it as a sign of a civilization whose inventiveness and creativity was in decline. Rather than ponder the alleged and actual fallacies of these approaches, I take them as important and revealing both for and of their time. They grant a significant insight to the scientific and technological landscapes against which the work of men like Song was created and evaluated.

As this is a fascinating and important subject in its own right, I address it in the epilogue. The book itself, however, moves the original process of knowledge production into the foreground. It reincorporates Song's documentation on crafts into the complex writing campaign to which it initially belonged and delineates his efforts in the synergy of social, political, and philosophical concerns in which it was created. In approach, the study concentrates on Song's ways of gaining, assessing, and presenting knowledge. It evaluates his discussion through his own taxonomy of knowledge, which was also that used by his contemporaries. This draws attention to the original structures of Chinese intellectual discussions on nature and material inventiveness, their internal coherence and their potential to make sense of the world. Song's life and work are presented in their original shadow theatre, with the flamboyant light of late Ming concerns cast on it. This book is an invitation to appreciate the divergent tones and variegated sets of figures of scientific and technological thought that the historical world has to offer.

The Crafting of the 10,000 Things illuminates Song's efforts in six chapters. Within all chapters, sections start with an original quotation taken from one of Song's writings. Short vignettes, furnished with the figures and settings Song employed in his magnum opus, then provide glimpses of the minutiae of seventeenth-century arts and crafts. Each vignette guides the reader through the textures, colors, smells, and sounds of Ming times, into the world in which Song lived.

The first three chapters deal with the contemporary affairs that shaped Song's writing, for I see knowledge-making as an embedded practice. Chapter 1 investigates Song's private affairs, the social implications, family surrounding, education, political events, and historical circumstances that affected Song's efforts. This lays bare the character of a scholar who was an ardent servant to his ideals, a well-educated literatus directing his brush with the soft hand of a gentleman, a frustrated minor official spurred to put his thoughts in writing by the promotion of a military official to a high civilian rank. The inquiry into this contemporary political scandal and Song's view of contemporary politics explains his motives to start writing. Laying out the schemes of his writing campaign, this chapter exposes the comprehensive scope of Song's approach to knowledge and shows that Song was struck by the chaos of his era and took it as an "affair of honor" to bring order to the world.

The ideals and ideas shaping this affair of honor is the issue of chapter 2: it examines Song's way of discussing knowledge and order and how he outlined these issues as a matter of *qi*. Sketching a historical perspective on Chinese usages of *qi, li,* and *yin-yang,* this part of my study also marks out the intellectual and sociopolitical implications of Song's choices, how his era and later generations read them. Thus Song's cosmological viewpoint, his concept of truth and belief, the role of man, and heaven and earth are brought on to the stage.

Chapter 3 delves into the changing social and political landscape of craft professions and scholarly activity in the Ming dynasty. With urbanization and state intervention the world of crafts diversified: subsidiary tasks expanded to become full-time undertakings and crafts such as textile work, once performed by women in private households, moved into the public sphere and became a male occupation.[8] Institutional changes and the commercialization of society provide the backdrop against which this chapter discusses the contemporary documentation of craft manufacture that supplements our view of Song's ideas about craftwork. I delineate this era's intellectual and political debates on experience, talents and skills, education and training, and Song's perspective on the social standing of the craftsman, the scholar, the farmer, and the merchant. Song saw the function of their work in the production of knowledge as essentially affected by the circumstances into which man was born: customs and habits, made up by man, determined his opportunities and social role. "Public Affairs" divulges Song in this context as an innovative traditionalist, a scholar who venerated scholarly skills and distinguished talents in terms of intelligence rather than morality, an elite representative who considered craftwork to be an object of inquiry, an epistemic thing, without recognizing craftsmen as a social group. Putting crafts together with farming Song redrew the boundaries of knowledge fields in premodern Chinese culture.

"Things and affairs" is the term I use following Song's notion of "the inception of things and accomplishment of affairs" to identify the approach of seventeenth-century men of letters such as Song to natural things and phenomena, material objects, working processes, textual sources, man, heaven, the state, society and the individual, the mind and the body. How Song dealt with these things and where he located "things and affairs" in his writing is the major issue running through chapters 4–6.

Chapter 4 focuses on how Song presented the "making of" knowledge, and how, when he included things and affairs into written culture, he approached observation and used it rhetorically. Investigating the role Song assigned to theory and practice and their application and how Song employed the mundane or the specific, I describe how this seventeenth-century scholar generated evidence and facts, and what universal validity or rationality meant to him. This investigation starts with an analysis of the rhetoric of both word and image and then elucidates the implications of Song's references to the classics and historical actors and his usage of general schemes and styles.

The subtle interlocking of erudition and empiricism in Song's writings is also the theme of chapter 5: it marks out the limits Song sets to both sensory experience and reasoning in the knowledge-making process. The schemes of his systematic terminology are described and I engage with his strategic usage of experimentation and quantification, shedding light on the concerns that shaped his application of numbers and interest in the details of processes and things. Song ascribed a cosmological meaning to both details and numbers: for him they were more than just descriptive elaborations. Burning wood and boiling water, Song measured volumes to substantiate the proportional relation between *yin-yang* forces in transformation. Song analyzed the principles of growth and decay using references to common sense and everyday events. In the process he found the world of observed facts inconsistent with his universal principles. This chapter shows how Song carefully adjusted his schemes and analyzed observational method to argue the world into rational schemes.

The purpose of chapter 6 is to reveal the original structure and actual dynamics of Song's knowledge presentation. It takes a topical approach, portraying Song's inquiry into sound and silence, his interest in explaining the world of the audible *qi* as one that aimed at verifying the universal validity of his world of *qi*-principles. Traditional views on sound and *qi* are described and Song's position therein is defined. In particular, this chapter engages with the physicality of sound in Song's concept of *qi*, his idea of the human voice, and hearing ability. Sound is identified in its cosmological importance in relation to concepts of resonance and harmony.

Particularly in the last two sections my study draws extensively on the fine research recently published on both visual and literary rhetorical strategies in various fields of Chinese literature, philology, law, popular writing, tracts on natural philosophy, mathematics, and the arts. Throughout this book I place Song's efforts in conjunction with the works of other Chinese philosophers and writers, before, during, and after his time. Doing this I avail myself of the growing literature on late Ming culture that delineates this era's politics, literature, and philosophy, the commercialization and commodification of society, and economic and political developments. When applicable, I compare Song's writing with the work being produced in Europe in the seventeenth century, pinpointing the subtle differences or similarities that mark intellectual approaches in this era in both cultures (in the hope that this comparison could be extended to other cultures in the future).

A keen and meticulous observer of physical work emerges, a scholar who recognized technologies, and natural phenomena as issues of how knowledge was made and where understanding came from. His literary effort takes shape as a complex and coherent study on "things and affairs," part of a body of writing on observations and personal experience. Chinese culture grouped these together as documents of personal experience, of what one had "seen and heard" (*jian wen*). They were known as jottings or "brush notes" (*biji*), the term preferred by the scholars themselves. Song's approach to knowledge thus appears intimately involved in Chinese natural philosophy and deeply informed by traditions reaching from the Song dynasty to late Ming dynastic intellectual culture. At the same time, it also defied his time not only in its choice of content and disregard of moral and ethical ideals: Song's work ultimately threw a new light on the play of Chinese civilization—a light that his contemporaries shaded with great care.

Understanding the manifold sociopolitical implications that shaped Song's efforts, his life and work, also throws a new light on the transmission of Song's oeuvre, one that illuminates alternating fashions and reader's concerns. This is the scene added by the epilogue: the perspective of the aftermath. Using Song's most intimate peers and friends, Tu Shaokui (1582?–1645?) and Chen Hongxu (1597–1665), as representatives of literati culture, it describes the importance of literary sponsorship in the seventeenth-century world. It goes on to explain how various factors affected the reception (or rather the non-reception) of Song's literary legacy. Song's minor position, his uncommitted stance on loyalty at a time of strong Ming loyalism, political prosecution by the Qing and last but not least the physical availability of Song's oeuvre, all played a role in the shadow play of Song's work on the historical stage.

KNOWING "THINGS AND AFFAIRS" IN PREMODERN CHINA

A seventeenth-century Ming scholar making knowledge could draw on a long and refined discourse about the relation between practical endeavor and theoretical speculation. There was a huge diversity in views about the values that needed to be assigned to each issue or how themes could be fruitfully combined, for the benefit of the state, oneself, or the people. One possible anchor was the so-called ancestors or sage-kings. Epitomizing the genesis of Chinese civilization(s), these ancestral kings signified both moral and theoretical concerns and artisanal knowledge, skills, and the understanding of things and matter. Thus practical and theoretical endeavor were given equal prominence in the construction of Chinese cultural identity. This allocation of things and affairs to the beginnings of Chinese civilization lent credence to the various knowledge fields that were relevant to man's world. It assigned them importance in the identification of man's relation to heaven. Philosophical discussions about whether man created or imitated, produced or fabricated things reflect the Chinese literati's interest in testing man's role and heavenly influences in nature.[9] Asking how things came up or affairs arose they struggled for an approach to knowledge in the making, a deeper understanding of how knowledge could be gained and what this knowledge should be.

Among the manifold schools of thought that discussed such topics in the early periods of Chinese culture, a select few prevailed. These left behind a refined corpus of texts and thought which representatives of the elite throughout the following centuries used to place different forms of knowledge in a purposeful relation to the society and state; those who understood the principles and ruled, the aristocratic and later the bureaucratic elite (*shi*); those who worked the land and sustained food production, the farmers (*nong*); those who manufactured things and provided goods and devices, the craftsmen (*gong*); and those who traded what others produced, the merchants (*shang*). The men of letters who authored these texts often claimed that they applied a consistent methodology and that they had attempted to find universal principles and moral causes. Farmers may provide for worldly necessities, military men may conquer the world, and craftsman produce its assets, but it was the scholars who systematized and classified the world. These scholars presumed that craftsmen, farmers, and merchants made things and initiated affairs without such knowledge, relying merely on their experience and physical skills.

Inherent in this broad categorization—the exemplification of an idealized state—is the idea that those who approached the world theoretically were to be distinguished from those who approached it practically and that the theoreticians ruled over the practitioners. And yet, the ideal promulgated in Chinese classical writings stressed the necessity of all forms, with boundaries negotiable as needed. Thus the scholar should also be a farmer. Owning the land, he had to cultivate it to supply the people with food, and he was obliged to devote his mind sufficiently to agriculture to meet this demand. He had to understand the merchant's potential and control him. Craftsmen had to be supported to provide for social and state needs and in case state power was threatened, the scholar also had to master strategic military tasks. Performing all these duties, the scholar engaged with nature and gained knowledge about it. He had to ensure the supply of goods and devices and, although he may have avoided physical involvement, tradition and classical texts obliged him morally to manage all kinds of craft tasks and mundane issues, providing for the people and sustaining the state. And thus craftsman work could also become part of his duty.

By the late sixteenth century, the time when Song was born, an autocratic state had made scholars its elite. Elite status had to be acquired and was then recruited. It was not inherited. A constantly growing pool of men thus memorized canonical texts, read, and wrote. Each erudite individual was forced to increase the sophistication of his learning. This was one issue characterizing late Ming scholarly approaches to knowledge. Another was the Ming dynastic state's policy of forcing its civil servants into close contact with practical matters of statecraft. Developing a system inherited from the Mongols, who had ruled China under the reign name Yuan for almost a century (1271–1368), the Ming rulers tied craftsmanship and artisanal work much more tightly to the state than their predecessors. The ancestor to the Ming, Zhu Yuanzhang (1328–98, reign *hongwu* 1368–98) established a network of state-owned manufacture under the control of the central government and local administration and enforced taxation and administration of raw materials and products. He furthermore adopted the Yuan practice of assigning a hereditary status to households, distinguishing between civil, military, and craftsmen households for tax payment (all households were in principle allowed to participate in exams or receive scholarly training). Craftsmen households were levied to provide service as a tax. Zhu dragged artisanal work into the focus of scholarly pursuits, obliging scholars to deal with issues they generally considered below their status and preferred to avoid. Acquiring a state position hence required a sophisticated training in philological skills and philosophical matters. But once the career was achieved, the state busied its servants with the rather more practical tasks of organization and control throughout all fields of knowledge. Secluded from the realities of life in private studies and schools, most scholars who aspired to a state career hardly ever had time to worry about issues such as hydraulics or agriculture, much less craft manufacture or labor, even if they wanted to. Their education secured their career, it did not offer any preparation for the execution of their daily tasks.

While this ambiguity characterized other dynasties, for example, the Song states (960–1279), the Ming rulers increased the pressure by making the field of craft enterprise explicitly a state concern, a "public affair." This interest in crafts was not only driven by the state's sheer demand for material goods or the luxurious desires of the court: Silk and porcelain were pertinent to the Chinese economy as well as its political power. As tributary goods, they traded in peace and loyalty. Scholar-officials of the Ming were aware of these implications, as they coped with the challenge of exercising state control over the craftsman. At the same time, it was in their interest to foster scholarly education as an authorization for political leadership; the state system implied that the right to control practical work rested on morals and the authority of knowledge. Growing commercialization and market forces, the sheer presence of material culture, added to the manifold factors impacting Chinese views on fields of knowledge at that time. The writings of high-level officials such as Qiu Jun (1421–95) exemplify the growing importance of crafts and also demonstrate one way in which Ming scholar-officials dealt with these ambiguities: Qiu discussed the utility of craftsman work as an issue of ritual and control in *Daxue yanyi bu* (*Extended Meaning of the Great Learning*, 1487), a political handbook addressed to upper-level officials in the central and provincial government.[10] This high minister of rites (*libu shangshu*, highest rank 2a) objected to the increasing level of skills possessed by the craftsmen of his time, and he advised the emperor to regulate the degree of artifice. This is one indication of how officials picked up the gauntlet craft know-how presented to their role in the state and their identity as scholars. Defending their role as leaders, the scholar-officials exercised control and defended their leading position by taking the role of managers and claiming that knowing was prior to doing. This is how they pushed the craftsman politically and socially to the side.

Research on early modern European cultures has highlighted the relevance of "hybrid" figures— artisans, sculptors and painters, apothecaries, engineers and practitioners of other occupations that stood at the intersection of practical work and theoretical knowledge—to the changes in Europe's seventeenth-century cultures of knowledge, emphasizing that these people often approached their products as a manifestation of their knowledge and, on this basis, also developed and articulated an artisanal epistemology in their writing.[11] Such hybrid figures also existed in the Chinese arts: painters, sculptors, and even more so practitioners in fields such as palace architecture and landscape gardening.[12] In contrast to the European case, however, the Chinese "hybrids" preferred to define themselves as scholars (or were defined as such in records), for social, political, and intellectual reasons whenever they took part in literate discourse, even if they had actually come from an artisanal background.

Artisanal knowledge and experience thus may have informed many authors. But from a historical perspective, the figure of the artisan is conspicuous by his absence in written discourse. What effect the rhetorical insistence on scholarly status actually had on the seventeenth-century quest for knowledge and its implementation is open to debate, especially as many of these social-occupational identifications (i.e. of a person as a scholar, or laborer, artisan or merchant) are also historiographically biased and not necessarily indicative of the personal view of his tasks. In general, however, we must assume that, similar to other cultures, what premodern Chinese artisans left were their works, not writings about their efforts. What kind of knowledge and ideas do these objects display? An understanding of a culture's and an individual's ideas about the forms and fabrics of labor, the product technology and its nexus are necessary to decipher the knowledge displayed in objects. If the craftsman is free in his work and able to decide what he wants to do and which materials to use, it is his knowledge that the objects display in design, form, and combination. And yet, even then his work is limited by many outside influences, such as the availability of resources and the nature of demand. Authorities can control the work of artisans to the point that individual and object are almost detached from one another. This happens, for example, when the working process is broken down into step-by-step tasks that obstruct the craftsman's view of the whole. The artisan's hand still shapes the object, but it is the mind of the manager that is visible in its composition and design. Art historian and Sinologist Martin J. Powers has found traces of such social control in the decoration of ritual vessels from the Han period (206 BC–220 AD). He demonstrates that the complex designs and variety of form were based on standardizations and predefined pattern units. Powers interprets this as a sign of a new social order, in which the officials found the means to control the craftsman's knowledge in order to manifest the power of the state.[13]

Not all crafts were open to such methods of control, but we can assume that the scholar-officials of the Ming favored managerial techniques of segmentation and moral regulation whenever possible. Ritual discussions reflect the existence of such viewpoints in the Ming period. In this era, men like Qiu Jun subtly contrasted craft work produced adhering to the rites (*li*) and "within heaven's natural order" (*tian shi*) with "excessive artifice" (*gong qiao*), those objects characterized by complicated patterns and splendid craftsmanship distracting the gentleman from his duties and awakening his cupidity.[14] Such reflections indicate a societal ideal in which the official and ruling class wanted the artisan mainly to serve utilitarian needs. He may shape the object, but the credit was given to the one who put the mosaic together: the expert scholar who had the knowledge about the aim and purpose of work, not the one who actually "performed" the work.

NATURE, CRAFTS, AND KNOWING

Coping with the tension between ideals and realities, the seventeenth-century world in which Song compiled his view on mundane tasks, crafts, and technological achievements engaged in a discourse on the sources and ways of knowing. By that time a growing network of state-owned local and private schools had produced a mass of literate people, far more than required to staff the bureaucratic structure. Searching for their identity, a place and task in society and state, these people discussed practical knowledge and theoretical inquiry, the extension of knowledge into fields of concrete matter and nature. Most people did not give up their scholarly aspirations, though, contributing their mite to the massive growth in literature, ranging from examination guides to household compendia, classical reference books to novels. This spread of literary knowing was further supported by the explosive expansion in the printing and publishing field. Indeed during the late Ming, officials of all ranks, exam candidates, unemployed and impoverished scholars, privateers, rich literate merchants, landowners and gentry, men and women, all wrote, published, and read what interested them most. Scholars compiled books on hitherto largely disregarded topics. Those interested in practical issues, work, nature, medicine, and physical materials during the Ming and Qing dynasties often had no position in the state apparatus and stood outside highbrow elite society. People like Wang Gen (1483–1540), a self-educated salt merchant and follower of Wang Yangming (1472–1529), or the physician Wang Qingren (1768–1831) provide evidence of an enormous stratum of society that took part in intellectual discussions of that period although they were not scholars or state officials.[15]

The atmosphere of this period was hence lively, with people both consuming and producing new approaches to knowledge, especially in written culture. The commercialization and commodification that mark Chinese society during the seventeenth century, a growing interest in art and an increase in the culture of the collector gave rise to new views on artifacts, the manufactured object. A remarkable number of sophisticated essayist studies emerged on topics such as coining, bronze vessels or the technique of producing lacquer wares. Others concentrated on the manifoldness of the natural world, the various shades of colors of goldfishes, or the grafting of peonies.[16] Scholars compiled them for themselves, for their friends, or for wider distribution. Rich estates housed huge private libraries, assemblages of classical literature and extraordinary writings, as well as obscure prints and curiosities. Some, such as the merchant Hu Wenhuan (fl. ca. 1596), made their private collections widely available and thus fostered the dissemination of knowledge about things and affairs that was traditionally not considered part of scholarly ways of knowing. Song's literary legacy, as part and product of this vibrant and innovative culture, had to compete with many others for the audience's attention.

The consequences of this increase in literacy and in sheer textual material were manifold. Kai-Wing Chow credits the progressive increase in printing enterprises with the increasing denial of the tutor-pupil linkage and suggests that examination aids "significantly reduced the degree of dependence on the teacher's personal instructions for advanced students."[17] Indeed facilitated access to the written word was one pivotal reason for the rise in discussions on the proper sources and methods that marks this era's intellectual life. Another incentive to this discussion was the fact that the Ming state had made a specific corpus of text interpretations, the Cheng-Zhu school, mandatory for the state examination. Cheng-Zhu learning synergized the thought of several eleventh-century thinkers, Zhu Xi (1130–1200), Cheng Yi (1033–1108), Cheng Hao (1032–85), and Shao Yong (1011–77). Throughout the Ming, scholars increasingly opposed this limitation, arguing that candidates no longer aspired to a true understanding of the philosophy of these great thinkers and within this context pursued text studies only to achieve a post in government and not with the aim of true knowing. A huge debate flared up, with philosophers questioning the purpose of bookish learning and the role it should play in the revelations of the mind. Should one study texts or observe one's environment? Which things and affairs were worthy of investigation and which kind of knowledge could be drawn from it? Those oriented toward literary pursuits suggested that a thorough understanding of the inception of things and affairs, their origin and relation provided the answer, and thus they documented and classified the multiplicity of their world. Those practically involved in politics, the scholar-officials, declared that knowledge had to fulfill social rather than private concerns. The intellectuals questioned attitudes and the ideals at the heart of knowing. An influential group led by Wang Yangming, for example, proposed that as knowledge was innate, and only became evident in action, such issues were indeed all contained in one's mind.[18]

Because the Cheng-Zhu text corpus was made a mandatory part of the state examination curriculum, it represents a major reference point in the Ming scholarly tradition, with opinions decidedly for or against it. Beyond this general identification, however, orthodoxy was a fluctuating concept in Ming times, with scholars finding individual positions and establishing intellectual groupings quite flexibly around the issues of the moment. In view of the growing social and political insecurity, discussions by the end of the Ming increasingly circled around practical problems, assembled under such headings as "pragmatic learning" (*shixue*) or "statecraft" (*jing shi*). Others chose for the same purpose the motto of Wang Yangming's school, namely, "knowledge and action" (*zhi xing*). Referring to these terms, these men indicated their ideological roots. Beyond that, however, most of them insisted on diverse interpretations over the reign of one truth. For many scholars of this era integrating mind and things, subject and object, presupposed that the principle in things had to be individually researched and assessed.

Within this diversity, we can discern a refined group of scholars from Song's time and the previous generation who turned their attention to nature and material inventiveness under the major premise of *qi*. Wang Tingxiang (1474–1544), Luo Qinshun (1465–1547), Liu Zongzhou (1578–1645), and Tang Hezheng (1538–1619), all claimed to probe the underlying principles of their world.[19] Polymaths a generation after Song, such as Huang Zongxi (1610–95), Wang Fuzhi (1619–92), or Fang Yizhi (1611–71) would extend and refine much of these thoughts. All these men investigated a huge range of topics that correspond to categories in science and technology, such as light and sound, magnetism and hydraulics. While the history of science and technology has paid much attention to their ideas, it is often overlooked that not only the themes but also their methods developed: Wang Fuzhi, for example, introduced a refined methodology of argumentation and also explicitly stressed physical experimentation (*zhi ce*) as a means of gaining knowledge.[20] The same can be said about Fang Yizhi's *Tong ya* (*Supplement to the Erya*, 1666) and *Wuli xiao zhi* (*Notes on the Principle of Things*, compiled between 1631 and 1634 and published in 1666). Fang's studies embraced agricultural issues as well as medicine, mathematics, and crafts as concrete matters, which he saw combined with issues such as morality, governance, literature, philology, and minor skills: "all of them divulge the heavenly way (*tiandao*) in human affairs. The result of thorough understanding (*tong*) is the principle of human nature and the principle of things: probing leads to enlightenment (*ming*). Concrete things are the way; this is the general principle of things" (translation mine).[21] Bringing to maturity a trend that began in Song culture and had flourished since Shizong handed state affairs over to Yan Song in the *jiajing* period (1521–67), by the mid-seventeenth century Fang promoted concrete studies and "minutest comprehension" (*tongji*) as the prime source for knowing.[22]

A dismissal of the concerns of formerly accepted theoretical discourse and new approaches to knowledge production characterize this era. Yet, the examples mentioned above display another unifying element: ideas about the knowledge-gaining process may have flourished, but in fact most scholars ruled out the craftsmen as a knowledgeable person. These men debated concrete things as part of scholarly knowledge and as an issue of epistemological awareness, intricately relating it at one point in their discourse to philosophical, moralistic, and ethical ideals. I assume that the expression of humanistic aims had a highly authorizing function, in the same way that teleological suggestions were important to European approaches to nature at this time. I maintain, however, that this should not be taken as a qualifying statement: to assume that Chinese approaches to scientific and technological issues were primarily humanistic is in its absoluteness as fallacious as assigning a purely naturalist interest and investigative frame for the Europeans of this era.

Positing Song in the context of other Chinese authors who articulated their notion of the material realm, I suggest it is possible to trace some points of continuity throughout the seventeenth-century intellectual world. An important point is that cosmology in the late Ming dynastic intellectual approach held a similar position that scientific theory in all its shades now occupies in our contemporary recognition of knowledge about nature and man, that is, a culturally specific stance on practical and theoretical knowledge interaction. It was this cosmology that made Song look at the body of craftsman work, mundane tasks, and material inventiveness. And it is this background that made him perceive the craftsman's active work as the extension of universal principles and the scholar as the one in a position to observe just this—what nature was, what things revealed, and how affairs developed. For Song, this is where practical knowledge was located and was the goal of theoretical speculation.

When we look at *Works of Heaven* this way, we can see that Song set clear markers for his colleagues, beyond the subtle reference to encyclopedic categories in his preface. The very words: the "works of heaven" (*tiangong*) and the "inception of things" (*kaiwu*) confirm that Song intended to embrace a larger context than the simple documentation of craft procedures. He saw the revelation of universal principles within the performance of crafts and technology; it manifested a cosmological order that man had to comprehend in order to be able to quell the chaos that ruled his era. The terms Song addressed in his title refer to classical Chinese texts. *Tiangong* is presumed to be a quotation from the *Shujing* [*Book of Documents*]: "The work is heaven's, it is man's [duty] to act for it (*tiangong ren qi dai zhi*)."[23] The second expression, *kaiwu*, is a quotation from the *Yijing* [*Book of Changes*]: "Alas, change (*yi*) opens things (*kaiwu*) and accomplishes affairs (*chengwu*), encompassing (*mao*) the way (*dao*) of heaven and earth (*tiandi*), that is, everything in the realm. It is like this and that is it."[24] On the basis of these two classical references most modern readings interpret Song's title as describing a relationship between heavenly and human endeavor in which man makes use of natural resources provided by heaven. While this notion is implied, Song's idea of what this interface between human and heavenly action entails differs fundamentally from our modern perception: man was not his focus. Instead he was interested in the knowledge displayed by the "works of heaven" and in the "inception of things." He believed that by describing the eighteen selected crafts man could learn the principles that would bring about order to his world.

Scholars of Song's generation were reminded by the first part of Song's title, the "Works of Heaven" of a legendary dialogue between the administrative clerk (*sifa*)—later chief speaker—Gao Yao and Emperor Yu. This was set down in a classic of Chinese erudition, the *Shujing/Shangshu* (*Book of Documents, Documents of the Shang Dynasty*), an approved source of authority for Ming dynastic claims to knowledge.[25] By referring to this dialogue, Song was primarily exhorting the emperor to act virtuously and appoint virtuous officials. A commentary on this passage by Cai Shen (1167–1230), interpreting the *Book of Documents* for the *dao* of the human mind, clarifies that scholars of the thirteenth century understood this passage in regard to the relationship between heaven and man. Cai Shen's view was a keystone for the civil service examination during the Ming. Especially in the late seventeenth century scholars referred to him to emphasize that officials and the emperor had to take up their tasks and not forget one single thing, however subtle it may appear at first sight. "The emperor has to recognize his tasks and duties. . . . *Tiangong* has the meaning of the works of heaven (*tian zhi gong*). The gentleman (*renjun*) acts on behalf of heaven whilst ordering/organizing (*li*) the things. Everything governed by the officials is in any event a matter of heaven. If one is careless toward only one single duty or neglects [one item], then [he] destroys heaven's works."[26]

Song's application of the term ignored the moralistic implications of prevailing Ming interpretations of the Cheng-Zhu school and instead identified heaven as a realm contributing a structural prerequisite—a guide to pragmatic action. The gentleman (*renjun*, generally addressed as *junzi*), as the commentary says, was the one who enacted order, organizing and governing the myriad things in the state. Song thus targeted his criticism at the officials, whose focus on morals and consequent negligence toward other essential issues had resulted in chaos. Using the expression "the works of heaven," Song omitted a crucial part of the original phrase—that "it is man to act for it." Although we can assume that scholars recognized the full reference the abbreviation still emphasizes the "works of heaven," not man's role in the making of things. In Song's view, man, applying things and divulging matters, made up his own world and yet remained marginal to the procedures of nature and material inventiveness, the actual order of the universe. For him, the relation between heaven and man was not defined in terms of sharing tasks or responsibility in the making of things. He understood the relation as one in which man had to understand the "works of heaven." As his text *On Qi* reveals, heaven, together with all other things, rested on the paradigm of *qi,* which brought forth things and affairs. In Song's approach to craft knowledge, man's role is thus solely to appreciate cosmological principles and act in accordance with them, not to be "the maker." Within this scenario, heaven is a natural entitlement, just another manifestation of *qi,* and *qi* is what provides unity.

Taking up the issue of *qi,* Song was embedded within his time, following a subversive intellectual trend within a society that thought itself shaken to its foundation by the contemporary crisis. His example shows that, despite having received much attention in recent decades, the intellectual effects of the political, social, and economic changes in this era are still hardly understood. The commercialization in printing may have brought forth numerous publications that were savored for the moment and have since vanished without a trace. Song's legacy and its transmission history demonstrate that some publications survived in the twilight zone of intellectual discourse. Private scholars stuffed their study rooms and bibliophiles their libraries with books reflecting scholarly ideals, as well as private interests, social status, political ambition, and financial possibilities. The surviving examples suggest that the number of texts produced was large and the reactions of intellectuals to them diverse—a scattered landscape of individual reactions rather than a unified or linear narrative of knowledge in the making. From this perspective Song's legacy is exceptional, offering a chance to reveal one of the multitude of moments and influences on knowledge production. This defines the great significance of Song's written record about the knowledge of his time and marks his place in the history of science and technology and China: studying Song and his works gives a view of a unique history of knowledge in the making.

CHAPTER I

Private Affairs

The end of Spring is approaching. I am going to have a break and travel in the Qian mountains.
An Oppositionist's Deliberations, Preface, 3

In the introduction to his essay, *An Oppositionist's Deliberations,* Song Yingxing described a picnic with his friend and senior peer Cao Guoqi (n.d.) on a balmy late spring day in 1636. Equipped with a generous supply of wine, they set out to find a shady spot, intending to while away the day with nothing more arduous than rhyming games. They had just reached the outskirts of the Fenyi County seat, where Song held an appointment as a teacher, when an out-of-breath errand boy caught up with them. He handed over an "imperial bulletin" (*dibao*), rudely interrupting a rare private moment for two assiduous officials of the late Ming who had, for a moment, laid aside their duties of administering the people and managing the state. This was how Song Yingxing learned that a minor military official named Chen Qixin had succeeded in obtaining a high capital appointment merely by "establishing a discussion" (*litan*) about the ills of the state. Song took the news personally and hard, describing Chen's elevation as "a freak event (*qishi*) that could happen only once in a thousand autumn seasons."[1] His lifelong ambition to pass the metropolitan exam and his efforts to establish an honorable career now looked meaningless; his ideals had been rendered ridiculous. At the age of fifty-one, Song felt truly humiliated or, worse, dishonored. He aired his grievances in an essay written in the style of a memorandum to the throne. The title of his treatise, *An Oppositionist's Deliberations,* imparts its prime agenda: a sociopolitical reflection on the demise of the late Ming reign.

Published on May 8, 1636, *An Oppositionist's Deliberations* is a typical report from a late seventeenth-century minor official, a man with no political influence bemoaning the ills of his time: apathetic officials and a corrupt state, impoverished farmers and immoral scholars in a profligate society suffering from profusion ad nauseam. Voicing the widespread frustration of this period, Song's treatise illustrates a world consumed by the chaos "made by man." Also true to type, it was not the only expression of Song's dissatisfaction and was shortly joined by several other publications. From our modern perspective, it is the publication following Song's political diatribe that appears to fall so decidedly outside the frame of seventeenth-century Chinese literary conventions: The *Works of Heaven,* a text dealing with crafts and technologies in an exceptionally comprehensive manner.

Broadly speaking, in the seventeenth-century Chinese world, scholars and craftsmen practiced two distinct forms of gaining and making knowledge. Making things and settling affairs, craftsmen gained their knowledge through experience and tested it by trial and error. Much of what craftsmen did remained embodied or tacit knowledge and was not verbalized or written down. Chinese scholars pursued textual research, deduced meanings, developed theorems, and compiled commentaries about problems of philosophy, philology, or politics. Scholars could attain high social status and political influence; those who got their hands dirty usually belonged to the lower echelons. Engagement with issues such as the hazards of candle production or the muddy business of molding clay bricks was not beneficial to one's social standing or career, and consequently it had no place in elite literati pursuits.

This high contrast black-and-white image of Chinese knowledge culture featuring scholars sitting aloof in their study rooms physically and socially detached from craftsmen working in their bleak workshops is torn apart by Song's writings. His work connects the scholar to the craftsmen's world, proving that, just as many studies have shown for seventeenth-century Europe, distinctions between practice and theory in premodern China were never quite as clear-cut as this caricature implies.[2] Song's approach is particular to its time and place, a radical and comprehensive attempt to situate technologies and crafts within Chinese written culture. To determine exactly what is particular about Song's efforts requires careful analysis of its original temporal and cultural context. This chapter proceeds from the perspective of Yingxing's life. It examines the social, cultural, and physical conditions that Song inhabited and introduces the work he produced in those conditions. Questions to keep in mind as we proceed are: What influenced Song in his view of knowledge? What circumstances informed and formed his interests, this man who embodied the ideal of a dutiful and morally upright literati-scholar in *An Oppositionist's Deliberations* and who wrote sagaciously and extensively about crafts and technology in the *Works of Heaven*?

This chapter examines Song's childhood and family background. It searches Song's education to determine how it prepared him for his future efforts and observes the concerns that characterized his adult years. In this realm I question whether poverty or people in his social circle raised his interest in matters lacking status in Chinese culture. I also ask if Song's endeavor could have been affected by external influences, in particular by information on the West imparted through Jesuits or their works. Delving into the Chen Qixin affair that motivated Song to write *An Oppositionist's Deliberations,* I speculate on Song's state of mind on the eve of publishing his works and reveal his ideals and ideas.

The description of the picnic and his dramatic response to the court appointment launched Song's writing campaign in 1636. The almost novelistic tone used to introduce this fiercely critical political tract may surprise the modern reader just as much as Song's emphatic reaction to the appointment of a military official: Why did Song relate the details of his picnic and why was he, a minor official, so concerned about an event at court? What made this appointment so freakish when the dynastic state was suffering attacks by northern barbarians, harassment by pirates, and social unrest embodied by farmers' uprisings? Examining his use of stylistic devices and placing them against their historical backdrop unfolds the layers of motives and ideals that packaged Song's engagement with crafts and technologies. It shows Song's interest was intricately tied into a wider approach to knowledge encompassing issues such as sound production and meteorological phenomena, putrefaction processes, social responsibility and economic politics.

Short narratives, like the one used by Song to open *An Oppositionist's Deliberations,* were common overtures to early seventeenth-century literary outpourings. Typically, a libation of wine would make the author tired and emotional. In the ensuing state of catharsis he would compose a number of rapturous poems, a memorandum to the emperor, a novel or some sort of essay.[3] Any or all of these writings could become part of the scholar's literary life work: a comprehensive testimony to the individual mind belonging to the genre of miscellaneous private jottings called *biji,* literally "brush notes." By the end of the Ming period widespread education augmented the number of literati and these writings multiplied. More and more literati were discontented with their fate and the tone of their writing became increasingly critical. Major or minor, employed or unemployed, scholars related their desperate efforts to carry on their duties in spite of the decay of state and society. Those in service often announced their abdication or withdrawal from government service, while those with no appointment praised their reclusion because it allowed them to maintain ethical ideals in what they called a depraved and corrupt society. And while some won inner peace through literary dalliance or indulged in their hobbies, others made pointed comments on moral behavior and righteous statecraft or pronounced the values they thought should be pursued.

Song epitomizes the minor official searching for his social and intellectual identity in the seventeenth-century Ming world. He belonged to the group of men who identified themselves as "scholars" (*shi*), reflecting the values of a world in which social distinction and political power were grounded in academic training. The term "scholar" in seventeenth-century China did not designate a profession or an occupation as it did in medieval Germany and the emerging academia of seventeenth-century Britain and France. In the seventeenth-century Chinese intellectual world, "scholar" meant individuals who shared an educational background and personal commitment. It was a self-identifier as well as an indicator of social status for men (and a few women) who worked in or outside state or private institutions and frameworks of knowledge production. A Chinese scholar, Song had mastered a defined canon of texts and pursued extensive literary studies to best serve the state: by passing the official exams at the highest possible level and entering public service. By the sixteenth century, the number of aspirants, however, far exceeded the needs of the state. If their higher-level exam results were insufficient, some scholars, like Song, only achieved the lowest level of service, and that long past middle age. Some kept trying to pass the highest metropolitan exam until the end of their lives. Victims of the ambiguities of social insecurity and economic prosperity, the huge surplus of educated men were forced to find new occupations. They became ghostwriters, novelists, teachers, and doctors; the independently wealthy engaged in gardening, peony growing, fish or poultry farming; they explored foreign lands and seas and compiled travelogues. The rich assiduously collected arts and crafts, and others concerned themselves with the natural world, gathering knowledge about plants and minerals. Many of the Chinese scholars implemented and expressed surprisingly individual perspectives on the boundaries between the domains of knowledge such as natural philosophy or agriculture, banausic culture, trade, and the world of crafts. In sum, the diversity of interests and range of careers was equal to that of contemporary scholars in Europe, but the institutional and social framing was much less explicit.

Against this backdrop, the documenter of the nuts and bolts of eighteen selected crafts moved onto the scene in stereotypical costuming—a man in a full-sleeved round-collared cloud-pattern satin gown with a scholar's cap on his head. So attired he published a sequence of six texts in a mere two years: *A Return to Orthodoxy*, a philological study; *An Oppositionist's Deliberations*, a sociopolitical treatise, in 1636; *Works of Heaven*, on crafts; *On Qi*, a cosmological treatise with a discussion of sound and meteorology, the *Talks about Heaven*, in 1637; and *Yearnings*, a collection of poems, in 1638. Rendering material inventiveness and natural phenomena from an unrecognized perspective, Song thus made his mark on premodern Chinese literature and Chinese seventeenth-century approaches to nature, technology, and practical work.

THE MING DYNASTY AND THE SONG FAMILY

> Since the inception of historical documentation and the standardization of carts (i.e., their axle widths) and characters, officials have kept order and preserved the peace.
> *An Oppositionist's Deliberations*, chap. *shiyun yi*, 5

The world in which Song lived was vibrant and challenging. In 1587, the year our protagonist was born, the Ming dynasty could look back on a long, stable phase of enlightened rule that had brought two hundred years of prosperity to most regions under its control. Established by a man of humble birth who had spent part of his life as a monk, the Ming state system was dedicated to material needs. A tax quota system regulated the demand and supply of agricultural goods and the state-owned manufacturing system within sectors such as silk and porcelain. The ancestor's provisions brought stability to the Ming state, enabling its rulers to avert the negative effects of enormous population growth and to protect the country from outside invaders. Prosperity and peace also encouraged the growing commercialization of goods by the late sixteenth century; the stable environment nurtured literacy and stimulated the diversification of intellectual life on a large scale; in the year Song was born Chinese culture was blooming, materially and culturally.

The historiographers of the subsequent Qing dynasty would mark 1587 as the moment when prosperity went out of control and morals became lax, when indulgence and the lust for luxury goods debauched the Ming world. Ray Huang spotlights this particular year in his account of late Ming political history.[4] Everything was blighted by a succession crisis provoked by the Emperor Shenzong (1563–1620, reign name Wanli 1572–1620), after he declined to designate his first-born son Zhu Changluo (1582–1620) the future emperor. The Wanli emperor's noncooperation had been preceded by his gradual withdrawal from government. His move lay bare the conflict between himself and his highest officials, to whom the emperor reacted by not acting. This paralyzed the autocratically organized Ming state. As decision-making became more a matter of chance than a trustworthy political process, good men became hermits and left the political stage, while the demon puppets—eunuchs, corrupt officials, and scholars of low morals—took control and men of mere military training such as Chen Qixin achieved high rank.

The gradual loss of political control and an economic recession would accompany Song's generation throughout their lives while the ambivalence of a society torn apart by material prosperity and social insecurity frustrated their ideals and encroached on their professional careers. Song may have been particularly sensitive to the atmosphere of decay, because his family's micro-history seemed to follow the dynastic macro-political events: The family genealogy (*zupu*) reveals that the family prospered and declined alongside the Ming. Like the founder of the Ming, Song's ancestors were of humble origin. The Song family lived in Beixiang village in Fengxin County, part of the Yuanzhou prefecture in the southern province of Jiangxi. Situated in a hilly landscape with a moderate monsoon climate, the family estate stood on fertile ground in a region that harbored excellent natural resources, both mineral and organic. Even a moderate-sized estate could yield a profit. Throughout the early Ming period the farmer forefathers of Song Yingxing had made good use of these resources. Their fortunes increased and, as the family advanced to the level of wealthy landowners, it began preparing some of its sons for careers in public service. This was standard practice for landed families of this period who hoped thereby to acquire social prestige and ensure their newly gained riches through political influence.[5] The Song family proved to be among those who managed to climb up the slippery social ladder. By the mid-Ming period Song Yingxing's great-grandfather Song Jing (1477–1547) had attained the position of chief censor to the imperial censorate (*duchayuan zuodu yushi*, rank 2a) at the court of Emperor Shizong. This was the heyday of the Song clan from Fengxin County.

But the Song family could not maintain these dizzy heights. Not one of Song Jing's four sons achieved an official appointment. His grandson, Song's father Guolin (1546–1629) did not display any scholarly talent whatsoever. Rather he indulged in military strategy and dedicated his life to his filial duty to secure the perpetuation of the clan. At least in the latter, he showed prowess, for he married three wives who gave birth to four sons. Apart from Yingxing, he fathered the elder Yingsheng (1578–1646) and two younger sons, Yingding (1582–1629) and Yingjing (1590–?). As far as we know, Song Guolin had daughters, too. The genealogy listed female descendants only when their marriage cemented social relationships. Several of Song's aunts, cousins, sisters, and granddaughters were married to local worthies, some to the family of Song Yingxing's boyhood teacher Deng Liangzhi (1558–1638), the retired administration vice commissioner (*buzhengshi si canzheng*) of Guangdong Province, ranked 3b).[6]

This was the family's political and social standing on the eve of Song Yingxing's life. The advancement of Song Jing, Yingxing's great-grandfather, into a high court position had raised their hopes. They now expected continuous participation in state politics, fortune, and a superior social position. Two generations later the hopes turned out to be based on a bubble, as even Song Jing's success had been a case of too little, too late. The official Ming annals, which tend to formalize the major career steps of high court and state officials, depict Song Jing as a loyal official who had worked his way up laboriously via various provincial positions. Song Jing eventually reached the height of political power, a position at the court, ranked 2a (the highest was 1a) at the late age of seventy-one under Emperor Shizong. An awareness of the political situation of this time highlights the vulnerability of Song Jing's position. Although Shizong's forty-five-year rule had helped stabilize the Ming state, he was self-aggrandizing and inscrutable. Indeed, he often dismissed and executed officials at a whim whenever he deigned to take a personal interest in state politics. Most of the time, he left mundane politics to his grand secretary Yan Song (1481–1565; *daxueshi, shoufu,* bestowed with several additional high titles *shaozhuan, taizi taibao,* rank 1a), who came from the same prefecture as Song Jing. Song Jing's and Yan Song's shared provenance would have been seen as a good basis for Song Jing to at least consolidate his social and political position. But his ideas often conflicted with those of his countryman who was seen as the archetype of Ming official corruptness and moral depravity in the later literature, particularly in the famous Chinese opera and shadow play "Beating Yan Song" (*Da Yan Song*). Standing true to his honorable ideals, Song Jing seems to have refused to participate in the dull game of political networking that his colleagues used to create "safe havens" through kinship, regional affiliations, or teacher-disciple relationships.[7] He may have kept his integrity, but within a few months Song Jing was sidelined. His rank was not high enough and his relationships too vulnerable to prove useful for his descendants. The Song family enjoyed a very brief spell in central state politics.

On the local and familial level, though, it was another story. Song Jing's appointment to a high state position made the Song family a big fish in the small pond of village life. Since the eleventh century the quota of bureaucratic participation at the highest state level from Song Jing's county Fengxin, and its prefecture Yuanzhou, had been consistently low, and dropped dramatically toward the end of the Ming period.[8] The local notables saw their influence in central government diminishing, as did the people in the neighboring county of Taihe (T'ai-ho), described by John W. Dardess.[9] Metropolitan degree holders thus became particularly important to the construction of local identity, defining the status of the family on the local level long after their death. From the sixteenth century onward local gazetteers (*xianzhi, fangzhi*) inserted the major degree holders of the Song family, from Song Yingxing's great-grandfather Song Jing to his elder brother Song Yingsheng into their biographical section.[10] As Joseph Dennis exemplified in his study on how the Wanli-edition of the *Xinchang xianzhi* (*Local Monograph of Xinchang County,* 1579, in modern Zhejiang Province) had included genealogical materials of local worthies, this was a common way for the local elite to market family lineages and create the basis for continued cooperation and intermarriage.[11] We can thus assume that contemporaries listed the Song clan among the eminent local worthies. Yingxing's brother Yingsheng furthermore promoted his great-grandfather to the level of local hero, embodying the values of loyalty, diligence, and high moral integrity. The family genealogy honored their ancestor Song Jing as an ardent and exacting scholar, a paragon dedicated to the exposure of corruption and disloyalty.[12] The local sources reflect this image, suggesting that Song Jing bestowed a glowing reputation on his clan. He and his family exemplified on the local level the better times when integrity was still worshipped and when the good guys could still take over from the bad guys. Similar dedication was expected from Song Jing's descendants. It was in this climate that Song grew up.

CHILDHOOD AND EDUCATION

> Every civil servant, from the biggest to the smallest, even the insignificant education officials responsible for no more than a hundred miles, every single one is recruited by the examination system.
> *An Oppositionist's Deliberations,* chap. *xuezheng yi,* 33

Only one successful ancestor Song's family tree does not make it a well-established, venerable scholarly lineage. The clan was, however, respectable, and on the local level they were considered part of the elite affiliated to state power. This image is supported by the Song family genealogy, the only source we have for Song's familial background and early childhood years. The genealogy is therefore an important source, yet, when we draw information from it, we must take into account who wrote it and why. While it gives insights into the lives of family members, it has at least one hidden agenda. By the seventeenth century its compilation and recompilation had become part and parcel of a set of practices used by the gentry to establish and maintain lineages and substantiate social and political relationships. The Song genealogy emerges as an instrument used to construct and preserve the family's identity. While it mentions that Song Yingxing's generation had to suffer some hardships, it gives no proof that the family demonstrated any affiliation toward or interest in agricultural work or even craft knowledge. Rather, throughout the generations the family anchored its identity in erudite pursuits, apart from some minor excursions into military affairs. And while this identity utilized poverty as a rhetorical theme, it made no concession to practical tasks: only scholarly ideas and ideals shaped Song's early life.

By and large the family genealogy corresponds to what Keith Hazelton describes as the southern type.[13] Chronologically arranged, it assembled biographies of individual family members, usually only the males. The biographies themselves followed a standard formula that included the topoi and tropes relevant to the individual and his time. A family member would take up the duty of biographer or would be assigned the task of adding biographies of the recently deceased to bring the genealogy up to date. Later editors copied the contents of previous versions, but some also adapted contents to emulate a social ideal or to avoid damage to the family's reputation. Since the integrity of family lineage was important in Chinese society, black sheep were disguised rather than fully excluded.[14] The Song genealogy was updated sporadically, as were many other family genealogies in the Ming period. Researching biographies was time-consuming and expensive, as was the compilation of the entries and the genealogy, which had to be done to a high standard of accuracy. Editorial history often reflects the needs of the time: social, political, and financial circumstances affected its production. By the 1630s Song Yingxing's brother Yingsheng had brought their genealogy up to his father's generation. In the case of Song Yingxing's biography, traces of influence or bias in the recompilation are only partly visible because his biographer has not been clearly identified. Pan Jixing speculates that Song Yingxing's grandnephew Song Shiyuan (1649–1716) could have been the author. Pan suggests that the recompilation took place in 1668, which would mean Song Shiyuan was twenty-one years old, unusually young to take on such an important family responsibility.[15] As information on the lives and thoughts of the later editors of the Song family genealogy is only available in the genealogy itself, their attitudes and the way in which they may have manipulated the contents remain somewhat hidden. The genealogy constitutes the sum of various individual and collaborative historical efforts. Therefore, Song's biography within it provides a testimony of a familial rather than an individual agenda.

Families with scholarly ambitions in the late Ming and early Qing periods usually shunned anything in the genealogy that might imply nonscholarly efforts, except possibly some modest remarks on their humble efforts to administer agricultural tasks, as these demonstrated social responsibility.[16] There was also a strong element of denial toward family involvement in trade although by the end of the Ming this attitude was slightly undermined by merchants intruding into elite realms and striving after official posts. Still, most scholarly families tended to downplay any such linkages in their genealogies as well as any reference to craft work, for both were considered disgracefully lower-class activities.

Looking at the Song family genealogy from this perspective proves that its editors, and thus the clan members, stood by their erudite ideals. The absence of references to crafts is entirely in line with seventeenth-century Chinese standards. The absolute lack of any trace of any familial link to practical endeavors, agricultural issues, or technological projects, not even as a personal interest, is, however, striking. As the high provincial official Chen Hongmou (1696–1771), described by William T. Rowe, vividly exemplifies, the Chinese elite of this era considered agriculture and tasks such as constructing waterways to be the duty of a gentleman.[17] They promoted this by taking it as a subject in their literary efforts, often claiming they were personally active in agricultural matters (for the social implications of agriculture and crafts, see chap. 3). The Song genealogy presents a lineage of men who avoided any implication of involvement with practical work and insisted on the unsullied image of eagerly studying men who managed on their moderate estate with no ambition apart from exemplifying moral rectitude on the local level.

Family genealogies selected information to embody a common ideal; they also employed a common set of topoi in the individual biographies. One of the recurring themes throughout all periods was that of the impoverished yet utterly dedicated scholar and his scholarly family. Another was that of the promising young talent and extraordinary thinker. The Song family genealogy employs both. Song emerges as the model of a Chinese scholar, pursuing a state career and academic interests with great dedication. Ever since the Song dynasty scholars had made poverty respectable. The eleventh-century statesman, poet, and historian Ouyang Xiu (1007–72), for example, advertised his humble background and officially proclaimed that poverty was nothing to be ashamed of while making his way into high court positions.[18] On the contrary, it demonstrated a high standard of ethics. Perhaps because Song's generation experienced growing literacy in an atmosphere of political decay, scholars constantly embellished this topos, declaring that excellent scholars inevitably lived impoverished, yet erudite lives because of their ethics and high ideals. In the scholarly rhetoric of seventeenth-century China, humble living conditions indicated that the family pursued its ideals despite economic hardship. It must be stressed that if authors mentioned agricultural tasks in such biographical compilations it was not in order to indicate a low standard of living. Tilling the fields, managing hydraulic projects, or constructing bridges indicated moral integrity; these tasks proved that the man or his family was part of the social and political elite and the duty to look after the common people was taken seriously.

The Song family genealogy states that in the late 1570s, shortly before Song Yingxing was born, the family estate hosted twenty tenant households with a population of approximately 200 persons, thus placing it in the middle-level land-owning gentry. Then disaster struck, according to Song Yingxing's eldest brother's revision of the family genealogy compiled in the 1630s. A fire devastated the family residence, and the costs for reconstruction resulted in such serious hardship that Song's natal mother was forced to take on the duties of a servant, as the household could no longer afford domestic help.[19]

While Yingxing's eldest brother emphasized his family's hardship, his general depiction did not depart one iota from the scholarly ideal: the father retained his integrity, the mother served her family humbly, and the sons, studying diligently for a scholarly career, displayed filial piety. Further documentation in the genealogy reveals that the family recovered sufficiently to provide all four sons of Song Yingxing's generation with a thorough and sophisticated classical scholarly education. Their uncle Song Heqing (1524–1611) undertook their early schooling. Heqing was an upstanding but probably quite frustrated scholar who had passed the palace examination (*jinshi*) in sixth place in his youth. This achievement would usually have led to a position at the court or at least in the capital. Heqing was, however, only appointed to vice magistrate (*tongzhi*, rank 6b) of a subprefecture of

Zhejiang Province, Anji zhou. The shamefully low position of Heqing was possibly a result of his father's (Song Jing) ongoing criticism of the politics of the influential grand secretary Yan Song and his son Yan Shifan (?–1565).[20] In any case, Heqing soon returned home. Once Song Yingxing and his three brothers reached their teens, the family engaged private tutors, investing considerably in their future careers. Like most of his contemporaries, Song Yingxing thus spent his life up to the age of thirty-six in his family's private, yet austere, classroom preparing for an official career.

How does the family genealogy depict Song in detail? Did he display any signs of an unusual interest in practical things? The biographical note in the family genealogy is the only source we have about Song Yingxing's early years. I quote and analyze it here in full. Its subtleties are crucial for our interpretation of Song's efforts, although the description is for the most part entirely in accord with the familial identity and depicts in standard terms a true Chinese scholar. And yet, while the biography goes on to paint him in the standard colors of a talented, traditionally oriented young man, it still gives the portrait of an individual, not a stereotype: a Chinese literatus who—though his achievements were more or less ineffectual—was a potential candidate for an official career, an unusually capable intellectual dedicated to the philosophical theorems of *qi* thought.

In the conventional manner, Song's biography skips quickly over his childhood years, alluding briefly to his youthful precocity and some characteristics of his mother. Further references are also more or less standard, albeit indicative of a certain amount of freedom of thought: Song Yingxing

> showed the first signs of his many talents at a young age. His protruding, bushy eyebrows were envied by his peers. At a young age he was already capable of composing clever rhyming prose, and he was gifted at the entwined *bagu* style. He demonstrated an independence of thought (*jiaoba*) that startled elders. He always studied together with his older brother Yuan [Song Yingsheng]. Their teacher in the study hall required them to recite at least seven sections of an unknown text every morning. One morning Song Yingxing overslept while Yuan Kong [Song Yingsheng] was preparing and memorizing the assigned texts. His teacher reprimanded Song Yingxing [for arriving late and unprepared]. Then much to his teacher's surprise and disbelief, Song Yingxing recited the entire lesson by heart without faltering. When the teacher asked him how he had managed to do so, Song replied, kneeling in respect: "While my elder brother was learning his lesson this morning, I (Song Yingxing) was still in a deep state of sleep, but was dreaming with open ears to my brother's

> discourse, which is how I memorized the text." The teacher encouraged the boy's remarkable talents, which subsequently increased.

His biography emphasizes that Song studied the various classics and necessary reading with great enthusiasm and a sharp scholarly mind:

> Without exception, he uncovered every important relationship and connection and the original sense within the writings of the Zhou-, Qin-, Han-, and Tang-dynasty periods, and the period of the Five dynasties, [he studied Zuo Qiuming's] *Commentaries of Master Zuo Traditions* [on the Chronics of the State Lu, ca. third century BC] and *Guo yu Discourse from the States* [ca. fourth century BC], as well as those of the Hundred Clans (*bai xing*); he was able to trace his way through every unseen passageway and chasm of these texts, classifying and arranging them according to their hidden meanings. He was very gifted and therefore undertook extensive studies.[21]

Viewed through the lens of the genealogy, the world of Song's childhood and education can be seen as a marble, veined with the traditional ideals of Chinese text culture. Frederic E. Wakeman, Jr., illustrates the various steps of an education that shaped young talents like Song: an above-average apprentice began learning to write characters at the age of five, had memorized *The Four Books and Five Classics* (*Sishu wujing*) by the age of eleven, mastered poetry composition at the age of twelve, and thereafter studied the so-called eight-leg *bagu* essay style with rigid rules for the compilation of argumentative treatises.[22] The texts specified in the biography were a traditional part of the Ming dynastic educational curriculum, with one slight difference: when pointing to Song's interest in the writings of some prominent northern Song philosophers, the texts reversed their standard order. Usually the four schools would have been listed chronologically in two pairings, identifying each philosopher by his local origin: *Lian, Luo, Guan, Min,* referred to Zhou Dunyi (1016–1073) from Lianxi, the two brothers Cheng Yi and Cheng Hao from Luoyang, Zhang Zai (1020–77) from Guanzhong, and Zhu Xi from Minzhong (i.e., Fujian), respectively. As suggested by Pan Jixing, the fact that the biography names Zhang Zai first can be interpreted as a sign of Song Yingxing's respect for his philosophy.[23] While this respect reflects a popular trend at the end of the Ming dynasty, the slight deviation is an important indicator of Song's ideals: reordering the references is like reordering the colors in a child's marble—combining the same elements in different ways changes the outward color of the sphere. In this subtle manner the family genealogy substantiates that Zhang Zai's tenets vitally influenced Song Yingxing's cosmological notion of *qi,* a reading whose manifold implications are discussed in more detail in chapter 2.

The family genealogy reveals another important nexus, explicitly mentioning that Song was a talented composer of the "entwined [*bagu essay*] style," an important issue of his early training. Eight-leg essays were highly formalized, applying a strict outline to an essayist style. The assigned "theme" was often a quote from the classics. Many late Ming scholars complained that the exercise had become perfunctory by the end of the Ming. Benjamin A. Elman suggests the tone of the end result tended to be schoolmasterish rather than reflective of mature poetry or reasoning.[24] Song in his writings also harshly criticized the emphasis on the eight-leg writing style, which he called a "misleading and useless feature of official education."[25] Song thus condemned a practice that the compilers of his genealogy or its later editors identified as one of his major talents. The genealogy also describes the young Song Yingxing as a connoisseur who could grasp even subtle implications in the classics. It appears that either Song or his heirs would have liked to make some sort of contribution to future generations and deliberately stressed that he did *not* fail the metropolitan exam because of a lack of talent in, for example, compiling *bagu*-style essays, but because he *himself* considered this skill worthless.

According to Wakeman, the selected group of pupils who could maintain the rigorous schedule without flagging would generally attempt to pass the prefectural exams when they reached fifteen. Students almost invariably failed at the first sitting. But, by dint of repetition, fortunate pupils were able to acquire the *shengyuan* degree, the next step in the career line, at the relatively tender age of twenty-one. Most applicants did not pass the prefectural level until the age of twenty-four, while the average *juren,* the base qualification for an official low rank position, was achieved at the age of thirty-one. The average *jinshi,* the title for all candidates who had passed the highest examination in the capital, was thirty-six years old (see fig. 1.1).[26]

Judged within this scheme, Yingxing belonged to the less than brilliant, but better than average candidates when he took the triennial examination for the second *juren* degree in Nanchang in 1615 at the age of twenty-eight and managed a third-place ranking. His elder brother Yingsheng, who went with him at the advanced age of thirty-seven, only achieved sixth place.[27] From then on Song Yingxing and his eldest brother Yingsheng, who had the same birth mother and with whom Yingxing had an intimate relationship

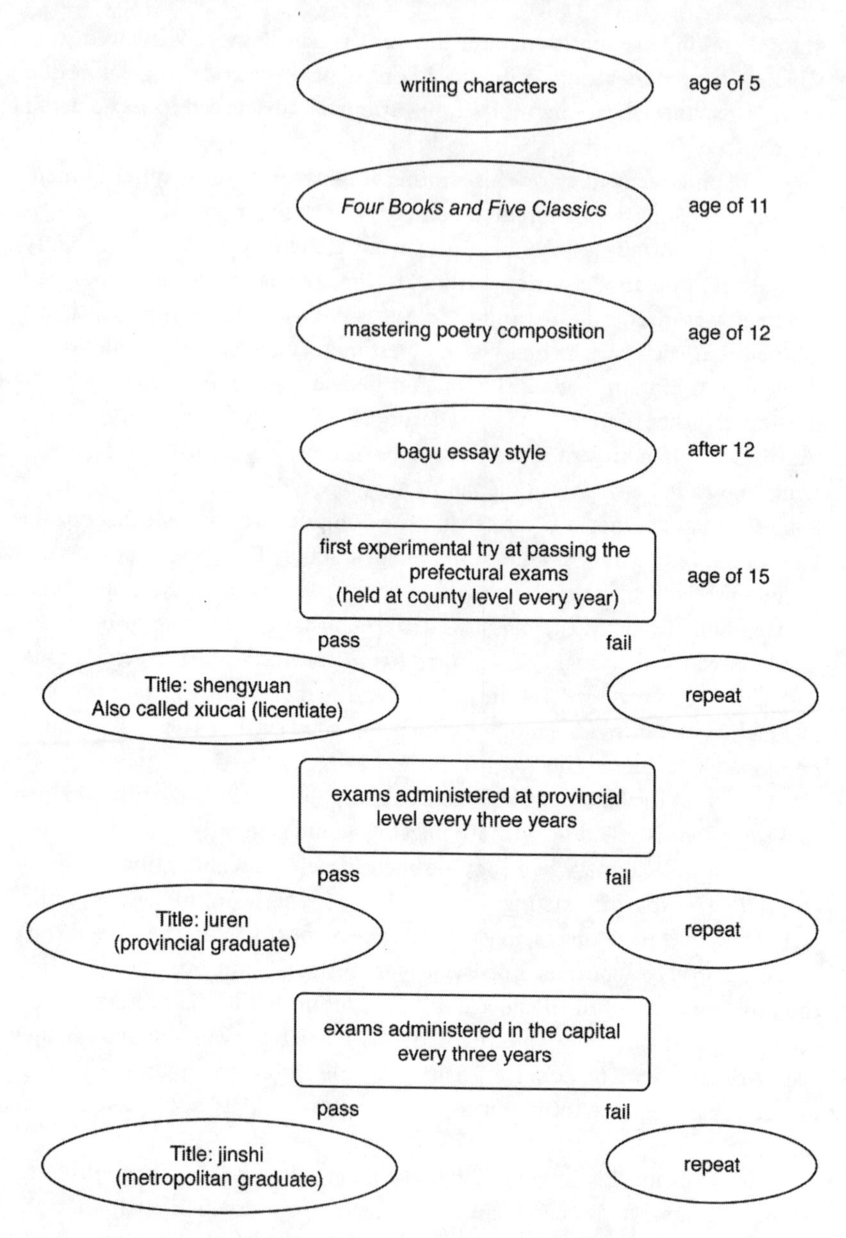

Fig. 1.1. Flow chart of the imperial examination system during the Ming and Qing dynasties.

throughout his life, prepared for the metropolitan exams that would open the doorway to a court or governmental career. The third brother, Yingjing, followed several years later, but was disqualified because he made use of taboo characters in one of his literary essays. The youngest son of the family, Yingding, finished the traditional curriculum but refused to take any form of higher degree exam.

Compared with his brothers, Yingxing seems indeed to have been quite gifted. Although as the third son he was not the first choice for educational promotion, the family again decided to invest heavily in his scholarly pursuits, enabling him to join his eldest brother at the exclusive White Deer Grotto Academy (*Bailudong shuyuan*). It was one of the oldest and most famous Chinese academies with an undeviating tradition dating back to the tenth century (some even claimed the seventh century). By Yingxing's lifetime, scholars used it for study and exchange, learning the ropes of intellectual debate and political socializing.[28] Yingxing seems to have profited from his stay intellectually but failed to make the social ties that would help foster his official or scholarly career. He mainly consolidated existing connections to his local friends Chen Hongxu (1597–1665), a bibliophile son of a high-ranking official, and Tu Shaokui (1582?–1645?), an influential provincial official who would later help Song finance his publications.

Much of the school's fame derived from the philosopher Zhu Xi (1130–1200), who had reestablished the academy in the year 1180. The Ming made his interpretation of the classics mandatory in the curriculum for the state examination, an orthodoxy that persisted until the abolition of the examination system in 1905/6. By the sixteenth century the academy had become a conglomerate of institutions involved in teaching, collecting and preserving books, and developing curricula for civil service applicants. It was crowded with diligent pupils, lofty scholars, and low-paid teachers, who provided a willing audience for visiting celebrities such as the founder of the school of mind (*xin xue*) Wang Yangming, a man who inverted Zhu Xi's emphasis on the "investigation of things" into an internalized search for knowledge and morals. By Song Yingxing's time, disciples of Wang had made the school one of their centers, discussing and extending the master's notion of innate moral knowledge (*liangzhi*). As discussed in more detail in the following chapters, Song's thoughts on talents were deeply influenced by the lively contemporary discourse circling around Zhu Xi's ideals and Wang's theory that knowledge and action were one. The visit of the renowned author of travel writings Xu Xiake (1587–1641) in 1618 may also have affected Song's line of thought, but none of these factors left any concrete evidence in Song's life or work or any other historical document.[29]

While we know that Song spent some time in the academy, we can only speculate whether, after having passed the provincial level, he ever actually took the highest exams. We know that his brother Yingsheng traveled six times to the capital to take the exams, and it is probable that Song followed suit. Yingsheng's failure (and Yingxing's if he did indeed take the challenge) may have been due to personal ability as much as external circumstances; the quota of attendees during this era constantly increased, which reduced the chances for each individual dramatically. Furthermore each of the six examination events were overshadowed by profound political crises, all of which facilitated lobbying and political machinations.

In 1616, the first occasion when the two Song brothers could have sat the metropolitan exams, the Manchu declared themselves successors to the Jurchen and founded the Later Jin dynasty (Hou Jin, 1616–35). They thus openly contested the Ming's entitlement to rule China. The regional exam candidates from the southern regions on their way to Beijing in northern China had to pass through a region devastated by peasant uprisings and foreign invasions. These incidents occurred far more frequently in the North of China; Song Yingxing's birthplace remained largely spared from violent acts during his lifetime, as James W. Tong's study of Ming dynastic uprisings and revolts illustrates.[30] Such a journey may have had a strong impact on Song Yingxing, forcing him to confront the desperate situation. In Beijing, scholars and statesmen were occupied with a foreign threat of another kind in 1616. The Jesuits, who had entered Ming realms of power shortly before Song was born in 1583, had tried to bring their religious ideas into Chinese society, building up communities of converts and influential proponents.[31] They offered their services as astronomers, military technicians, and mathematicians. Matteo Ricci (Chinese name Li Madou, 1552–1610) had succeeded in making his way into the court in 1601. By 1616 some Jesuits had adopted an aggressive strategy of proselytizing the Chinese elite, prompting eminent Chinese politicians and intellectuals to rise up in opposition in what came to be known as the first anti-Westernization movement in Chinese history (the Nanjing Missionary case). If he was a candidate for the metropolitan exams, Song may have been among the crowds in the streets of Beijing during this period. From this source Song may have heard rumors about European ideas of the world as a globe, their expertise in military weaponry, and probably also their interest in technological matters, such as hydraulic engineering, porcelain, or silk production. Yet all of this would become no more than a fleeting shadow in Song's life and work.

The subsequent three occasions when exams were held in Beijing were in 1619, 1622, and 1625. All three were darkened by the successive loss of imperial power to the eunuchs, which culminated in the temporary empowerment of the eunuch Wei Zhongxian (1568–1627). In 1625, for example, the court was in a state of stasis. The denunciation of Wei Zhongxian by the firebrand Yang Lian (1571–1625) had turned out to be a momentous misstep ultimately leading to the massive purge of a group of dissenting scholars who had assembled under the academy name Donglin.[32] In his writings published in 1637, Song sympathized with this group who had taken up the cause of moral standards and efficient state control.

Song's elder brother Yingsheng made a final unsuccessful attempt in the year 1631. In the same year the shepherd and apprentice ironworker Li Zicheng (1605–45) raised an army of rebellion, precipitating the final collapse of the Ming.[33] The death of their father and thus the official mourning rituals disallowed both brothers from participating in the capital exams of 1634. By this time the whole of northern China was already feeling the pressure of pillaging bandits and invading borderland tribes. The Manchu invaders still hovered ominously on the horizon, establishing their own dynastic reign and proclaiming themselves successors to the Ming dynasty. Factional alignments had stopped large sections of government dead in their tracks, with the true "gentleman" (*junzi*) preserving an ideal in privacy that the "small men" (*xiaoren*) in court had already abandoned. The *junzi-xiaoren* discourse about morality and integrity that Song joined originally mainly addressed the scholar's obligation to serve state and society. But by Song's time it had also become an issue of social class as the merchants attempted to approach the status of gentlemen and the gentlemen found themselves engaging with traders.[34]

It was quite usual for aspirants in Song's generation to continue to try to pass the palace examinations up to the advanced age of forty, dedicating most of their adult life to classical studies. This, indeed, was verifiably the case for Song Yingxing's eldest brother Yingsheng, who gave up at nearly sixty years of age, accepting the position of an assistant district magistrate (*fuxianling*, rank 6b) five years after his final examination attempt. In 1634 Song Yingxing accepted the position of a minor teacher in the Fenyi County school near his native county and part of the same prefecture. If we believe his biography, he did so not out of financial necessity, but for altruistic reasons. He had served four years in this position when he was rewarded with an appointment in 1638 in Dingzhou district to the slightly higher position of local judge (*tuiguan*), rank 7a, which he left in 1640 for reasons unknown. Song Yingxing stayed in his native hamlet during the next two years before accepting a position of a subprefectural magistrate (*zhizhou*), ranked 5, in Bozhou County in Nanzhili Province (modern Anhui) in 1643. We know this from a note written by his peer Liu Tongsheng (1587–1645), a retired historian of the Hanlin Academy (*Hanlinyuan xiuzhuan*) who helped Song resign from the position.[35] The local monograph of Bozhou gave Song a brief, standard mention, noting only that he sponsored the local library.[36] This suggests that Song had at least nominally taken up his duty.

According to the genealogy, his other two brothers married several wives and spent a leisurely life at home. The youngest son, Yingding, is said to have indulged excessively in wine, women, and song. Indeed, the males of the family in Song's generation seem to have lived in comparative ease and relative wealth. Apart from Yingsheng's dramatic account of the fire, the family gave no impression they were anything other than southern Chinese land-owning middle class with advanced scholarly ambitions. Almost all family members before and after Yingxing's time became scholars and landlords. A few took low military positions, which contemporaries may have judged a warranted choice in these times of crisis despite the implication of scholarly failure. No branch of the family appears to have benefited financially from Yingxing's accumulated knowledge about practical handicrafts (i.e., they did not make a living as craftsmen or earn money through the sale of Yingxing's book), nor did they become involved in any sort of project resulting from technical production; rather they were complacent in their position as middle-level landowners of moderate circumstances. A search through the registries of traders, handicraft rolls, and other local sources for any signs of a possible practical engagement by Yingxing or any other member of the Song family yields no results.

The historical record confirms that the Song family identity was geared to erudite subjects, rather than new fields of knowledge, and certainly not to the level of manual work or material comforts. This was the picture of themselves they handed down to future generations. Within the framework of his family's ideals, Song Yingxing's profile looks scholarly from every angle, a precocious youngster and conservative Chinese would-be official. His self-image as a scholar was damaged and the last remnants of his hopes for an official career were destroyed by the news of Chen Qixin's appointment. This proved to be the last straw. Looking at Song in his adult years, we find the stage set for the scholar at the picnic to take action in the only way he knew—with a pamphlet on the ills of the world "made by man."

DRIVING FORCES—THE APPOINTMENT OF CHEN QIXIN

Originally the village leader (*lizhang*) had a good reputation and was considered an honest and just man. But one morning [state politics and obligations] turn him into a bandit, and an unscrupulous thief.
An Oppositionist's Deliberations, chap. *luanmeng yi,* 45

The news that induced Song's frenzy of writing was announced in late April 1636. The gazetteer quoted directly from Chen's memorandum to the throne, *Lun tianxia san da binggen* (*A Discussion of the Three Big Roots of Illness under Heaven,* n.d.). It explained that Chen, who was originally a military selectee (*wuju*), had managed to submit this special memorandum to the emperor by way of a clever maneuver.[37] Barging his way through a crowd of petitioners waiting for the emperor to pass through the East gate of the imperial palace, Chen had provoked an uproar. He had managed to attract the emperor's public attention and thus guaranteed that his request would not be overlooked.

Chen's proceeding contravened the formal rules, as memoranda had to be submitted to the responsible ministry. It was not, however, totally unheard of, for in times of crisis men often had recourse to both self-humiliation and showmanship to attract the emperor's attention and hand in petitions directly. In his memorandum Chen lamented that those who had chosen a military career were prohibited from nomination for a civil service position. He went on to argue that the lack of skilled military administration had significantly contributed to the current crisis of the Ming state now threatened by rebellion inside and Manchu barbarians outside. Indeed, the civil service–oriented Ming state significantly hampered the career possibilities of trained military men, and most court officials did their best to enforce these restrictions. Envisaging the threat on his northern border and the inability of his civil officials to deal with it, Emperor Sizong (1611–44, reign Chongzhen 1628–44) was sufficiently impressed by Chen and his ideas to hear him out. Although the court officials were infuriated both by Chen's message and his method of attracting attention, the emperor appointed Chen to a high-level metropolitan position, thus bypassing all standard procedures of official examination, recommendation, or rewarded effort.

An Oppositionist's Deliberations was published a mere four weeks after Chen's appointment as the supervising secretary for personnel affairs (*like jishizhong,* rank 4). In the preface Song claimed that he had written his essay in the course of a single night—"finished by dawn (*jiechao*)"[38]—a common trope to express passionate cogency that underlines Song's personal engagement with the affair. In doing so, Song was in good company. Chen's appointment caused quite a stir among central government officials and influential people such as the supervising secretary of the Office of Scrutiny for Rites (*libu zhushi,* rank 6a) Jiang Cai (*jinshi* 1632) and the Confucian philosopher Liu Zongzhou, as well as Song's ally Jiang Yueguang (1584–1649), then commissioned to lead a delegation (*shi,* rank 4a) to Korea.[39] The fervent Ming loyalist and commoner Yang Guangxian (1597–1669) is reported to have carried a coffin through the streets of Beijing in protest against the "illegal" appointment of Chen, demanding the impeachment of the Grand Secretary (*shangshu,* rank 1a) Wen Tiren (1573–1639), as he considered both to be exponents of immoral government.[40]

Why did these highbrow scholar-officials react so strongly? As Chen was a military selectee himself, he could be rightfully accused of private ambition. In general, this alone would have sufficed to check his move. But in Chen's case the emperor had taken action and, after carefully listening to Chen's suggestion, had made him his personal protégée. In more than one respect Chen's criticism was far-reaching. The ministers, court officials, and clerks of Ming China, in fact all who deemed themselves scholars, felt Chen had dismissed their education, values, and ambitions. Their reaction was aggravated because Chen—in order to emphasize his call for the appointment of trained military personnel—refused the suggestion he sit the exams as a token of compliance. Chen's appointment rocked the very foundations of scholarly culture. Scholars saw it as an attempt to replace them with military personnel.

From the reign of the Han dynasty onward (second century BC), it was a cultural convention in Chinese scholarly tradition to regard the appointment of trained military men in times of crisis with great suspicion, especially if the power of the elite was based on civilian force. The bureaucratic state established by the Song in the tenth century promoted scholarly ideals, such as erudition, as markers of elite identity. The Ming state returned to this custom in order to distinguish itself from its predecessor, the belligerent Mongolian Yuan state. From the time of its establishment in 1368 up until its decline in 1645, the Ming system explicitly barred entrance to the civil service to those who had chosen a military career. A classic example of the reserved attitude of the Ming toward the military, especially if they were successful, was the opposition to the military official, General Qi Jiguang (1528–88), who fought the Wokou pirates on the Chinese coast.[41]

Chen's memorandum claimed that the exclusion of military men from governmental tasks was one of the political system's most critical deficiencies at a time when the Ming was under serious military threat. The civil servants' reaction was hostile because this was one more front on which they saw themselves attacked. They saw eunuchs flattering the emperor until they achieved influential positions; generals criticizing the scholar-official's strategic capabilities, merchants intruding into scholarly spheres; scholars without positions attacking their integrity, while on the border tribes such as the Manchu threatened to overthrow the state itself. In actual fact ad hoc appointments of this kind were no longer affairs "happening once in a thousand autumn seasons" as Song presumed in the preface to *An Oppositionist's Deliberations.* Yet, Chen's appointment came at a critical point and was made an exemplary case, a manifest instance of bad government.

For Song, the Chen Qixin affair had touched a particularly sensitive nerve. Throughout his life, he had kept silent about the great ills threatening the state. Even in 1627 when the Donglin party movement, the group of idealistic reformers opposing moral laxity, was smashed, Song held his peace although it concerned him on a personal level: in the year 1625 his friend and former classmate in the White Deer Grotto Academy, Jiang Yueguang, had been impeached as a Donglin member by the eunuch Wei Zhongxian; their subsequent confrontation is well described in the Ming annals.[42] The factional alliances and struggles of the Donglin, who spoke in the name of Confucian righteousness against the eunuchs and their allies, accompanied Song's life from the 1580s onward, culminating during the reign of Emperor Xizong (1605–27, reign Tianqi 1620–27) at and beyond the court. And although historical resources verify the widespread influence of these events, Song did not publish anything at this time, probably because he was still hoping for a career in the central government. The news of Chen's appointment in 1636 obviously came at a sensitive moment from a macrohistorical perspective and for Song in particular. In 1636 the Manchu reinvented themselves as a new dynasty, the Qing, and thus asserted their claim to rule the Chinese empire. Chinese scholars thus feared their own military would take this as a pretext to take over control in the Ming state. For Song the moment was critical because he had just given up his hopes of a court career and accepted a low-ranking appointment as a county teacher. The salt in the wound was that Song actually agreed with everything Chen said. They both felt the officials were incapable of handling the situation and the contents and execution of the civil service examination and the recruitment practices for officials were in desperate need of reform. Although

Song was against a military state, he felt that Chen's memorandum pinpointed a central reason for the decay that marked his era: officials and scholars amplified philosophical discourse and neglected their actual tasks. Song was also outraged by what he saw as Chen's selfish motivation. But for him Chen's appointment itself was not the worst part of the situation; even more damaging was the consequence that Chen's proceeding had rendered any literati discussion on the central issue of reforming scholarly minds and the state null and void.[43]

Song's subtle nuances in *An Oppositionist's Deliberations* reveal that he was not enraged because a military man had been appointed to the civil service; nor was he bothered that Chen called for the abolition of the examinations. He was enraged because Chen offered a valid critique that his colleagues dismissed for the wrong reasons. In Song's eyes the Chen affair had left things in a real mess, with attitudes deadlocked in a way that left few options open and none for himself: chaos had reached its climax. *This* was the event that only occurred once in a thousand autumn seasons. Song was acutely aware that whoever offered the same critique, even if he suggested an alternative feasible solution, and however accurate and moral his proceeding might be, would run into the same predicament. This was for Song truly the final blow. He realized that in the eyes of his contemporaries he was not only in a similar position to Chen, but that his goals had been the very same—changing scholarly attitudes—and would be dismissed in the same way. This blasted all his hopes of obtaining an influential position by the usual means—he was now a man with nothing to lose.

Song's choice of words as the weapon with which to make his last stand is understandable. Song was a scholar and identified himself exclusively within this framework. He was a minor teacher in a remote village and nevertheless saw himself, his activities, and his thoughts as deeply connected to central state politics and the fate of the Ming dynasty. But why did he choose the context of technological endeavor? There is no evidence of family involvement in or personal enthusiasm for craft knowledge, so we must dismiss theories involving Song's practical experience. Looking at the cultural, social, and physical conditions that formed Song, I propose Song's personal margin between theory and praxis was ultimately founded upon a basis not to be found in modern or historical definitions of technological endeavor. We need to take a step back, take a fresh look at Song's endeavor to document craft knowledge, and think anew our view of Song's effort. We need to investigate the peculiarities of his exceptional writing campaign within the context of the sociopolitical crisis, Song's personal failure, and his view of both.

SONG'S WRITING CAMPAIGN

> After a heated discussion, I returned to the silence of my room, lit the lanterns and worked feverishly until dawn.
> *An Oppositionist's Deliberations,* preface, 4

The mists of dawn had not yet lifted when Song set down his brush. Exhausted by his effort to express his beliefs and anger, his hopes and anguish in the ten thousand characters that now blackened the white sheets at his desk, he may have perched on the window seat and let his view drift over the verdant hills and the fields awaiting planting in the plain of Fenyi County, where Song had taken an official post. Up to his fifties he had dedicated his life to the study of classics, preparing for the metropolitan exams and a career as a scholar-official. He had verified his scholarly proficiency and almost finished his study, *A Return to Orthodoxy.* As a compromise, he had accepted the position as a teacher, only to then find himself forced to waste his time on the badly brought up, inept sons of lowbrow parvenus. Seeing the emerging chaos, Song started to "complain loudly and discuss his ideals and concerns about the state." But as he also remarked himself in *An Oppositionist's Deliberations,* his essay, written by the light of the lantern of true scholarly duty, was predestined to end as no more than a "rumor in the streets."[44] It might have been at this point, on this very day, having finished the last word of *An Oppositionist's Deliberations,* that Song, deprived of all his dreams, realized that there was something he still wanted, something he needed to accomplish regardless of the cost. He may have squared his shoulders and set off to finish what he had started—to publish his complete works of which *Works of Heaven* would meet the most resonance.

After he finished his first essay, events happened very quickly. Song obviously felt an urgent need to see his complete works in print. The rapid succession of Song's publications gives us an impression of the fury and energy that drove him. As far as can be established, Song's publishing activity was concentrated within a period of two years, between 1636 and 1638. Then came silence. His schoolmate Chen Hongxu mentions that Song in 1644 and 1645 wrote two other pieces, both were manuscripts and remained in private hands.[45] A careful look at Song's publication history shows that he controlled the sequence of publication of at least four of his works. *An Oppositionist's Deliberations* came out in the third month of the year 1636, and *Works of Heaven* followed almost exactly one year later in the fourth month of 1637. Only a few weeks later in the sixth and seventh month *On Qi* and *Talks about Heaven* were printed (see appendix 2). The sequence of Song's publications looks like a methodical campaign. This implies that although *Works of Heaven* would ultimately be the only work to achieve prominence, it was never meant to stand alone. It was deliberately placed within a set.

The strong relation between the various writings of Song is also evident in the technicalities of the printing. Song had all his writings printed in an identical style: All publications have the same style of characters. The prefaces are made up of six lines per half-page with thirteen characters per completed column. The main texts are arranged in nine lines per half-page and twenty-one characters per completed column. Modern research shows that the same paper was used and the same sort of ink. Pan Jixing suggests that, pieced together, this information strongly indicates that the same publishing house issued all of the texts.[46] All of Song's treatises not only harmonize with each other in their printing, they all display the same internal organization, classification, and arrangement of "chapters" (*juan*) and the subsequent numbering of the subchapters. Song classified each of the three main sections of *Works of Heaven* as a chapter, and then called the entire text a chapter. This was not a mere slip of the brush: Song was telling us that *Works of Heaven* (published in three chapters) was only part of the message.

Song embedded his technological writing within a thematic set to tell his contemporaries where it had all gone wrong and where the truth could be found. Song's interest in technology was part of an artful design. His epistemological environment carefully framed knowledge with regard to its purpose and value. Conceptualizing his testimony within the seventeenth-century world, Song thus employs traditional Chinese ideas of content organization and draws on encyclopedic categories that manifested the idea that knowledge expressed the relation between heaven, men, and earth. To him technology and crafts were expressive of this relation.

Song revealed his thought processes on the framing and internal organization of his writings and their purpose quite clearly in the preface of *Works of Heaven*. He stated that he had arranged his writing according to a leitmotif (*yi*): "The chapters are divided into before and after (*qianhou*) purporting the idea of 'holding the five crops in high esteem but considering gold and jewelry as less significant (*wan gui wugu er jian jinyu zhi yi*).'"[47] Song's arrangement of content in *Works of Heaven* was thus not based on a hierarchy of technology, but on a very traditional maxim—sustenance before luxury. But his maxim did not simply evaluate fields of knowledge in a traditional scholarly manner. It implied that Song organized knowledge about things and affairs from the roots to the end (*benmo*) in relation to each other to bring order to the world.

This general, yet essential, principle, "from the roots to the end," was widely applied by scholars. Authors addressed it directly in the genre of "tracts and registers" (*pulu*), and in compilations delineating the "origin of things" (*wu yuan*). Another important group that referred to this principle were authors with an encyclopedic approach. From a modern perspective, it looks as if the authors in these genres all pursued controversial goals: *Pulu* compilers aimed at systematically exhausting (*bu*) a specialist theme such as goldfish breeding or peony plants.[48] They gave a tremendous amount of detail on one issue. Those tracing the "origin of things" concentrated on giving issues a history and place, thereby integrating them into a lineage.[49] Encyclopedists collected and embraced knowledge as a whole. And yet, all three had two things in common: first, all intended to present a completed nexus (which could be from the perspective of either comprehensiveness or detail). Second, all of them defined lineages of knowing (in terms of homologous, or hierarchical, sequential or causative relationships). Song also pursued these two aims.

Another statement of Song's in the preface to *Works of Heaven* is particularly significant for understanding the conceptual arrangement of Song's oeuvre in its entirety: He explained that he originally intended a much broader scope but "as for the two chapters 'observation of heavenly phenomena' (*guanxiang*) and 'musical intonation' (*yuelü*), [after some deliberation I decided that] their ways (*dao*) are too refined. Consequently the characters [which were already carved] in the printing plates [of the table of contents] have been expunged."[50] Song, then, intended to compile a book dealing with heavenly phenomena and musical intonation in addition to the technological fields of the *Works of Heaven*. The three chapters were ordered in the standard way for works in premodern China: first (*shang*), middle (*zhong*), and later (*xia*). When Song originally considered including two further chapters, he was not thinking of making the number of eighteen topics up to twenty. Song's statement strongly implies the sections on musical intonation and heavenly phenomenon originally formed another two chapters. This would have made *Works of Heaven* a work containing at least five chapters.

Pan Jixing has suggested, based on his investigation of the original print of the first edition, that the printing plates (*zi*) for the missing two chapters had already been prepared when Song decided to omit them. The table of contents of the first edition of the *Works of Heaven* shows that several columns were retouched after having been cut. The corrections of the characters of the first edition of *Works of Heaven* verify that Song Yingxing had originally intended to include these parts.[51] While we do not know exactly why Song finally omitted them, the exclusion was a relatively spontaneous decision made shortly before the actual printing of *Works of Heaven*.

The titles Song chose for the omitted chapters are another indication that he originally had a larger framework in mind, namely, that of an encyclopedic effort: *yuelü* referred to music and harmonics and *guanxiang* to the observation of heavenly phenomena. They were valid variations of encylcopedic categories used, for example, by Shen Gua (Shen Kuo, 1031–95, the Chinese polymath of the tenth century discussed by Fu Daiwie) in his seminal private jottings, written during his retirement at Dream Brook, in present day Jiangxi Province, *Mengxi bitan* (also pronounced as *Mengqi bitan*) (*Brush Talks from the Dream Brook*, 1088). Song's note that he originally intended to include two additional chapters indicates that he thought of the documentation we perceive as craft knowledge or technological knowledge as part of a greater whole.

Looking at Song's writings as a whole raises some interesting issues about Chinese categories of knowledge and their identification within literature. Drawing on Fu Daiwie's recent research on key concepts and categories of knowledge in late imperial China, I suggest that had Song been successful in convincing his contemporaries of his approach, bibliographers would have categorized him under the rubric of syncretists (*zajia*): Syncretists make up a subsection of the main Chinese bibliographic category of matters (*zi*) where the Chinese intellectual world assembled exceptional thinkers who had a comprehensive, sometimes encyclopedic approach to fields of knowledge—an approach that aimed at revealing essential truths about the relation between heaven, earth, and men.[52]

Within Chinese written culture we find acknowledged organizational schemata of great collections of written knowledge that were encyclopedic in nature, *congshu* and *leishu*. Both forms display the fundamental knowledge categories within which Chinese scholars operated and according to which they organized their knowledge: from celestial phenomena (*tian*) to insects (*chong*), historical occurrences (*shi*) to administrative institutions (*guan*), ritual (*li*), music and harmonics (*yuelü, lülü*), metrology, or the observation of phenomena (*guanxiang*).[53] I suggest that Song in the seventeenth century followed categories that had been laid out by the Tang dynasty (618–907), and considered their underlying issues as did Shen Gua. In a way similar to Shen Gua's, Song reacted to the trends of his time, including new fields under new perspectives. Shen Gua examined subjects such as mathematical astronomy, the construction of city walls, the zither, musical composition, and whiskers.[54] He discussed the organization of the central state and the heavens, ritual and man's role in the world.

Song does not explicitly refer to the full set of issues that make up an encyclopedia in terms of *leishu.* Yet, taken together, the contents of his work touch many issues that belong in this category. In *An Oppositionist's Deliberations* Song was concerned about institutional settings and the administration of the state; language, naming, and identification was his theme in *A Return to Orthodoxy. Works of Heaven* made issues such as the ventilation of mine shafts and lime processing a subject of discussion within literary discourses on human-heaven interconnectivity. The contents and approach of *On Qi* and *Talks about Heaven* are very close to what Song may have intended to discuss in the discarded chapters on heavenly phenomena (*guanxiang*) and music (*yuelü*). *On Qi* contains no less than nine paragraphs discussing the manifestation of *qi* in acoustic phenomena (*qisheng*). It deals with musical instruments and meteorological phenomena elucidating the harmony of the universe in *yin* and *yang.* One can easily imagine that Song's basic assumptions about the role of *qi* in the creation of sound were meant to be part of an earlier proposed treatise on music and harmonics and that he left out ritual discourse because of its misleading falsification by his contemporaries. The six paragraphs of *Talks about Heaven* deal with observation of the movements of the sun as a visible body. The temporal and thematic proximity of these treatises begs the question: Were the original chapters on "musical theory" (*yuelü*) and "astronomical phenomena" (*guanxiang*) to some extent used in, incorporated by, or even wholly identical with the treatises published later? Fu Daiwie suggests that Shen Gua's *Dream Brook,* which he identifies within the so-called literary genre of private jottings, can be looked upon as an attempt by an individual scholar to fix the knowledge of the world: some fully accomplished this, others just lodged the claim. Both may be likely alternatives in Song's case, but in either case I suggest Song deliberately spelled out his original plan, signifying to his colleagues the scope of his efforts and that his *Works of Heaven* was part of a universal approach to knowledge.

Within this context Fu also points to the flexibility that private jottings offered "as quasi-alternative prototypes of an encyclopedic endeavor."[55] The development of *biji* is still discussed. But it seems clear that by the eleventh century, when private jottings turned into a genre, authors attested to a shared understanding of writing style mainly by the way in which they collected and recorded information and neither by stylistic/formal nor topical restriction.[56] As a relatively free form of written discourse, private jottings enabled seventeenth-century scholars to usher in new approaches to knowledge and arguments by assembling fresh observations within traditional schemes. Traditional encyclopedic compilations in general only drew from written sources, reorganizing everything from short quotations to full treatises and monographs. Private jottings, which were also grouped under the title writings on "observations" (*jianwen*), allowed scholars to include personal experience or quote from memory rather than referring to specific texts. If we consider Song's oeuvre in the contextual classificatory scheme of original Chinese tradition as private jottings and Song a syncretist (*zajia*), we realize that *Works of Heaven* contains an important layer of meaning apart from its technological content. As Fu Daiwie has remarked: jottings actually contain "non-trivial social messages and ethical lessons," which are for the author "equally important as those acute observations" that interest us.[57] I would go one step further and say that, for Song and his contemporaries, the nontrivial messages and ethical lessons were more important. Discerning the implied social and intellectual messages that Song delivered together with the technological contents of *Works of Heaven* gives us a heightened awareness of how Song perceived knowledge.

Each individual piece of writing divulges one particular facet of Song's mind. *A Return to Orthodoxy* verifies Song's firm consolidation in traditions of written discourse and his professional and social identity as a county school teacher; *An Oppositionist's Deliberations* links him with ideals of the scholar official who dedicates himself to the state and society and who is torn between his ideals and his contemporary world characterized by political unrest and social disorder. His poetry collection *Yearnings* sets out his aesthetic ideals and shows his development from an ardent aspirant for civil service to a compromising scholar discontented with his personal fate yet a fervent defender of scholarly values. In his treatise *On Qi*, one finds clues to Song's viewpoint on Buddhist thought as well as manifold linkages to the ideas of the *qi*-iconoclast Zhang Zai. *On Qi* also gives a picture of the meticulous observer and careful analyst of natural occurrences who follows the lead of the mind. His *Talks about Heaven* confirm that he was a consistent theorist in his aim to divulge the fallacies of contemporary "Realpolitik," a morally loaded political attitude that he identified as "the Confucian's (*ru*) irrational approach to heaven."[58] And finally *Works of Heaven* elucidates him as a man with a free rein, finding order in the things and affairs that made up his material surrounding.

If we look at Song's efforts one by one, his beliefs may seem to us as eclectic as his character. Combined they reveal that Song's thoughts all circled around the issue of human-heaven interconnectedness. Scrutinizing Song's work with this issue in mind, we can see that Song's approach to craft and technology was formulated within Chinese traditions and enmeshed in the issues that governed his time. In seventeenth-century China, Song was a thinker who brought forth a congruence of knowledge and action that rested on the understanding of universal principles. When Song referred in the second part of the title to the "inception of things" (*kaiwu*), he was indicating that he located this universal principle in the change (*yi*) of *qi*. How he understood this and what it implied is the subject of the next chapter.

CHAPTER 2

Affairs of Honor

At the height of order, thoughts and ideas are in disorder. When disorder is at its height, thoughts and ideas are in order. Indeed, this cycle has been the leitmotif for many generations.
An Oppositionist's Deliberations, chap. shiyun yi, 5

Song grew up in southern China where boats were commonly used for transportation and travel. He noted that this meant he could identify the various types by their silhouette, unlike "those who spend their lives in mountainous country or on the plains (who) only saw a simple scull or two or a chaos of rafts on the river." In the section on boats and carts of *Works of Heaven* he described the tax boats (*ke chuan*) whose prows, long and slender, rose above the other craft as they delivered revenues and wares at regular intervals on canals and the rivers of the Yangtze and Han rivers. Oars stabilized the boats when heavily loaded with grain, salt, or silver. Above deck they were fitted with a row of up to ten cabins, each just big enough for one person to sit or sleep in. Travelers preferred more spacious accommodation. Around Suzhou city, Song explained, voyagers used the "wave-riding boats of the Three Wu region" (*San Wu lang chuan*) with windows, passageways, and larger cabins. Officials passing through Fujian Province chose the "mizzen sailboats" (*qingliu chuan* or *shaofeng chuan*) comfortably appointed with large living quarters so they could travel with their family and entourage. Knowing the types, Song argued, was important; knowing the proper nomenclature imposed order on the chaos. And if chaos was on the agenda, it was the duty of a scholar to put thoughts in order.[1]

Chapter 1 has shown that although Song was an outsider who grew up far beyond the metropolitan, courtly, and elite worlds of Nanjing and Beijing, he was, nevertheless, passionately engaged with them. Deeply concerned about the chaos of his time, in his literary efforts Song tackled politics, morals, and society and explained the procedures of ship construction and cart making, the operation of grinding mills, the phenomena of sound, metrology, and decomposition processes, the movements of the sun and moon and eclipses. Looking at Song's interests within the intellectual atmosphere of his time, we ascertain the singularity of his character: a man who foraged through a continent of subject matter to contest what he saw as the irrationalities of his world. But what purpose did he assign to this knowledge and how did he authorize his comprehensive viewpoint?

This chapter places Song within the main intellectual debates of his era, scrutinizing his approach to knowledge through *qi*, his view of man's and heaven's role in the world and in the making of things and affairs. A complex intermeshing of politics, ideology, religion, and philosophy must be disentangled in this process. The central issues for Song were the question of morality as a guide to behavior, the responsibility of the intellectual to think independently, and the duty of the scholar to address chaos and impose order on a world gone wrong. On this basis we can see that in his writings Song juxtaposed the chaos of his time with the order inherent in things and affairs on heaven and earth. In *An Oppositionist's Deliberations* Song reacted to the moral standards and premises of his era. He identified and attacked what he saw as crucial weaknesses or misunderstandings: the fatal impact of the "isms"[2] on politics and social developments, belief in an omnipotent heaven, and cyclical change or fate. Attempting to quell the chaos of his time, Song promoted the key to order: *Works of Heaven* exposed the order inherent in things and affairs; *Talks about Heaven* divulged the rationality of heavenly patterns within the universal order of *qi*. *On Qi* unraveled the subtleties of *qi* workings on phenomena such as sound or meteorological phenomena.

In this chapter an overview explains the significance of *qi* in Chinese epistemology and provides background information on its close intertwining with morals and ideology in the discourse of Song's time. It shows that in a scholarly manner, Song identified knowledge and its origin by examining intellectual lineages of *li* and *qi*. He conceptualized *yin* and *yang*, the five phases (*wu xing*) and the relationships between them as cosmological principles shaping the world and human creative activity. Applications of the past and the sage-kings were the cultural devices he used to assign man his role: man had to allocate the appropriate sources for knowing. Intellectuals had to think independently to impose order on a world gone wrong. Examining where Song located himself within his time and its theories lays bare the levels of surface chaos and underlying homeostasis exposed by his theory. It reveals the ideas and ideals that drove seventeenth-century Chinese intellectuals when they wrote down scientific and technological knowledge.

Song lived in a world in which knowledge systems revolved around the question of how cosmic and natural forces affected human affairs or, conversely, how humans could live in harmony with the cosmic order. He ordered nature, phenomena, and events along these lines, considering heaven, earth, and human endeavor integral and contingent parts of one entity. Heaven was an impersonal agency, complementary to a supporting earth. Together they formed the framework of a realm in which man lived and gained knowledge. Chinese intellectuals approached this realm in terms of "that which is so of itself" (*ziran*), "all under heaven" (*tianxia*), the subtly working forces of *yin-yang qi,* and "heaven and earth" (*tiandi*). These concepts varied in their epistemological or sociopolitical emphasis, but all of them were equally grounded in the orderly schemes of *dao, li,* and *qi.* Hence, order and its implementation informed thoughts on man's or heaven's role, nature, culture, and creation, rather than the teleological implications of a godlike figure that lies at the heart of early modern European assumptions on the opposition of the categories: natural, cultural, artificial, super- or unnatural, or the nature-culture divide in modern thought.[3]

Searching for order, Song found reliability in natural phenomena and the production of material objects. With the compilation of the *Works of Heaven, On Qi,* and *Talks about Heaven,* Song directed his colleagues' attention to agriculture, crafts, and technology, the patterns of heaven, and natural phenomena. He argued that these issues revealed an order his colleagues had overlooked. Moral behavior, Song urged, was the natural consequence of understanding and following the universal rulings of *qi.* If we thus understand rationality as an issue embedded in *qi,* Song's universe was rational, based on knowledge and facts.

Song's view ran counter to a late Ming trend in which scholars imposed moral categories of "heaven" onto human nature. Officials of Song's generation attempted (unsuccessfully) to take refuge in this argument after the Wanli emperor cunningly avoided cooperation during the last phase of his reign. In the atmosphere of increasing chaos and sociopolitical insecurity caused by imperial inactivity, the ubiquitous association between heavenly patterns and man's action, fate and morality became increasingly popular, heavily influencing this generation's actions and its approach to knowledge. Many suggested with a reference to the Song iconoclast Zhu Xi that "man's mind" (*renxin*) was bound through personal merit to the "mind of heaven" (*tianxin*).[4] Some felt that all things and affairs—whether it be now natural occurrences such as weather, or earthquakes, or state politics, personal fate, man's abilities and talents—were impervious to man's influence. Heaven thus became the arbiter of man's fate, while fate became the cause and effect of moral behavior. Book dealers did a roaring trade in moralistic guidebooks. The classic *Taishang ganying pian* (*Tract of Taishang on Action and Response*, ca. 1200) became a best seller.[5] It described mechanisms of retribution in man's relation to heaven. Putting this theory into practice, scholars often left affairs deadlocked on moral grounds, paralyzing state and society.

Song's disapproval of contemporary "Realpolitik" led him to inquire into all kinds of thought to reveal the truth about how heaven and man were connected. In *An Oppositionist's Deliberations*, he claimed to have thoroughly investigated historical examples and contemporary thoughts on human nature (*xing*) and morality. He found them all lacking. A passage in the preface to *On Qi* is central to understanding Song's intellectual motivation to approach knowledge in a comprehensive manner. Authorizing his interest in *qi*, Song argued that because they followed their "emotional mind" (*renqing*), his colleagues were blinded by cultural bigotry, ideological bonding, and immoral inclinations. As emotions were by inclination "obstinate" (*gu*), the scholarly inquiry was superficial and they had become vulnerable to indoctrination. Song explicitly criticized the representatives of Buddhism (*fo*) and Confucianism (*ru*). He argued they had never really grasped the essential arguments and ideas of their opponents—not even the great eleventh-century thinker Zhu Xi, who many leading Ming intellectuals embraced as their doctrinal idol. This, Song believed, had brought about the chaos of his time:

> As for human nature (*xing*), well, the Buddhists say it cannot be explained straightforwardly. Whenever they ascend to the pulpit (*deng tan*) and expound this issue in preaching, they use mysterious allegories and miscellaneous methods. They are not capable of saying what they want to say in a direct and unpretentious manner. For this they are attacked by the Confucians (*ru*) as if they were enemies or thieves and their writings and principles are dismissed as heretical (*duanyi*). Zhu Xi once proceeded from the Buddhist canon of *Sishi'er zhang jing* (*Sutra in Forty-two Articles*). His words plainly and austerely recapitulated its contents. He ought to have known [how to] read the sutra (*nei dian*), but he never attentively studied the meaning between the lines![6]

Many of his colleagues in power believed their scholastic skills, their eloquence, and their sophistication in essayistic style made them superior to other cultures, with different languages or other ideals; "hence they overlook that Buddhist thinking is rich and inscrutable in spirit. Precious things are, however, never on the surface. Those who brand [Buddhism] as barbarian have only skimmed over the characters, disregarding its proper meaning."[7]

With his defense of Buddhist ideas in this passage, Yingxing superficially seems to join people like his contemporary Li Zhi (1527–1602), who defamed Confucius and denounced the Ming school of Confucianism even as hypocritical. But while Li Zhi then turned his eye to Buddhism (and actually considered himself a true Confucian in substance), Song's discourse on morality eventually concludes in the dissociation of all generalized doctrinal identification. In this radical abdication of his era's doctrinal frameworks, he must also be distinguished from contemporaries such as Jiao Hong (1540–1620), who expressed their wariness toward doctrinal premises by trying to integrate the three major teachings of Buddhism, Confucianism, and Daoism.[8] Song's search for truth and knowledge was far more radical: he truly ploughed a lone furrow.

Song attacked as irrelevant the issue of ritual and morality in discussions on human-heaven interconnectedness. This fits our modern conception of a "scientist": a man who suspects indoctrination, challenges contemporary thought, and systematically searches for a rational order in the world that surrounds him. Song's methodology was, however, that of a seventeenth-century Chinese scholar concerned about sociopolitical and intellectual chaos. His *Talks about Heaven* exemplifies this point: in it Song does not calculate or measure. Rather he focuses on cosmological and sociopolitical issues when investigating heavenly and earthly phenomena, eclipses, catastrophes, and the movement of the sun and the moon. For this purpose, Song pointed out that historiographic traditions had formed a faulty cause and effect correlation between morality and the ascent and descent of dynasties. He could prove with the very same historical examples that heaven was unrelated to human morals. Heaven was a phenomenon driven by rational patterns of *qi*-logic. Song's argument freed man from the reign of a moralizing heaven and at the same time obliged him to accept his duty to study the heavens as an exemplification of the order of the world and then put this knowledge into action.

Looking at the patterns of the heavens, Song said it was man's understanding and interpretation of the heavenly patterns that was effective, not the patterns themselves. This, he reminded his colleagues, was the original point of departure for any assessment of the relation between heaven and man. He admitted, however, that the association of human-heaven interconnectedness and morality involved one important truth: if the ruling house was negligent in its duties on earth, that is, those of the state, and failed to read the heavenly patterns correctly, chaos was nigh. His colleagues, however, used the heavenly patterns as an excuse for apathy, waiting for their expression instead of discharging their tasks. Song's vivid imagery in *An Oppositionist's Deliberations* depicted the disastrous results of the state officials' procrastination: "people dying at the hands of bandits and oppressors one day and when [they survive] it is only to die at the hands of soldiers on the next day."[9] Song was not exaggerating: millions of people did suffer during this period, and millions starved in the streets or died in military campaigns or rebel clashes.[10] Instead of facing the situation, court officials and scholars endlessly discussed which stage of the cycle of dynastic change they found themselves in at that moment or who was to blame for the emergence of chaos and disorder. Song called for action, for an emancipation of ideas, such as a predefined cycle of events or fate. By publishing *Works of Heaven* he urged his colleagues not to be fooled or discouraged by anything that failed to stand up to thorough inquiry: it was time to direct their attention to the "inception of things and accomplishment of affairs" (*kaiwu chengwu*). Only this would unravel the attributes of human-heaven interconnectedness and reveal that the universe rested on the order of *qi*.

KNOWLEDGE IN TERMS OF *QI*: UNIVERSAL RULINGS AND RATIONALITY

> Talking about it at length I am so grieved, I can only weep bitterly. I do not know where to begin and where to end. There are so many men who worry about the state's strategies (*guoji*). It would be a blessing if the emperor would lend an ear to the grass and reed cutter's gossip (*churao*)!
> *An Oppositionist's Deliberations*, chap. *lianbing yi*, 30

How were the Ming literati to deal with the wealth of detail and the bewildering variety of things and affairs that characterized their capricious world? One answer was to define the epistemological boundaries of literati learning and anchor oneself within it. Late Ming scholars did this by delineating their view of *li*, and *qi* and identifying eligible fields of scholarly inquiry. They signified their stance by affiliating with such classics as the *Book of Changes* or the *Liji* (*Book of Rites*) or their philosophical interpretations.

Some substantiated their interests more concretely by singling out specific quotations or technical terms. Those who propagated the "investigation of things and broadening knowledge" (*gewu zhizhi*), for example, referred to the ideas of the eleventh-century Zhu Xi, who drew this passage from the *Daxue* (*Great Learning*, one of the four books canonized during the Song dynasty). Zhu considered *li* the governing principle and proposed "seeking moral principles within one's moral cultivation" as a way of dealing with the expansion of literati learning.[11] By the sixteenth and seventeenth centuries Zhu Xi's interpretation of the "investigation of things and broadening of knowledge" was challenged by Wang Yangming who claimed that *li*, the structural principle, was all in the mind. Connecting to the *Great Learning*, both thinkers justified their interests in "things and affairs" and redirected attention to philological studies and morality. Others favored "practical learning" (*shixue*), which relocated Zhu Xi's ideals closer to statecraft. Ideals diversified considerably until the eighteenth century, when another practically oriented group of statesman and civil servants emphasized even more economic issues and the military. They distinguished themselves from their late Ming colleagues by referring to another term from the *Great Learning*: "ordering the world" (*jing shi*).[12]

The viewpoint on *qi* and *li* marked out an intellectual affinity, whereas references to classics functioned as mottos, specifying joint aims or methods. Scholars could still plough different fields or disciplines: one engaged in botanical studies, while another inquired into philological issues. Some scrutinized material objects, others heavenly bodies to "order the world" or "broaden their knowledge." They all hotly debated the methods of gaining and maintaining, preserving, and passing on knowledge. A large proportion of scholars dedicated their literary energy to reordering and newly classifying the huge range of new and old topics of intellectual inquiry. Some gathered knowledge in encyclopedias or compiled diverse philosophical essays. Others refused any literary engagement, concentrating on action and practical performance. Some insisted on text-based studies and the authority of classical sources, others contemplated the mind. Still others analyzed all of these issues and their relation to each other, reading books and observing the world inside and outside the individual mind. Benefiting from the freedom and challenged by the insecurities of their time, intellectuals became hard-edged and daring, but they never considered their choices arbitrary.

Qi and *li* are two basic concepts with a long tradition in Chinese philosophical approaches to natural phenomena and material inventiveness. Thinkers saw *qi* sometimes opposing *li* and sometimes complementary to it. Combined, *li* and *qi* explained in the regulative sense the problem of existence and in the normative sense how things and affairs "ought to be." For Song *qi* was an intellectual rather than a materialist issue (in the modernist sense of the word). Song's choice of *qi* was ethical-political as well as metaphysical. This makes him part of a long historical trajectory of Chinese thinkers who, speculating on nature, felt themselves in opposition to contemporary political and philosophical ideals. It also makes him part of a seventeenth-century trend in which exceptional thinkers favored *qi* over *li*, thereby opposing the Ming school of Cheng-Zhu learning. Cheng-Zhu learning synergized the thought of the eleventh-century thinkers Zhu Xi, Cheng Yi, Shao Yong, and Cheng Hao, and promoted Zhu Xi to its most dominant figure. As part of the curriculum for the state examination, Cheng-Zhu learning represents a major reference point for Ming scholarly tradition. Put very briefly, the Ming school of Cheng-Zhu learning presumed that *li* manifested a universal principle, and *qi* was the particular, existing only a posteriori to *li*. Late Ming scholars increasingly challenged this distinction of *qi* and *li*, expressing their dissatisfaction with state, scholarly, or societal developments. A number of individuals broke away from the accepted school of *li* thought and gave priority to *qi*. Song Yingxing belonged on the fringes of this group as he insisted on the complete irrelevance of *li*.

Approaching Song's effort from the perspective of *qi* studies gives due regard to the epistemic setting of seventeenth-century Chinese knowledge culture, its terminology and assumptions about the study of natural phenomena and material inventiveness. It opens perspectives on questions of intellectual authority and reveals that claims to truth and universality in this intellectual world were the effect and result of both long tradition and the individual approach. I consider "knowledge" and "understanding" here in terms of David Bloor's definition of "those beliefs which people confidently hold to and live by," and "which are taken for granted or institutionalized, or invested with authority by groups of people." It is against this background that I display Song's particular view of natural phenomena and technological contents.[13]

As mentioned in chapter 1, Song located his field of knowledge in "the inception of things and the accomplishment of affairs" (*kaiwu chengwu*), advertising it in the title to his work on crafts. Identifying the world as one of change and transformation, he can be grouped with a whole range of intellectuals who used this quotation from the *Book of Changes* to represent a comprehensive perspective with which to approach knowledge. A branch of this discipline concentrated on assembling the origins of natural phenomena and cultural achievements (*wu yuan*). Ordering the world, they gave things and affairs a place and a history.[14] This approach was sidelined by historiography, but it can be traced continuously in Chinese culture from the seventh to the eighteenth century, reaching a quantitative peak of literary production at the end of the seventeenth century. Another important term under which scholars approached natural phenomena and material objects has already been mentioned above: the "investigation of things and broadening of knowledge." Benjamin A. Elman discusses the new heights it reached in the seventeenth-century Chinese world when Western *scientia* entered Ming court culture and there was an increase in the popularity of Wang Yangming learning with its concept of "innate knowledge" (*liangzhi*).[15] The "investigation of things and broadening of affairs" was related to *li*, whereas "the inception of things and accomplishment of affairs" was referred to by thinkers of *qi*. How they related to each other is open to debate, but what is clear is that both approaches were a reaction to the growing variety of things and affairs intruding on the literati world.

The cornerstones of *qi*-concepts were *yin-yang*, the five phases, and resonance (*ganying*). The following short historical survey unravels their major threads and the stipulations influencing Song's individual interpretation. A dramatic disruption in the third century BC is central to the multiple cosmological interpretations in Chinese writings regarding *qi*. The first Chinese emperor, Qin Shihuang (259–210 BC, reign 247–210 BC), unified the region between the Yangtze and the Yellow rivers in the year 221 BC at the expense of intellectual variety: he banned most schools and burned many documents of the various former feudal states during his fifteen-year reign. Generations of Chinese scholars have blamed Qin Shihuang ever since for this break in the line of intellectual transmission and for leaving a landscape of fragmentary wisdom to the subsequent Western and Eastern Han Dynasties. Thinkers and statesmen of the second century, such as Liu An (179–122 BC) and Dong Zhongshu (179–104 BC) could refer only to fragments of earlier texts. They used excerpts from these formerly independent cosmologies and employed *qi* to systematically erect a philosophy that linked state, humanity, and cosmos. Liu and Dong established the basis for a new and influential line of thought. Around AD 350, *qi* was identified in the realms of physical conceptions as opposed to the mental levels of existence. Liu's and Dong's deficient text sources gave later scholars the opportunity to label their approach a reconstitution of the originals and identify previous interpretations, supplements, and commentaries as false or misleading.

From the very beginning *qi* discussions spanned a large topical and ideological field. In the course of time thinkers stressed different qualities of *qi* according to the focus of their interest. In the field of investigating one's body and natural surroundings, the ensemble of *qi, yin, yang,* and the five phases developed its own content and context. It combined the physical and the metaphysical, the body (*ti*) and heart-mind (*xin*) into one force and explained the world, rationally and poetically. Describing both its sense and function, Nathan Sivin explains in his chapter on "the fundamental issues of the Chinese sciences" in *The Way and the Word* that *qi* "bridged the transition from humanistic thought to state cosmology and then to distinct physical sciences . . . "[16] *Qi* indeed became a distinguished and fundamental medium in fields of knowledge that we identify as scientific: astronomy, alchemy, life sciences, medicine in particular, and meteorology. The terminology itself, however, resisted exclusive or selective stipulations. *Qi* remained eligible in all realms of human thought.

Two key terms in discussions on *qi* were *yin* and *yang.* Most intellectuals understood them as interacting forces of one *qi* operating in the universe. They created things and accomplished affairs either by complementing each other or as oppositional forces. Their synergy explained the beginning of all being, its variety, and its continuous ever-changing existence. This definition was given by Liu An in the *Huainan zi* (*The Master of Huainan,* 120 BC).[17] The concept of *yin-yang qi* embraced a unitary vision of heaven, earth, and human beings. It made all things subject to the same basis, *qi*. Therefore all things resonated with each other, including man with the world and vice versa. Harmony resulted from the balance between *yin* and *yang* forces. As complements, they joined to constitute the void (*xu*); opposing each other, they formed the earth (representing *yin*) and heaven (representing *yang*). In this sense the phenomena of *yin-yang qi* manifested a universal order that nature displayed and man enacted whenever he engaged in material production. Man had to understand and follow this universal order to bring prosperity to the world and be moral. Ignoring these principles created disaster and chaos in man's world, political life, and society. Song drew his moral interpretation of *qi* thought from the *Discourse from the States* and the *Commentaries of Master Zuo Traditions,* the latter being chronicles of the Lu state. Compiled around the fourth century BC, the *Discourse from the States* illustrates a political disaster in the period of the western Zhou dynasty (traditionally 1100–711 BC). In this era, the region that was later unified under Chinese rule was politically fragmented. The text suggests that the claimant to legitimate rule, the Zhou state, was doomed to collapse because it disregarded the order of *yin-yang qi.*[18] Song considered his era in a similar situation and his use of quotations from the *Discourse from the States* implies that he was following a tradition that saw sociopolitical and metaphysical issues as one and the same thing.

These classics were, therefore, an important point of departure for Song's epistemological approach to *qi.* As a mandatory text in the examination system, the *Commentaries of Master Zuo Traditions* established the notion that *yin* and *yang* were in congruence with the first two of the six heavenly *qi* forces (*yin, yang,* wind, rain, darkness, light), which made up the basic fabric of existence that "descend and produce the five tastes, go forth in the five colors, and are realized in the five notes." *Yin* and *yang* were foremost, as they actively produced rain and wind, darkness and light and obscurity as well as all other phenomena on earth and in heaven. Sage kings "carefully imitated (*ze*) the heavenly relations and analogies" in forming the ceremonies by which man ordered his world.[19] The six *qi* stood in relation to the general axiom of the five phases of *qi* (*wu xing*), which were water (*shui*), fire (*huo*), earth (*tu*), metal (*jin*), and wood (*mu*). The five phases were based on cosmological approaches reaching at least as far back as the fifth century BC. Such views conceived change in the light of a "Supreme Ultimate" (*tai ji*), which, through the medium of *yang* and *yin qi,* set in motion the five agents of cosmic evolution. These phases then yielded the production of the "myriad things" (*wan wu*) in the world. By Song's time the concept of the five phases had matured to a sophisticated theory used by scholars to explain the relationship between man and cosmos, human body and environment.[20] It had also developed into a major classificatory scheme to explain the particularities of physical phenomena as well as the interdependence and correlation of these processes. The relationship between these *qi*-phases had become essential to understanding and describing the order of the universe.

The concepts of the five phases and the relationship of *yin* and *yang* in the context of *qi* display a complexity that withstands any easy translation into modern scientific theories. Comparing the reception of *yin-yang qi* in Chinese thinking with the Greek concept of atoms or to notions of generative energy or a mechanical balancing of natural forces is illustrative, yet it is ultimately unsatisfactory. What then was *yin-yang qi*? Chinese culture understood *yin* and *yang* basically as types of *qi,* not as substances. For Chinese scholars, *yin* and *yang* could not act as independent forces. They were mutually reliant parts of one body of *qi* governed by a cosmological principle. Nathan Sivin describes "*yang* and *yin* as scientific and medical concepts [which] were precisely that: x and y. They were abstract foundations upon which a metaphysic could be distilled out of the multiplicity of physical situations, a metaphysic that remained applicable to all of them . . . *Yang* and *yin* are best considered the active and latent phases of any process in space and time. 'Latent,' 'reactive' and 'responsive' are better English counterparts of *yin* than 'passive,' since *yin* not only accepts *yang* stimuli but responds to them. This response is as important as the stimulus in bringing about change."[21] Each intellectual took his individual stance on *yin-yang qi* and wove an intricate network of functions and characteristics. As mentioned earlier, scholars following Zhu Xi prioritized *li* as the leading principle of order. For this group, *qi* emphasized the unity of things, while its duality of *yin* and *yang* expressed differentiation and individualization on the anthropological as well as on the cosmological level. This interpretation allowed the world to be in an incessant state of change between *yin* and *yang* that created all things and affairs and still was unified as a phenomenon of *qi*. Zhu Xi thus suggested that "*yin* and *yang* are one identical *qi*. The retreat of *yang* is creation of *yin*; it is not that as soon as *yang* retreats, a separate *yin* is created from it. One can look at *yin* and *yang* as one or two facets (*yi*). As two facets it can be distinguished as *yin* and *yang*, [seen] as one it is simply waxing and waning."[22] Song agreed with Zhu Xi that *yin* and *yang* were two facets of one *qi*. But he denied Zhu Xi's allocation of a minor status to *qi*. He assumed that the unity and its two facets both composed the cosmological system and made it operate. Song hence saw the world as an organism of *qi*, a dynamic system in flux that yearned for the harmony of *yin* and *yang* in *qi* and existed as a continuum of change.[23] Song further argued that because of the universality of *qi*, all things and affairs were interrelated and because of the complementarity of *yin* and *yang*, everything acted and reacted to each other in synchronization, and therefore returned to harmony. Interaction was spontaneous and not caused by an external agent. Following traditional interpretations, Song defined this nexus as the tenet of "stimulus and response" (*ganying*). He authorized his notion of this functional feature with a reference to *The Master of Huainan*. According to Charles Le Blanc, for *The Master of Huainan* "stimulus and response" (which Le Blanc also translates as resonance) was a rational device with which to understand the universe as a totality, man being part of that totality.[24] Song's view derived from these sources, and yet, as the following chapters will show, he considered "stimulus and response" a fundamental behavioral feature of the cosmos, not just an explanatory model.

In sum, Song considered *qi* as omnipresent throughout time and space; hence cosmological order did not have an end or beginning. Things and affairs, tangible and intangible, material or spiritual phenomena were manifestations of the interaction of *yin-yang qi,* which constantly tended to keep the system in balance, either by complementary junction or by balanced opposition. Song suggested, because all things and affairs originated in *qi* they were also contingent on it. This ensured both harmony and responsiveness. All observed things, from ceramic pots to the heavenly bodies, consequently became an exemplification of the order of *qi,* fundamental to the functioning of the whole.

Song's propagation of *qi* as the ultimate basis of everything was based on a tradition of thought at least two thousand years old. His approach did not appear *de novo.* Our interest lies in tracing precisely where he followed his predecessors and where he deviated from the traditional path. This divulges the intellectual constructs of trust and reliability Song implemented. Using Stephen Shapin's identification of trust in the knowledge-making process as the subtle result of a historical-social process with a personal dimension, I place Song within the community thought style, distinguishing actual commonalities from historiographic constructions, both in classical Chinese historiography and modern historical research.[25]

The Historical Reception of *Qi* and Thinkers of *Qi*

Intellectual lineages are one of the many threads historians follow when they research how knowledge was made and authorized at a specific time and place. They contextualize the thinker with his time and culture. Lineages must, however, also be understood as a subtle means to construct myth, tradition, and the identity of community; they are products mirroring a combination of historical and contemporary, communal and individual interests. From the historical perspective, linkages and commonalities also emerge that the individual himself may have deliberately rejected or unconsciously neglected. Therefore it is helpful to start with a clear view of the historical actor's categories: When and how did the protagonist see (or propagate) himself as part of a trend? What were his actual motives to associate or distance himself from others and which were their social or political implications? Did the protagonist identify specific affiliations or general rivalry between contemporary or historical groups of thinkers and where did he position himself?

In seventeenth-century Chinese culture, scholarly attitudes toward intellectual bonding were strongly influenced by a diversification that was a reaction to the state's attempts to bring the scholarly world in one line that followed Cheng-Zhu learning. This step led to much discontent among scholars. Some scholar-officials welcomed it as a way to disseminate righteous learning. Others found that the institutionalization distorted Cheng-Zhu learning or that it suppressed other fruitful lines of thought. They were concerned that it obstructed the scholarly quest for knowledge. Followers and opponents established schools and academies and promoted scholarly circles and literary communities. The commercialization of printing was conducive to these developments: the wide availability of written knowledge challenged former ideals of the pupil-teacher relationship, enabling self-teaching or retreat from public life without the loss of academic connections. In the late Ming intellectuals surfaced who rejected any intellectual linkage or institutionalization of thought. A select number of this group saw themselves as thinkers who addressed knowledge in terms of *qi*: Song belonged to this group. In lineages of thought with regard to Song, one of the major *qi* thinkers deserves special mention: Zhang Zai, who formulated his philosophy of *qi* in the eleventh century. Zhang's legacy exerted a significant effect on the following generations of *qi* approaches, because all Chinese intellectuals up until the twentieth century lived under the dominating influence of Zhang's notion of *qi*, although they did not all mirror it.[26] Even today, historical views on *qi* thought are obliged to refer to this lineage, if only perfunctorily. This tarnishes our view of the individual's intellectual orientation and his view on *qi* lineages.

For late Ming intellectuals such as Song, Zhang's theory was deeply embedded in state-promoted Cheng-Zhu learning. After 1368, when Zhu Xi's interpretation of the classics had become mandatory in state examinations, Zhang Zai's theories had also received imperial sanction, but only in Zhu Xi's reformed and abridged interpretation. Whether one favored the original, examined the discrepancies between Zhu and Zhang, or rejected Zhang Zai altogether, any view on Zhang Zai's theory had both sociopolitical and intellectual implications. Rhetoric was subtle, as original thinkers walked a tightrope between sociopolitical necessity and personal conviction. For Zhu Xi, Zhang Zai's ideas served mainly as a supplement. He synthesized Zhang Zai's interpretation with the concepts of the principal northern Song thinkers, the Cheng brothers Yi and Hao, Shao Yong, and Zhou Dunyi. Drawing on Zhang, Zhu described the formation of the world in stages from the original unformed *qi*, to *yin-yang*, the five phases and on to heaven, earth, and the myriad things. And yet, while he set forth a comprehensive philosophy of cosmic and human creativity on the basis of Zhang Zai's *qi* theory, Zhu Xi considered *li* superior to *qi*. In accord with the Cheng brothers, Zhu Xi saw *li* as a transcendental universal that existed outside *qi*. For Zhang Zai there was nothing outside *qi*, with the same *qi* going through a continuous process of condensation and dispersion to form the universe.

Zhu Xi's theory elevated Zhang's interpretation of *qi* to the status of a tenet for its followers. But at the same time, it denigrated *qi*, by subordinating it to *li* as the structural principle: *qi* was an a posteriori issue to *li*.[27] By Song's lifetime, Zhang Zai came thus delivered in a subtly contaminated package of Cheng-Zhu learning. The incorporation of Cheng-Zhu learning into the scholarly curriculum almost completely subjugated other traditions of *qi* thought, for example, the Daoist conventions. This had the side effect that idiosyncratic approaches such as the one offered by the late Han thinker Wang Chong (AD 27–91) also received little attention. Any individual's approach to *qi* at the end of the Ming dynasty was confronted with a heavy theoretical impediment: it was forced into or taken out of a linear trend culminating in Zhang Zai's theory of *qi* and Zhu Xi's interpretation of it. Furthermore it was demoted to a marginal sideline of *Cheng-Zhu* learning.

Modern historiographic tradition has long mirrored this attitude, especially when looking at scientific fields of inquiry, ignoring breaks in the transmission and scholars' explicit abstention from making linkages to their predecessors. In the late 1980s, when Taiwanese and mainland Chinese scholars, historians, and philosophers, thinking of Song primarily as the author of the *Works of Heaven*, first glanced at Song's *On Qi*, they quickly incorporated him into a late Ming philosophical trend that connected practical learning with Zhang Zai's *qi* thought and saw itself continuing the tradition of the Song dynasty.[28] But in contrast to other late Ming and early Qing thinkers engaging with *qi* (and reacting in one way or the other against the newly popular Yangming school of the mind) such as Wu Tinghan (1490–1559), Yang Dongming (1548–1624), Lü Kun (1536–1618), or Fang Yizhi, Song did not openly refer to Zhang Zai either to legitimize his theory or disagree with him. Unconsciously or consciously Song must have relied on his predecessor; after all, Zhang Zai's theory was part of his educational background.[29] However, I suggest that Song's practice of never openly using Zhang Zai to authorize his viewpoint was intentional. It contained a message in itself: he insisted on a free rein to follow his own thoughts and create his own knowledge directly from the original classics or through personal inquiry into nature and material inventiveness.

Song entirely dismissed *li*. As this did not align with the Ming dynastic sanctioned view of *li* and *qi*, Song was making a philosophical as well as a political statement: he not only disagreed with Zhu Xi's interpretation of Zhang Zai, but even more radically with Zhang Zai, whose original thought also tackled *li*. Song thus rejected the Ming state orthodoxy, and its schooling traditions. By connecting back to the original texts, Song furthermore emphasized his discontent with his dynasty's idealization of Song dynastic thought. Song placed himself well outside the acknowledged *qi* tradition. When he refused linkages to other *qi* thinkers of his era, he claimed that the truth lay beyond the philosophical sophistication of his colleagues in the actualities of life and nature. In his quest for knowledge Song did not admit to any scholarly crossbreeding or shared ideas and concepts. As an intellectual, Song felt only he approached truth in *qi*.

How useful, then, is the practice of taking *qi* as a common denominator to align traditions or legacies of thought over broad time spans, or even to group thinkers within a generation, as historical surveys often do? What were the ties between *qi* thinkers beyond a common interest in *qi*? Curiously enough, from a historical perspective it is Song's intellectual attitude (not his approach to *qi*) and its sociopolitical implications that make him part of a lineage and indeed an archetype of *qi*-thought: starting with Wang Chong in the first century, continuing with Zhang Zai in the eleventh century, up until Song's contemporary, Wang Tingxiang, and the eighteenth-century intellectual Dai Zhen (1723–77), thinkers of *qi* were, as Li also remarks, suspicious of ideological bonding and opposed the leading doctrines of their times.[30] They stood, deliberately or under protest, on the margins of social, and political power. And while their outsider position may sometimes have been the result of their personal interest in the investigation of practical matters and natural phenomena, it seems that by the time of the Ming, a scholar's identification with *qi* was more an expression of political conviction than a philosophical concern.

Intellectually, many claimed to draw their knowledge directly from primary sources, the ancient classical texts, and they insisted on individual experience, appreciating "events known through seeing and hearing" (*jianwen zhi shi*).[31] Some historical *qi* thinkers were individualists to the extent of being intellectual recluses although few of Song's generation were that radical. Many were also much more subtle and selective in their abnegation of *li* and previous thought than Song. Yet, like most *qi* scholars, they all claimed to pursue a "pure" quest for knowledge untainted by career ambitions or ideological bonding. Indeed by the seventeenth century this intellectual attitude had become so popular within *qi* thought, that from a historical perspective Ge Rongjin and Zhang Qizhi even suggest these scholars formed a school of "practical learning" (*shi xue*) which they contrast to the Cheng-Zhu "school of principle" (*li xue*)—also called the "school of dao" (*daoxue*). The historical actors would have fervently opposed Ge's and Zhang's anachronistic reading: their epistemological attitude rejected any formalized or associated form of learning. This effectively negates any retrospective identification of a "school" (*xue*). The discrepancy between contemporary seventeenth-century attitudes and the historical viewpoint throws an interesting light on the construction of intellectual lineages in general and *qi* thought in particular: what late Ming *qi* thinkers had in common was only the term *qi*—they shared a common philosophical denominator. Beyond this they interpreted *qi* in such diverging ways that it is often hard to believe they were talking about the same thing.

Does this imply, therefore, that scholars like Song did not draw from each other and always started from scratch? Indeed, some associations that seem likely from a historical perspective collapse on closer examination or are revealed as slightly distorting the point. Dai Nianzu, for example, identifies Song's treatise *On Qi* as a manifesto for acoustical knowledge and thus places him in a line of various thinkers who contemplated sound and *qi*, such as the third-century eccentric Wang Chong.[32] In fact, Song neither knew nor would he have been concerned about Wang Chong's ideas on sound. Most of Wang Chong's writings lay dormant until their rediscovery in the nineteenth century.

In assessing the historiography of Chinese scientific and technological thinking by the standards of modern issues of science, a huge variety of shadowy and dubious similarities have come up that obstruct our view of the links between thinkers such as Wang and Song. Neither Song nor Wang Chong were genuinely interested in acoustics. Still they shared a legacy of thought when they discussed *qi* and sound phenomena: Both were primarily interested in identifying the relationship between heaven and man and approached this question within the rationale of *qi*.[33] Sound was in their view the natural phenomenon verifying that heaven did not interfere with human affairs. Both opposed irrational beliefs that made heaven the prime arbiter of human morality and mystified natural phenomena. They drew on a common basis of texts, even if they were separated by more than fifteen hundred years. Despite their different interpretations, they both located reliability and truth in ancient sources such as the *Book of Changes, The Master of Huainan* or the *Discourse from the States*. Only on this basis can a line be drawn between Wang Chong and Song. This line can be extended to many other thinkers, including some who were influenced by Western ideas: Wang Fuzhi, for example, approached *qi* in this manner, adapting new ideas effectively within indigenous traditions.[34]

These are some of the original factors that define a tradition of *qi* thinkers and Song's individual place within it. They reflect the agencies of the historical actors. In the Kuhnian prescription of historiographic approach, I consider evaluations of knowledge consistency historically and individually dependent, with no objective everlasting value.[35] An assessment using modern terms of science and technology helps identify discrepancies as well as alternatives; it does not, however, help us understand the original thought or the huge range of culturally specific scientific or technological knowledge and how it came about. Looking at the actor's agency is significant for the interpretation of the scientific knowledge contained in writings such as Song's. For example, Song's writing may lack—from our modern view—an explanatory context for the understanding of sound. But this was not Song's aim. His aim was to reveal human-heaven interconnectedness. From this perspective his account was both systematic and accomplished. Establishing *qi* as the ontological concept that explained how and why sound is produced, he employed sound as one of the many phenomena explaining what *qi* was in his view. The technical contents in the *Works of Heaven* have to be looked at in the same way. When Song detailed how to calcinate minerals, he did so to exemplify the performance and logic of *qi* and the ultimate principle of change behind it, not to explain why or how mineral stones could be calcinated. From this perspective Song's view of truth and knowledge become coherent and are in line with his era's claims to universality and objectivity. Thus we can, as Bloor suggests, distinguish the discriminatory, material, and rhetorical functions Song assigned to truth and knowing.[36]

Song's approach to *qi* exemplifies the epistemological ideals that shaped a seventeenth-century Chinese approach to fields we nowadays call scientific or technological. For Song, *qi* was a universal theoretical model that explained all being. It could comprise seemingly contradictory qualities, but it was nevertheless and above all, still *qi:* It could be, act as and have the quality of being energetic or at rest, it could materialize and suspend, fuse and disperse. And yet, in all cases *qi* was never discharged from its essential quality of being *qi.* Therefore Song never considered *qi* to be a yet-to-be-defined energy or matter conception. To him, *qi* was not material but the materiality of *qi* was behind everything. It was not energy, but explained energy and its effects. For Song, the universal validity of *qi* could furthermore not be fragmented and thus he selected a terminology that maintained this universality. In this regard Song's notion of *qi* was that of a completed scientific concept through which one could approach the world that surrounds us.[37]

The Role of a Scholar: Ordering the World of Heaven, Earth, and Man

Investigating *qi* was Song's remedy for a world of disorder and his authorization to engage in all kinds of knowledge fields. But how did he legitimize himself, a minor scholar with neither influence nor position, as the one to bring forth order? Song mentioned that at one point in time a great thinker—it is unclear to which period and person he refers—had proposed that chaotic seasons gave rise to a remarkable number of remarkable thinkers with orderly ideas, whereas "at the height of order, thoughts and ideas are in disorder."[38] Intriguingly, for Song, the flourishing of ideas, including his own, was the ultimate proof of decay. He would thus have agreed with the modern historian Timothy Brook who cites the *Works of Heaven* as a sign of the blossoming of ideas. At the same time, Yingxing would have fervently opposed Brook's historical identification of the period between 1550 and 1642 as flourishing.[39]

Using the correlation between chaos, order, and thought, Song expounded a historiographic construction: Over time, people had stretched this idea to its boundaries until they developed the belief in a cyclical reemergence of great sages who would help to order the world whenever society fell into decay. Thus they waited apathetically, instead of searching for solutions.[40] Most late Ming thinkers would have agreed with Song's skepticism about the fatalism of their time, as they also believed "that what creates [man's] destiny lies within ourselves."[41] But they set up an a priori construct of morals, which Song denied. For Song order was inherent in any natural or human activity: knowledge and action were hence congruent. Knowing this congruence caused morality, whereas morality could not, in Song's view, instigate knowing. Song's logic made man the arbiter who controlled his own fate and his world through "knowledge" alone. He thus presumed that "chaos and order are unavoidably instigated by human behavior (*renwei*)," namely, by man acknowledging and enacting or neglecting and thus going against the order of the universe.[42]

Song openly ridiculed those colleagues who waited for a sage, instead of taking needed action. "The Confucians say that in the correspondence of celestial events with the affairs of humans, solar eclipses are the most important. Using the eclipses [as a threat], these officials warn the emperor not to be selfish. If I take on this accordance with heaven I can only say: If it is true that there will be a eclipse whenever the emperor is weak and the officials are too strong, why is there none now?"[43] With this remark Song objected to his colleagues' belief they could read morality from heavenly patterns and tie their findings to human events without understanding the principles. Seventeenth-century court intellectuals insisted that disorder in one part of the relationship, whether heaven's or man's world, brought disorder to the other. This thought was rooted in the idea that with respect to morals and ethics, nature, heaven, and human society formed a whole. The emperor, the son of heaven, received his mandate from heaven and performed rituals in accord with heavenly principles. Early thought had given this relationship a sound and pragmatic basis; dividing the earth into areas in relation to its counterparts in heaven, astronomers during the Warring States period correlated events observed in the sky with political events happening in the various states then in existence. The heavens were always in order, yet man's world fell into chaos; such chaos meant that man had misinterpreted heavenly order or failed to investigate it thoroughly. The correlation between heaven and man was hence based on knowing patterns and fulfilling one's tasks. Morality was not in decay because of the occurrence of oddities such as eclipses. Heaven revealed that morals were in decay. It was, however, not connected to them. Thus even when no eclipses occurred, man's world could be in disorder. It also meant that eclipses were thus not bad omens per se; rather it was human disregard of the appearance of eclipses and comets that was dangerous.[44] Throughout the subsequent centuries both astrology and astronomy thrived on this fertile ground, attempting to provide continuity and unity between man and Heaven.

In his treatise *Talks about Heaven* Song blamed state officials for hypocritically putting heaven on a pedestal, against basic common sense. How could officials, mindful scholars, be so irrational as to suggest that heaven could not be reached or contacted, while on the other hand maintaining a huge state apparatus of officials, grand chancellors (*taishi*), and astronomical (*xingguan*) and calendaric officials (*zaoli zhe*), whose main task was to communicate with heaven and interpret its ways in the constellation of the stars?[45] Critique of a seeming paradox between ideology and behavior, knowledge and action, is a recurring theme in Song's works. Song considered this deliberate malice on the part of his colleagues. For him, it was an affair of honor, his obligation as a knowledgeable scholar, to reveal the truth about human-heaven interconnectedness in *qi*. Following his lead the following passages demonstrate how Song in *Talks about Heaven* tore off the morally tinted clothes with which his colleagues had dressed up their heaven. Devoid of any philosophical attachment, Song suggested that there was no spiritual component in the manifestations of Heaven: Heaven was a manifestation of the universal rationale of *qi*. To substantiate his claim, Song analyzed the heavenly phenomena in his treatise *Talks about Heaven*, artfully combining the physical sky with the display of ultimate cosmological principles to expose heaven as a phenomenon of *qi*.

THE TRUTH IN HEAVEN AND THE ORDER OF *QI*

> Since [I] am breaking a taboo when I divulge the secret of heavenly warnings and providence I will suffer from ridicule, thus I turn my back on Confucian speeches. However, I am not fearful in haste and agitation [to discuss such matters]. I wish to leave this writing for future generations, and I plead that knowledge is already in front of our eyes!
> *Talks about Heaven*, preface, 99

White and perfectly round, pearls are conceived, in Chinese tradition, as products of the night, created by the luminescence of the full moon, shining only by moonlight. Originally a product of the southern seas, pearls became increasingly familiar to the Chinese in the second century AD. Stories, legends, and fantastic notions of their origin spread throughout the country. The creation of pearls was assigned to sea beasts and dragonlike creatures. Harvesting the oyster's precious nucleus, pearl divers knew about the pearl's slimy origin. Assisted by a cord around their waist and weighed down with stones, they attached curved tin pipes to their noses and mouths and dived into the deepest sea to reach the best and most beautiful exemplars, pearls labeled poetically "bright moon" (*mingyue*) and "light at night" (*yeguang*; see fig. 2.1).[46] Each year, according to the gazetteers toward the end of the Ming, the boats had to travel farther and farther out to sea and the divers had to search ever deeper to find enough pearls for the growing demand.[47] Dipping and dragging stone-loaded net-bags over the sea bottom was less risky for the diver and in the short term it brought higher yields. Yet many coastal folktales recount the common people's view of this practice, reminding all who would listen that the balance of nature was delicate; vandalizing the oyster colonies would destroy the prosperity of all future generations. Song considered this point, too, warning his reader in the section on pearls and gems in the *Works of Heaven* that oysters should not be gathered too often. He explained that pearls were obscurities of the transformative forces of *qi*. The crafting of these materialized paragons of perfection out of nonmaterial qualities (*wu zhi*) of *qi* was slow and subtle. Thus, he submitted, they needed time to float around with open shells and absorb the *yin* quality light of the harvest moon.[48]

Heaven was for Song an observable phenomenon of *qi*, a sky-heaven working in a complementary relationship to the earth. Song presumed that heaven "brings forth the phenomena (*xiang*) of the two *yi*."[49] The two *yi* are the *yin* and *yang qi* evolving from the Supreme Ultimate. In Chinese cosmology this refers to the celestial bodies of the sun and the moon. Discrediting the Confucian creation of a ritual heaven, Song saw heaven and all its phenomena as products of *qi*: stars, the body of the sun, and the moon nurturing pearl production. The sun was accumulated *yang* and the moon accumulated *yin* (*ji*). Accordingly, Song explained, the pearl literally absorbed the *yin* quality of the moon's light to create "substance (*zhi*), where formerly no substance existed (*wu zhi*)."[50]

For Song heaven was hence the product of a rational, comprehensible order. It was important for humans and should be venerated because it could be observed and knowledge could be gained from it. It did not, however, actively intervene in human affairs. Verifying his argument, Song concentrated on structural issues, refraining from any computational or astronomical inquiry. Christopher Cullen has pointed out that astronomically speaking Song's approach lagged behind his time.[51] I suggest, however, that Song consciously avoided contemporary cosmological models or astronomical theory because he had quite a different agenda: he actively distrusted any kind of indoctrination. True knowledge, he claimed, was always universally valid, obvious and comprehensible to any open-minded scholar. He noted: "Heaven has a clear plan which brings out the phenomena (*xiang*) of the two, *yin* and *yang*, the sun and the moon, and fears only that mankind will take no heed of it. [The plan is clear throughout,] from the head to the tail, it has an origin from which it starts and an end to which it proceeds. It inspires mankind to observe and investigate. Hence, if the plan is clear, but the three systems of belief (*san dao*) are still seeking illumination about the earth's center and the four directions, then their stupidity and delusion is glaringly obvious."[52]

In their delusion, Song argued, astronomers following their theories on the heavenly patterns were no better than the clerics who mythologized heaven as a spiritual force when it was simply an observable phenomenon—the sky: "Those who discuss what is under heaven in confusion without reconciling their contradictions are as detestable as those who extend the heavens up to the thirty-three heavens [as Buddhist sayings imply]."[53] Believing in a world of *qi*, Song in the preface to *Talks about Heaven* openly attacked the three major traditions of Chinese cosmology, the theory of the "covering sky" (*gaitian*), the "enveloping sky" (*huntian*), and "the dark night" (*xuanye*), dismissing their structural approach as unreasonable. He opposed the earliest cosmology of the covering sky that saw all celestial bodies as a rotating umbrella-like heaven above the earth. Within

Fig. 2.1*a–b*. "Boat for the collection of pearls following the waters" (*yanshui caizhu chuan*). Double-page image. The divers are depicted with baskets for the pearls and apparatuses that functioned as snorkels. The sailor to the left tries, as the caption in the illustration explains, "to counteract whirls [by throwing] straw mats on them" (*zhijian yuxuan*). *Works of Heaven*, chap. 3, no. 18 *Pearls and Gems*, 56a/b.

没水採

this theory sunset and sunrise were considered optical illusions caused by the distance between observer and object. According to Sun Xiaochun and Jacob Kistemaker, the covering sky cosmology assumed its most refined form in the *Zhoubi suanjing* (*Arithmetical Classic of the Gnomon and Circular Paths*, approx. 300 BC). This divides the sky into six circular belts arranged inside seven circles (*qiheng liujian*), with the celestial bodies moving within these belts.[54] Song also denounced the conjecture of the enveloping sky. This model conceived the sky as a celestial sphere (*tian yuan*) surrounding the earth (*di fang*), which explained the motion of the sun and the stars by saying they were attached to the inner surface of the celestial sphere, which rotated once daily about an inclined axis.[55] Court astronomers used a celestial globe to illustrate the daily motions of the sky.

And finally Song was also dissatisfied with the darkness of infinite space (*xuanye*) model.[56] This school illustrated the heavens as an empty void and an infinite universe with planets made of condensed void. The celestial bodies of sun, moon, and stars were suspended, floating freely. The earth lay beneath the universe and was neither flat nor round but had infinite depth.[57] This theory was delivered to posterity only in fragments. Astronomers hardly used it, but Zhang Zai (and thus also Zhu Xi) addressed it in his philosophical approach. In this guise elements of the darkness-of-infinite-space school became relevant for Song.

Zhang Zai envisioned the earth as consisting of pure *yin qi* solidly condensed at the center of the universe. The heavens consisted of a buoyant *yang qi*, revolving counterclockwise. The fixed stars were carried around endlessly within this floating, rushing *qi*. Zhang Zai furthermore substantiated the universe as a space where *qi* was not able to accumulate, identifying this as the void (*taixu*), a stage when *qi* was in its original state (*yuan qi*). The earth was thus swimming in an ocean of *qi*.[58] Cosmologically the idea of a balancing polarity of heaven and earth as *yin* and *yang* provided Song with the conceptual frame to explain all things and affairs in the universe as complementary bodies of *qi*. Yet, it also limited his explanations to a rather amateurish and superficial analysis of heavenly features. Like Zhang Zai, Song illustrated the universe with the image of an egg—a parallel typically drawn in the enveloping sky theory. He took the heavens as similar to the white in a hen's egg enveloping the earth as an egg envelopes its yolk and suggested that like the egg white the heavens were a sphere embraced by a round-shaped exterior shell.[59] Like most of his contemporaries, Song was not interested in extending this analogy or in giving any further explanation on the shape of the earth or the heavens beneath the horizons. But he explicitly doubted his colleagues' ideas that there "could be a sun entering down into the earth [beyond the horizon] or moving to distant places when being fixed in form" or "that any body could be great enough to conceal a body like the sun."[60]

For Song, the universe constituted a bipolar space with accumulated *yin qi* forming the earth and concentrated *yang-qi* establishing the heaven with the sun at its center. In principle, all heavenly bodies, sun and moon, planets and stars, as well as phenomena such as winds, comets or the cloudy shades of the milky way were ephemeral, bound to the waning and waxing of *yin* and *yang*.

> The *qi* of the *yang* emerges from beneath and ascends. Its period is from the third until the seventh hour. Beclouded *qi* accumulates in the East. It coagulates and forms the sun. [One can] climb [the mountain] and gaze at the sun accordingly and see [this phenomenon]. And in fact the shape of the sun only forms during the sunrise. Concealed between the black *qi* golden glances smear [like silk threads]. It is violet and like rippling water. Wait for a while and it will become round. Once it is round, the day comes. From afternoon three o'clock until seven o'clock the *yang qi* diminishes step by step. If one climbs the mountains of Yadala one can observe [this process]. The white turns into red step by step and step-by-step the red turns into jade-green. It turns into confusion and disperses brilliance in ten thousand hollows. It vanishes, submerging in a moment like the ashes of an extinguished fire. How could one believe that there can be a form of the sun entering down [beyond the horizon] of the earth or moving to distant places?[61]

For Song the sun was hence newly created out of *yang qi* every day. This shows that he considered heaven, just like all other things in the world, to be subject to universal principles. It was a *qi*-phenomenon and thus not particularly related to man's action. Composed by alternating phases of *qi*, sun, moon, planets, and stars approximated stability only by periodically recurring to a similar state, yet never quite to the same thing: "To take today's sun as yesterday's sun makes as much sense as carving a notch into the boat at the place where a sword dropped into the water and believing that one could find the place again."[62] Song's dismissal of the astronomical models of the universe of his era was not for scientific or epistemological reasons. His major point in *Talks about Heaven* was that his colleagues misconceived the heavens because they assigned an ideological purpose to astronomical investigations instead of investigating for the purpose of "knowing." In this regard Song tarred all scholars with the same brush. Arguing within a contrasting cosmological model, he identified scholars from the West, by which he might have meant Indian and Arabian theorists as well as presumably the Jesuits, as "without knowledge" (*wu zhi*).[63] He claimed the ideological ideals of the Westerners interfered with the general understanding of the "heavenly way" (*tiandao*). He ridiculed Westerners (here specifying a particular Indian model of thought) who believed "that dawn depends on the distance to the sun, whereas being far away constitutes the night."[64]

Song was concerned with the primacy of his cosmological worldview, not astronomical method. In this regard he faced a similar conflict to that of the Catholic Church when they refused the new theories brought forth by Johannes Kepler (1571–1630) and Galileo Galilei (1564–1642) in sixteenth- and seventeenth-century Europe. From this viewpoint he also ridiculed the ideas of a global earth brought to China by the Jesuit missionaries, asking how they could truly consider "the earth to be a globe, located in the middle of the void/emptiness (*xuxuan*) with all things on it swarming around like ants on a carcass?"[65] On a global earth, he insisted, people would drop off, thus reassuring his reader of his experience of the world as a secure level place. Fearing that the rug was being pulled out from under him, he claimed it was irrelevant that the Westerners were good in mathematics or able to predict eclipses with great accuracy.

Song's denial of the concept of the spherical earth, placed him with contemporaries such as Wang Fuzhi, who, despite being connected to informed astronomical circles, speculated about the cosmos primarily as a framework of knowing. From this perspective Wang Fuzhi, like Song, located the earth in a universe of fluctuating features, suggesting that "since it is in some places level and in others steep, in some places recessed and in others convex, then wherein lies the sphericity? (. . .) Thus from the earth's inclines, irregularities, heights, depths, and vastness, it is clear that it has no definite form."[66] Both thinkers argued in this way because something more than scientific knowledge about the heaven-sky was at stake—their view of world order.

THE POWER OF HEAVEN—OMENS AND ECLIPSES

> For the Confucians (*ru jia*) solar eclipses are the most important element in the correspondence of celestial events and human affairs.
> *Talks about Heaven*, chap. *ri shuo* 3, 101

Enshrined as a heavenly stone, jade was a preferential cult object for imperial ritual and valued in grave furnishings. Song thought jade a product of the moon, classifying it, like pearls, as a *yin* item. Valued for its attributes of toughness, texture, sensuous properties, and attractive color, jade was a metaphor for human virtue and an expression of spiritual and earthly powers. The elite of the late-Ming dynasty employed jade for decorative purposes and fashioned it into all kinds of luxury utensils and art objects. Assigned singular properties such as preventing the human corpse from decomposition, it was an emblem of immortality and part of the search for eternal life; jade suits that encase the corpses of rulers for protection belong to the showpieces of early imperial burial sites. Imports from South Asian and Central Asian regions such as Khotan introduced stones featuring new shades and stupendous size, satisfying a growing taste for sculptural works. Song's generation was especially fond of animal figures, both real and mythical, and jade dishes with tiny, delicate inlays.

The quality of a stone was difficult to judge from the surface. To break open its crusty shell the stone carver sprinkled the stone with fine sand and then cut it with a spinning round iron disc that he turned with pedals. The sculptor then carved the jade into a rough form, choosing a position for the figure that necessitated the fewest possible cuts, carefully following the lines and gradients of the original pebble or boulder. Optimally a stone would be sawn from opposite sides and while lapidarists of earlier periods did not bother to smooth away the ridges where the saw scarfs met, craftsmen of Song's period carefully polished them away.

Manipulating hard nephrites, the artisans of this period demonstrated a high level of technology. As well as introducing many innovations, the masters among them emulated past ideals and willingly employed arduous methods from times gone by. They reproduced, for example, the *bi*-discs that the *Zhouli* (*Rituals of the Eastern Zhou*) regarded as a suitable offering to heaven, revolving a hollow reed between the palms of their hands for hours to drill holes into the middle of flat round jade dials.[67] Resembling in form a modern compact disc, the *bi*-disc manifested man's close relation to heaven. Many late Ming literati venerated them as a compensation for the loss of political power and social security that marked their era and its failing approach to the heavens.

Chinese rulers legitimized their rule from their knowledge of the heavenly patterns and their resulting assignment as intermediaries between heaven and man. Dynastic houses ritualized and institutionalized the fields of astronomy and astrology. Only specifically assigned state officials could use refined instruments to observe heaven and mathematically analyze the collected data. In the late Ming the mathematical astronomer Wang Xichan (1628–82) complained that the political emphasis given to human-heaven interconnectedness had bifurcated astronomy since the tenth century.[68] Technicians had increasingly concentrated on predicting phenomena, while scholars, deprived of advanced calculation and empirical methods beyond the naked eye, pursued theoretical speculations about cosmological patterns.[69] Indeed there was little contact between the fields. Scholars outside the court relied on common literary resources, the classics, and philosophical works for their cosmological argumentation, which they supplemented with occasional observation of the sky.[70] Song's view on heaven and his discussions in *Talks about Heaven* on eclipses should be seen in the tradition of this schism between cosmology and astronomical knowledge. Eclipses were observable to all, and they were challenging events that openly cast doubt on dynastic authority and capability. Scholars throughout the empire used eclipses to discuss the morality of rulership and eclipses also played an important role in scholarly speculations on cosmological structures. Half of *Talks about Heaven* is dedicated to this topic.

We know today that a solar eclipse occurs when the moon passes between the sun and the earth, obscuring the light of the sun. Identifying the moon as an accumulated *yin*, Song did not see the moving planets in the sky that the Chinese specialists of his era had noted. He understood the moon as another phenomenon of *qi*, a spatial accumulation of *yin* water *qi*. He identified the sun as an accumulation of the *yang* fire *qi*. Song described solar and lunar eclipses as the result of the exchange of *yin-yang qi*. Moon eclipses occurred, he explained, when the sun and the moon intermingled their two divergent energetic *qi* (*jing qi*) of *yin* and *yang*. As their natural dispositions corresponded to each other they willingly fused to the void: "The phenomenon is gloomy and retired. Its peripheral agency cannot be grasped or called by name."[71] In the case of a solar eclipse, Song also saw the essences (*liang jing*) of *yin* and *yang* fusing, explaining that the difference between the two phenomena lay in the spatial distance that they had to cross to then merge in fusion as an eclipse. Yet, from the viewpoint of a world of *qi*, solar and lunar eclipses were the same: in both cases *yin-yang qi*, longing for reunification in the void, generated an eclipse.[72]

Song defined eclipses by their diverse circumstantial parameters, arguing that solar eclipses in winter were different from those in summer because sun and moon moved at varying distances to each other. In winter, Song suggested, "when the courses of the sun and the moon are nearest to each other, the sun ascends above the moon."[73] A solar eclipse then occurred because the earth was directly beneath. In summer, however, "the rays of the sun are high above the gloomy moonlight; the distance between them is approximately a thousand miles. And at the direct opposite meeting of above and below the (*yang*) essence of the sun falls downwards and the moon (*yin*) essence raises upwards."[74] Song surmised that the essence of the pure *yin* and *yang* phases reacted according to the universal principles of *qi*, the *yin* and *yang* unified and reverted back to balanced *qi*. In his theory of *qi* Song saw eclipses as proof that the interchange of *yin* and *yang* was uniform whether it happened on earth or in the heavens. In his theory of heavenly phenomena, Song paid particular attention to the directional attitudes of *qi*. His descriptions create the image of a magnetic effect, in which an eclipse happened because an accumulation of one *qi* agency, *yang*, attracted the other, resulting in *yin qi* moving upwards to meet *yang qi* falling downwards. Both elements yearned to fuse with each other, but could only cross a defined distance. Thus *yin* and *yang* could move toward or away from each other, but eclipses only happened under particular conditions.

Song's ideas were systematic, based on a strict rationale of *qi*. Step by step he established the universe and the relationship between heaven and earth, heavenly bodies, and natural occurrences in *On Qi*, like the lapidarist cutting open the jade and carving a *bi*-disc. He trimmed away the crust of ideological purpose to polish out the true lines of heavenly order. Doing so, he addressed Zhu Xi, the leading philosophical authority who had stressed the moral implications of the heavens in relation to eclipses in a commentary to the *Shijing* (*Book of Poetry*), where he argued that an eclipse was a reminder to the kings of antiquity to be virtuous and rule benevolently.[75] Song accused Zhu Xi of being superstitious in his view of heaven,[76] evoking a number of historical examples to argue against correlating heavenly movements to man's actions in general. He even came up with a quantitative overview to demonstrate that often when a tyrant had taken over the country or an emperor was defeated there was no eclipse. He pointed to the absurdity of Zhu Xi's seeing the increase in eclipses in the period of the usurper Wang Mang (45 BC–AD 23, reign AD 9–23) as a sign of immoral behavior, noting that even more eclipses had occurred during times of benevolent rule.[77] During the sixteen-year rule of Han emperor Jingdi (188–141 BC, reign 156–141 BC), considered by historians a virtuous and enlightened ruler, Song found documentary evidence of nine solar eclipses. In contrast, during the interregnum of Wang Mang, a period of more than twenty-one years during which the state was impoverished and nothing was in the right order, only two solar eclipses were reported.[78] These statistics, drawn from historical record, refuted the idea of his contemporaries that portents were related to human action.

The theoretical approach to knowledge in the discussions on eclipses in *Talks about Heaven* is shaped by Song's ideological agenda: his aim is to produce a coherent theory of *qi* substantiated by personal observation and verified empirically by historical statistical data. Replicable facts were his method to debunk heaven's role as moral arbiter. Assigning heaven a moral role, he warned, was harmful to the state. It only encouraged credulity. Song accused his colleagues at court of shamelessness for "making the poor poets believe in something like a portent manufactured by heaven for man's world."[79] By no means did this imply, however, that heaven could be omitted: heaven was still essential because as an observable phenomenon it revealed the universal order of *qi*. Divulging heaven as a rational phenomenon, Song also identified man's role. It was man's duty to recognize and investigate this important source of knowledge and act in accordance. Song made use of an acknowledged rhetorical means in Chinese scholarly culture to model proper human activity: the sage-kings. As prominent figures of the past the sage-kings lent credence to knowledge. Emphasizing that sage-kings had invented agriculture and weaponry, textiles and dyeing, Song subtly argued that crafts and technology were indispensable within the relation of heaven, earth, and man and that truth and knowledge lay not only in the star maps of heaven but also in the mud of the mundane.

SYSTEMS OF VALUE: THE SAGE-KINGS, THE AUTHORITY OF THE PAST, AND MAN'S ROLE

> Heaven exhibits (*chui*) the phenomena. The human sages (*shengren*) follow (*ze*) them by applying the five dyes according to the five colors (of the Five *qi*-phases concept). Indeed Emperor Shun considered this very carefully!
>
> *Works of Heaven*, chap. 1, no. 4, 49a

One of the great secrets in producing dyes and paint lies in the mineral or vegetable basis of the components, in knowing how to combine them and the best medium in which to dissolve the pigments effectively. Considering the magnificently colorful art objects produced during the Ming, both artists and artisans must have experimented extensively with materials, oils, and mineral and plant extracts, thus demonstrating a keen curiosity about their inherent qualities and transformational behavior. Landscape painters of that period reached new heights of skill in depicting the infinite hues of the clouds and mists that Song observed in the sky.[80] Portrait artists captured the gaudy plumage of birds and the subtle tints of diverse human complexions. In terms of color, seventeenth-century Ming China also attained superb results when dyeing its textiles. Lavish silks, damasks, brocades, and satins with delicate textures in iridescent, opalescent colors, from aubergine to intense aqua enlivened the street scenes in the great urban centers. Robes and garments in every shade of blue embroidered with subtle multi-colored geometric patterns, peony and peach blossom competed for the viewer's attention. Sophisticated craftsmanship was required to produce the dyes for threads and fabrics in all the subtle shades consumers demanded. Color varied with regard to process and was dependent on the raw material used for the cloth and its condition. Lotus pink, peach-blossom pink, silver pink, and clear pale pink, all obtained from safflower cakes, could only be applied to white silk, while the less processed yellow silk would, as Song remarked, take on no color at all.[81] The *Commentaries of Master Zuo Traditions* stipulated the significance of color in man's world, as they displayed the cosmic order man had to follow. Man may distinguish himself from the beasts by dressing in garments and indicate his social rank through the quality of his clothes, yet it was only through the proper application of color that man showed his ability to bring his order in accordance with that of heaven.

In a hierarchal system of symbolism the five colors represented the cardinal points of the Chinese spatial worldview. As an epistemological feature they could be correlated to the primal forces of all being. The South was a red phoenix reflecting the nature of the tropical summer sun; the North was the cold arctic winter corresponding to two black reptiles, the snake and the tortoise. The East, a blue-green dragon, was correlated to the spring and the West, a white tiger, signified autumn and snow-covered mountain peaks. A yellow dragon occupied the center and balance-point of the Chinese world system. Here the emperor stood alone. In the Ming rituals on clothing he was allowed to dress in yellow and live under yellow-tiled roofs. The Ming official, who viewed the court and had to bow to him, dressed in blackish-blue. Colors and dyes furnished the ritual and representational performance of society and state. Apart from the technical considerations this moral basis justified the Ming state in making dyeing an integral part of the state-owned system of silk manufacture. Dyers stained the silken threads for the emperor's robes with an aqueous solution of boiled Venetian sumac wood and shampooed it with an alkaline solution of water-leached hemp ash. They mixed indigo and yellow berberine to dye the raw material for officials' clothing, ensuring that the severe blackish-blue gowns would stand out in the colorful throng. Color also marked the importance of buildings in the growing urban centers. Potters blended ochre, rosin and rush to glaze the tiles for the imperial palaces in the capitals of Nanjing and Beijing. They painted fired tiles with a mixture of pyrolusite and palm hairs to obtain the green color used for princely courts and sacred temple sites and accentuated roofs of administrative buildings by applying a shiny blackish coating. But while scholars and officials all knew the symbolic value of color, few of them would have given a second thought to the ingredients of dyes. Song in the *Works of Heaven,* however, drew attention to these details. He quoted the *Book of Changes,* pointing out that "heaven exhibits the phenomena and the human sages follow them."[82] In this way he reminded his colleagues that symbolic practice originated in a close understanding of universal order.[83] The sage-king Shun's veneration of color was due to his awareness of the nexus between crafts and heavenly order.

Crafts and technology, identified here as the intellectual and physical nexus of making things and accomplishing tasks, have a critical weight in the balance between man and the surrounding world. Man, exploiting the possibilities of raw materials and processes, produces unique objects and initiates events that would not otherwise exist. Cultures met this intellectual challenge in many ways, sometimes defining human efforts within and sometimes outside nature; some assessed crafts as a positive, others as a negative achievement; some considered them central, others marginal to their world: in each and every case the epistemological positioning of crafts was, however, expressive of a cosmological viewpoint and a common cultural stand on creation. It carved out the territory of man, and heaven and all other forces involved in the making of man's world and the knowledge contained within it.

Throughout almost all periods premodern Chinese literature features a refined philosophical discourse among its elite about the relation between practical endeavor and theoretical speculation, asking which value should be assigned to each or how they could be fruitfully combined for the benefit of the state and the people. Identifying man and nature as integral and contingent parts of one entity and not as self-referential issues, Chinese scholars discussed and assessed the origin of knowledge, its creation, and relation to moral action and practical statecraft and also the role of crafts. By reference to the sage-kings, literati anchored these concerns firmly within the genesis of Chinese civilization. They made clear that they addressed a cosmological order and were equally concerned about society, state, and self. The mythological figures of the sage-kings epitomized moral concerns and artisanal knowledge. They represented all issues relevant to the construction of Chinese cultural and historical identity: ethical behavior, skills, material inventiveness, military leadership, agriculture, arts and crafts, and the comprehension of natural phenomena. References to the sage-kings implied philosophical discussions about whether man had initially created or imitated, produced or fabricated things, thus specifying man's role and heavenly influences in nature.[84]

Association with the sage-kings and the past was an important method of assigning value to knowledge in premodern China. It had implications within chronological narratives of development. Within the history of science and technology the explicit recognition of and reference to advance or progress was long held to play an important role in the making of scientific knowledge in Europe. After generations of scholars had used Greek literature and Roman statecraft as a source to ground and authorize their knowledge, in the eighteenth and nineteenth century scientists increasingly advertised their efforts with adjectives such as modern, new, novel, and unique. In comparison, historians branded Chinese literature as comparatively static, in that it seemed to have consistently applied the past, its features and values as its point of reference up until the twentieth century. In actual fact, Chinese philosophy and scholarly culture often used originality, novelty, and uniqueness to advertise both products and ideas. Shifts in fashion in the practice of labeling issues as novel or old, unique or a continuation of traditions can be discerned. Song dynastic authors, for example, extolled their exam guides and literary works as unmatched and novel, while Ming dynastic authors seem to have avoided such claims. The social and epistemological specificity of such labels to a particular time and place in Chinese culture is still poorly understood. But it is notable that claims to novelty and originality were used as often as references to the past, both in propagating literary knowledge and material creations. Thus sixteenth-century book dealers carved "new revised version" on the covers of books that claimed to revive the ancient. Pharmacists claimed to have developed "improved medicines" and new unseasoned soaps, while literati doctors showed as much interest in discovering new herbs in and beyond the empire as in recovering secret old recipes or retracing lost ingredients.[85]

New and old were hence flexibly used to shape the present and set out for the future. A thorough survey (both in European and Chinese cultures) leaves us suspicious of apparent continuities within the rhetoric of fields and their chronology as such: The past served the present, it did not determine the present. The mathematician Isaac Newton (1643–1727), for example, authorized his task by including the recovery of lost wisdom of the ancients and undertook painstaking philological studies to support that enterprise. The English natural philosophers William Gilbert (1544–1603) and Robert Hooke (1635–1703) may have insisted on direct observation. But they also equally evoked antiquity and drew an analogy between political and philosophical decay. Song's scholarly tenor was similar to these innovative European thinkers. One got ahead by going back to the origin: progress through purification.[86] Both Chinese and European thinkers of the sixteenth and seventeenth centuries treated truly ancient texts as enormously valuable sources of truth about the natural world, and claimed that the originally pure sources of that ancient truth had been polluted over time. They used the past as a rhetorical tool to propagate manifold issues. For them continuation or revival were as important as progress or radical change.

Despite such cultural parallels, Chinese approaches do display idiosyncrasies and contextual bearing. The majority of scholars strongly emphasized the past in moral issues: They applied antiquity (*gu*) and history rather as an ad hoc guide to moral behavior.[87] This combination of high morals and the past was ingrained, and literati used it quite capaciously to authorize their views. They did not always specify what "antiquity" referred to. In Chinese discourses, rhetoric of the past were often expanded during times of crisis. Frederick W. Mote points to the invigoration of the movement toward the "reinstitution of antiquity" (*fu gu*) that marks the Song dynasty (960–1279), which was constantly threatened by invasion from the North. At this time "nothing could compete in value with or lend more authority to a concept, an institution or a model of behavior than to claim that it was authentically ancient."[88] This expansive and rather free use of the past was, as Mote observes, partly because "the weight of antiquity notwithstanding, the revered past was something vaguely ideal rather than literally attainable. The best minds of all ages were clearly aware of that. The best statesmen sought the spirit of the past in governing, not its literal re-creation."[89] Premodern Chinese intellectuals already critically reflected on this usage of the past by their colleagues. The historian Cui Shu (1740–1816) noted that Zhu Xi, Zhang Zai, and other major Song dynasty scholars tended to "add on to antiquity" quite generously.[90] In fact, by Cui Shu's lifetime, literati had successively added many figures to the time before the Yellow Emperor (*Huang di*). Joshua A. Fogel also points out that Chinese not only reinterpreted the past, they also deliberately introduced novel features: "While Confucius spoke of antiquity going back no further than the sage-kings *Yao* and *Shun*, by the Han dynasty Sima Qian (145?–86 BC) began his *Shiji* (*Records of the Grand Historian,* second century BC) with the Yellow Emperor; later historians went still further back to Fuxi [i.e. another legendary figure]."[91] Rather than submitting to the past, Chinese scholars thus invented it to suit their contemporary interests.

As the past was creatively extended, the sage-kings were assigned specific tasks; Shennong, the Divine Agriculturalist, represented agriculture and knowledge about plants; Huang Di, the Yellow Emperor, established state institutions, built palaces, and practiced weaponry; Da Yu, the Great Yu, exemplified hydraulic engineering. In socially significant fields such as agriculture, scholars could draw on a broad range of reference points, whereas other fields, especially those of everyday tasks and minor crafts such as tannery, were rarely addressed. The agricultural *Qimin yaoshu* (*Essential Techniques for the Peasantry,* fifth century) written by Jia Sixie, for example, opens with a lengthy enumeration of all sorts of worthy sages in antiquity: from the Divine Agriculturalist to ideal monarchs, sage-kings such as Yao and Shun to sages (*shengren*) such as Confucius (ca. 551–479 BC) and Mencius (second half of the fourth century BC) all recognized the importance of agricultural knowledge for society and state.[92] The *Nong shu* (*Book of Agriculture,* published 1313) by Wang Zhen (1290–1333) mirrored this attitude. The medical-pharmaceutical work, the *Bencao gangmu* (*Systematic Materia Medica,* published 1596) by Song's contemporary Li Shizhen (1518–1593), enumerated the lost works of an ideal antiquity and the involvement of cultural heroes such as Yandi as the forefather of the field.[93] Jia Xian (eleventh century AD) legitimizes the knowledge of his mathematical study *Huangdi jiuzhang suanjing xicao* (*Detailed Solutions for the Problems in the Nine Mathematics Chapters of Huang Di,* 1148) directly by referring to a sage-king in the title.[94] Mentioning the sage-kings was a means to authorize knowledge as worth knowing. In this sense the sage-kings represented the basics of Chinese civilization. They were a universal source for the authorization of knowledge and manifested everlasting truth. They were beyond the interpretative reach of later generations. Early texts mention every task the sage-kings found necessary to govern the people, the state, and their environment. These sparse references offered a huge spectrum of possibilities to which one could anchor all kinds of interests, from moral concern about ritual proceedings to an interest in the construction of carts and wagons.

As a seventeenth-century Ming scholar, Song had cut his teeth on these traditions. He used both the past and the sage-kings skillfully to validate his approach to craft endeavor. In keeping with his era he acknowledged the value of sage-kings in their assigned roles. But he also moved on, emphasizing their emblematic function for practical tasks in the first section on agriculture in the *Works of Heaven.* He stressed the irrelevance of whether "the clan (*shi*) of the Divine Agriculturalist actually existed or not, because it is the connotation of his title [i.e., divine (*shen*) and agriculture (*nong*)] which has endured up until today."[95] He affirmed their role as originators of Chinese civilization, and thus made their activity essential to man. He denied them any philosophical or moral characteristics. For him they conveyed praxis-oriented values.

Zhu Xi provides a counterexample to illustrate how Song's approach differed from orthodox thought. Zhu also applied the great sage-kings as emblems, but he outlined them as philosophical and quite human figures openly doubting that the practical abilities of the sage-kings could exceed man's competence. He presented the Great Yu as a paragon of ethical behavior beyond human capabilities but suggested that even a man like the Great Yu could not have handled a flood alone.[96] Technically, Zhu Xi's comments were meant to empower people to prevent floods themselves and to motivate officials to organize the necessary tasks instead of waiting for heavenly or sagely intervention. Yet, by questioning whether the Great Yu's practical ability had really gone beyond human limits, Zhu Xi deprived the Great Yu of his emblematic function for practical tasks.

In the first section of *Works of Heaven* Song blamed the ignorance of his colleagues regarding agricultural knowledge by pointing to the changes since the times of the Divine Agriculturalist:

> But what is the reason for the fact that the classification and explanation of the numerous varieties of grain had to await the coming of Hou Ji? More than a thousand years elapsed between the time of the Divine Agriculturalist and King Tao Tang, during which grain was used as food and the benefits of ploughshares (*leisi*) had been taught throughout the country. . . . The sons with the white silken breeches (*wanku zhizi*) regard the rain cloak and the leaf hats' [of the common] people as if they were criminals' caps (*lisuo*); erudite families (*jing sheng zhi jia*) apply the term "farmer" (*jiafu*) as a curse. Man eats steamed rice in the morning dawn and in the evening the rice is used as provision. The officials recognize the grain's flavor and yet have forgotten its origins! Well, [envisage this ignorance] it seems unlikely to me that the deification of the first Agriculturalist could be the result of human force alone.[97]

While Song's emphasis on agriculture was in full accord with his contemporaries, issues such as hammer forging were seldom considered remarkable enough to be mentioned by scholars in textual studies. In such cases, scholars could quite freely assign a historic figure or a past event to anchor their interest in the past and thus authorize it. In the introductory passage to the section on boats and carts Song asked cunningly if "a sailor who lives afloat on the ocean year after year and moves in accord with the ten thousand waves as if they were plain ground" was not in fact similar to a sage figure like Liezi who could "ride a cold breeze" (*yu lengfeng*)?[98] Comparing a philosopher, even a cryptic thinker of the hundred schools phase of the fifth century AD such as Liezi, with a humble sailor, Song mocked all contemporaries who argued that the actions of ancient sage-kings could not be repeated. Affairs only appeared divine or fabulous to those who did not master the situation.

For a seventeenth-century Chinese scholar the linkage between crafts and sage-kings was functional. It addressed the responsibility of the scholar as a sociopolitical leader and affirmed the primacy of scholarly knowing over all other skills. This sort of reference is common in almost all texts that refer to material inventiveness. And yet, once again, Song deviated from the norm by drawing attention to the sailor's performance as a manifestation of knowledge. Song underlined this epistemological claim through his exclusive employment of the sage-kings in the introductory parts of the sections in *Works of Heaven*, in which he delineated each subject within its perceived cultural achievement. This is in contrast to his colleagues who generally addressed this issue in prefaces and paratexts. Song's rhetorical practice indicates that, for him, the sage-kings were more than a source of authorization: they were signifiers of his structural argument. The sage-king's engagement of crafts carried his concept of man's role and the value of the offered knowledge for the state, society, and self—an argument that aspired to define the relation between heaven, earth, and man.

THE KNOWLEDGE IN CRAFTS

> The rust and the bottom of the crucible are brought together in the stove that separates the ores. In an earthen pot filled with fire, lead (*qian*) will transform (*hua*) first and flow out through the bottom [hole]. Copper sticks to the rest of the silver like glue. They can be forced to separate with the help of iron rods. All this is in good order, and thus the "works of man" (*rengong*)" and the "works of heaven" (*tiangong*) are disclosed.
>
> *Works of Heaven*, chap. 3, no. 14, 6b

The coherent scope and methodological approach of Song's oeuvre and his view of *qi* in both *On Qi* and *Talks about Heaven* indicate that Song saw heaven not as an equivalent to nature, nor an omnipotent force but an issue subject and integral to his notion of the world as *qi*. Song regarded heaven (in its combination with earth) as contributing a structural prerequisite to a world of *qi*. Man was part of this world and not outside it. Man's creative activity, crafts, and technological efforts thus enacted universal patterns in the same way as natural phenomena. Crafts displayed "the works of heaven." When Song reminded his reader "it is man to act in accordance," he urged him to understand and respect how change, natural phenomena, human affairs, the "inception of things" (*kaiwu*), and the "accomplishment of affairs (*chengwu*)" came about, not to perform crafts.

In *Works of Heaven* Song mentioned that "the helmsman rules over the entire crew. He is a man who possesses thorough knowledge, and a steadfast sense of duty, not merely a dash of bravery."[99] In the metaphor of a ship sailing the seas, Song saw the official as the helmsman whose duty it was to direct the ship of society, state, and self safely through stormy weather. Song, in line with Confucian orthopraxy, thus urged the scholar to take up action and arrange the world according to the order displayed in natural phenomena and material inventiveness.

Looking at this metaphor from our historical view Song was the man at the helm taking measures to pilot the ship against the prevailing winds back into serene seas. Despite his gloom and pessimism, Song assumed that man, in principle, could influence his fate if he took up the rudder and struggled against the strong winds that would lead the world into chaos.[100] Taken together his works demonstrate that he still saw a way to brave the storm: if only man acknowledged the sources for knowledge, order would come about. He urged his colleagues that it was in their hands to reestablish order. Song also called for practically oriented action as a way to make fortunate times return: "After all, heaven endows living persons [with the facility of] hands and feet and clever schemes of how to make a living. There are plenty of ways and means."[101] Song saw crafts, therefore, as a way to bring prosperity to society and state. In this regard, we can easily group Song with other contemporaries such as the philosophers Li Zhi, Wang Tingxiang, or his temporal successors like Fang Yizhi who all undertook the pursuit of scientific or practical knowledge for some social use rather than private good. Whether we now follow Robert Crawford's definition and call this attitude pragmatic Confucianism or classify it under the rubric of practical learning in the sense of Ge Rongjin, the conclusion can be drawn that most of them felt obligated to explain the absence of a philosophical discourse by presenting their central concerns *as* a philosophical discourse. This is true for well-known figures such as the chancellor Zhang Juzheng

(1525–82) as well as philosophers outside the mainstream, like Wang Tingxiang, who terminologically defined and philosophically aligned his ideas with contemporary conventions. By ignoring moral issues and insisting on a world uniquely defined in structure and essence by *qi,* Song challenged his time, a time when even extreme thinkers evoked moral and social obligations in their approach to practical matters. For him this moral obligation was the major fallacy of his era. He propounded instead the a priori knowledge revealed in nature and material inventiveness. He did not even argue this point. I suggest this was intentional. It was a conscious step away from the priorities of his contemporaries. As knowledge and action were congruent, the description of the actual thing and affair sufficed to reveal the order. Any discussion of merely philosophical concepts or abstract models would confuse the issue. Therefore Song' rhetoric was per se void of philosophical terminology. It was a hypothesis of a material and observable "objective reality" of *qi.*

Song noted in the final essay of *Works of Heaven* on pearls and gems, "With heaven and earth creating all things . . . the clear is the opposite of the turbid, moistening is the foe of drying, and treasures are at this place, while worthless things are over there."[102] None could be without the other, and only together they make a completed whole: "Heaven creates numerous things. If only one step is missing, a good bow cannot be accomplished. This is not by chance!"[103] This nexus identified the facts and facets that needed to be investigated. Often, he reminded his colleagues, these facts and facets were like precious jades concealed beneath a blackish crust, or pearls hidden in soiled and muddy mussels. Song's call to his colleagues to observe the mundane raises the issue of his approach to theory and practice within the wider world of Ming life, its social structure, material surroundings, and political system. Switching between two reflective surfaces, Song's work and Ming society, the following chapter spotlights the sociopolitical shades of Song's approach to knowledge and delineates his views on the role of the scholar and craftsman and the purpose and role of practical skills and theoretical acumen in the making of knowledge.

CHAPTER 3

Public Affairs

> Master Song says: the five *qi* of heaven create the five flavors. The king [Wu of Zhou] asked JiZi about the production of salt through saturation (*run*) and thus learnt its inherent meaning.
> *Works of Heaven,* chap. 1, no. 5, 66a

A valuable resource, salt was monitored by the state during most periods of Chinese history. Mechanisms of control included taxing the production tools or end product and commissioning production and trade. In the Ming period, the tax on this far-from-scarce commodity provided a steady source of income. Only Jiangxi, Guizhou, and Guangxi provinces had to import salt, as all other areas were self-sufficient. Salt was extracted from the sea, crystallized from salty lakes, drawn from brine wells, and washed from rocks or gravel. As one of the pillars on which economic prosperity was built, salt was a matter of dynastic politics and public policy, a "public affair."

The varied production methods for salt are described in *Works of Heaven.* When Song explored the causal relationships of technical processes, his interest lay beyond tax issues or economic considerations. In his introduction to the section on salt, Song argued that salt was a legitimate scholarly subject because it was essential for human nutrition. Apparently if man "was deprived of salt for a fortnight, he would become too weak to tie up a chicken and feel utterly enervated."[1] Following a cultural imperative, Song furthermore evoked a figure from antiquity, King Wu of the Zhou state (eleventh century BC), to authorize his interest. He explained that when King Wu questioned his first minister Zhu Zi about the production of salt, he learned that the five flavors—sour, bitter, sweet, acid, and salty were correlated to the five *qi*-phases, water, fire, earth, metal, and wood.

Realizing that general underlying principles related all things and affairs to one another, King Wu concluded that salt was related to the phase of water. By using the term "saturation" (*run*), the king emphasized the essential role of water in the production of salt, focusing on the addition of water rather than on its removal. Acting in accordance with these principles, the king could provide for his people and his rule exemplified benevolent rulership.[2] Implicit in Song's description of the event was his perception of the need for someone to remind the officials of his era of the causal relationship between understanding order, acting on it, and moral leadership.

Classifying the world into types of *qi* agency, Song reveals that he believed in the intent of physical matter and the order in natural processes within *qi*. The universality of *qi* explained the occurrence of a substance like salt in seas and lakes, on and under the ground. As all was *qi*, one and the same phenomenon could occur in various places. Song mapped out his cosmological view in his *Works of Heaven*, suggesting that the "public affairs" of Ming China such as salt, silk, and porcelain production, were all relevant if one wanted to understand (*zhi*) order. But what did it mean for Song that man had to "act in accordance"? Who could (and should) do this, and how? What did Song think of the hand, the craftsman who saturated the brine or gathered the salt after its crystallization from the fields, the meticulous performer of the work described in his book? How did the scholar Song, a theoretician rather than a practitioner, approach talents and skills and the tacit knowledge implicit in the bodily activities of laborers?

The answer to these questions reveals the sociopolitical implications of the relation between scholarly and craft knowledge in the seventeenth-century Chinese world and the complex dynamics of community construction and systems of evaluation affecting individual efforts. As Steven Shapin in his portrayal of the Royal Society in seventeenth-century Britain has convincingly argued, individual and communal assessments of the relation between knowledge and person, talents and social status were essential to every period's claim to knowledge and truth.[3] In this context the social identification of knowledge as theoretical or practical became part of a system of evaluation that affected how knowledge was gained. Some British aristocrats pursued a "scholarly" approach. Their high status ensured that their claims were trusted. Artisans, servants, merchants and other folk made assumptions on the basis of practice. Even if scholars appreciated artisan skills, they took it as given that craft knowledge required further substantiation. Each individual may have blurred these boundaries in the quest for knowledge, but few rendered them arbitrarily in their writing. Constructing communities and associations, mapping out professions and disciplines, British, French, German, and Italian natural philosophers, instrument makers, and cartographers laid claims to various forms of knowledge and acknowledged talents and skills. These viewpoints determined their approach to facts and the nature of knowledge they ascertained. For example, although René Descartes (1596–1650) promoted the observation of artisanal efforts, he believed that "the gestural knowledge of artisans could not lead by itself to the production of ideas, yet it could exhibit courses of action that guided the mind towards scientia."[4] Descartes described the methods of the blacksmith in detail in terms of his own theoretical approach. He focused on working methods and the smith's orderly performance, with no thought to the nature of skills or the blacksmith's experience. Artisans' works were in Descartes' mechanistic worldview "objects of study," perhaps even "epistemic objects," impersonalized, complex, but stable entities that acted as question-generating machines.[5]

In *Works of Heaven* Song's attitude is similar to Descartes's: he documented the eighteen selected crafts as if he were logging a record of proceedings with his eyes glued to materials and procedures. Only rarely did his attention stray to the artisan and acknowledge his part in the process. Like Descartes, Song approached artisans from the viewpoint of a natural philosopher. What made Song's view different from Descartes's was that the Chinese seventeenth-century world's conception of skills and social roles was not the same as the European model, and, not surprisingly, Song's conclusions differed from those of his colleagues in the West.

Theoretically a meritocracy, the social ideal in the Ming world held that anyone with sufficient intellectual ability could become a scholar. Scholars occupied the highest social rank and made up the political elite. Entering public service, they governed the state and maintained order. Agriculture, despite its earthy roots, was honored for its importance to state and society. When scholars, whether high officials or minor landlords, engaged in farming, discussed the breeding of plants, or praised the benefits of hydraulics, they demonstrated their moral integrity and high level of commitment to society and state. Still the elite usually shied away from craftsman work. Although such work had a recognized importance as the provider of useful goods, it smacked of vulgar and dirty labor. Artisans were generally illiterate. They gained their skills through experience and plied their trades without the benefit of theoretical learning. A lower rung on the social ladder was reserved for the dissolute merchants and fraudulent traders who, unlike all other groups in the Chinese societal ideal, only benefited from the work of others and produced nothing themselves.

By the seventeenth century this stereotyped view of the social distinction of tasks and fields of knowledge was increasingly challenged by the realities of life. The early Ming dynastic promotion of craft production, changes in agricultural methods, and increased population pressure led to urbanization as farmers moved into the new centers. Many of them now practiced full time the crafts that had once provided a subsidiary source of income. Economic interests of the state, trends of commercialization, and commodification had made merchants respected and influential in state politics and intellectual life. By Song's time the social group most affected, both quantitatively and qualitatively, was that of the scholar (chap. 1). Men of every social background studied the classics and dreamed of a career in civil service. Sons with a literati background, as well as the offspring of landlords, military, and merchants, studied the classics. Their education became their common ground, even if those who finally failed to enter civil service had to find other ways to occupy themselves or earn. They worked as teachers, publishers, or doctors. Often they spent their time with studies on plants, collected herbs or pursued practical things such as ship building and urban planning. Some moved easily back and forth between their hobbies and jobs, their administrative posts and academic pursuits; others made awkward forays across blurred social boundaries, engaging with village people and foreigners, traders and monks. The detailed picture that emerges within the literature of this time shows a finely differentiated assortment of individuals that dragged the mundane, virtuous, and convenient into the scholarly realm.

The Ming state involvement in production made the practical tasks of salt, silk, and porcelain production matters of political and public concern. This can be identified as an important factor in the emergence of new scholarly topics: Employing civil servants as administrators to manage lowly hands-on tasks, the state challenged the scholar's role of "men of letters." This intensified a long-standing intellectual discourse on "knowledge (*zhi*) and action (*xing*)" in which Song participated (chap. 2). Featuring ideas about human disposition and testing the boundaries of scholarly identity and learning, Song classified the world in two groups: intelligent and dumb, scholars and commoners. He presumed that only the scholarly mind realized the higher formalized knowledge inherent in things and affairs. Intelligence was necessary to reveal the underlying universal principles. Song argued that all nonscholars, farmers, and merchants, performed mindlessly. Throughout his writings, Song edited out the artisan, in his role as laborer and as craftsman (*jiang*), discussing neither his skills, nor his social role. This can be seen as a logical reaction from a scholar who felt superfluous in a world in which the state increasingly relied on craftsman skills for its manufacture and society longed for refined wares of silk, porcelain, or lacquer. Defending his scholarly identity Song emphasized that the performance of crafts revealed knowledge only the scholar was able to grasp. In this context Song also took his stance on morality, arguing that good or bad behavior was situational; it was the result of customs, habits, and circumstance. Thus, good behavior could be found in all social groups.

Seen from a sociopolitical viewpoint, Song was a child of his ambiguous times: he insisted on the superiority of scholarly talents, intellectually and socially, and gave academic achievement and literati studies the highest social rank. His attitude was a reaction to the increased importance of trade and the growing visibility of craft work in public life. Like many of his contemporaries he acknowledged the necessity for merchants, but did not approve of their scholarly ambitions. Some men of the generation that followed Song's, such as the scholar and merchant Wang Yuan (1647–1710), suggested that the traditional order of the four professions—scholar, farmer, artisan, and merchant—was no longer valid and thus should be dismissed. Richard John Lufrano shows that several others were "placing merchants before artisans in importance and added the category of soldier after farmer."[6] Song would have agreed with Wang on the importance of merchants and soldiers, but he would not have included artisans. Curiously, for a man who wrote about crafts, he dismissed craftsmen as a category. His schema twisted the usual categories and presented only two groups, intelligent scholarly (or military) leaders and inept commoners, who took on whatever practical tasks were required at the time, functioning as merchants, farmers, or soldiers.

CRAFTS AND THE MING STATE

> For the construction of city walls and houses of well-to-do people oblong bricks laid lengthways are used. They are fitted snugly, regardless of cost.
>
> *Works of Heaven,* chap 2, no. 7, 3b

One of the most famous sights of modern Nanjing is the view of its fully intact city walls. Built five centuries ago by more than 200,000 workers and skilled craftsmen they were intended to protect the first Ming capital. We know the name of almost every official manager and many artisans who made a brick in this wall, and where and when it was fired. This data was carefully inscribed on each and every brick.[7] A prodigious source for modern historians, the practice of marking was introduced by the founder of the

Ming, Zhu Yuanzhang, for purely practical reasons. It provided information for tax purposes, it allowed the maker to be traced to guarantee quality, and now and again it might have ensured fair payment for its maker. The Nanjing bricks exemplify the close affiliation between craftsmanship and state enterprise that shaped Song's world. The next part of this chapter sketches a general history of Ming imperial involvement in technology and uses details of the silk manufacture in Suzhou city and porcelain in Jingdezhen city to illustrate the ways in which Chinese written culture reflected on crafts. An overview of the history and structure of Ming imperial production sites and the incentives defining the relation between scholar, state, and artisan in the late Ming will assist the contextualization of Song's work.

From the very beginning of his rule the first Ming emperor kept an eye on craftsmanship, closely linking it to the state. He adopted the system of hereditary household registration first introduced by the Mongol rulers of the Yuan dynasty. This bound all male family members and all future generations to one profession. Zhu initiated a network of production sites under state control. Potters, minters, carpenters, and weavers had to perform set term levy service in state-administered workshops. Production included ships, carts, and weapons for military use, ritual articles for state performances, trade and tributary goods handed over to the enemies and allies of the state, as well as luxury items to fill the imperial treasure chest. Workforce and working cycles, raw materials, and end products, all had a fixed imperial quota.

Previous dynasties had been content with establishing manufacturing sites for the imperial household and purchased additional wares on the free market, skimming off the cream of common artisanal production. The first Ming ancestor's interest in agriculture and craftsmanship was a considered response to the methods of the Yuan state. The Mongolian Yuan dynasty in the thirteenth century had given material development a higher priority than scholarly pursuits, but the top priority went to warfare. This meant that the Yuan rulers favored crafts that stimulated economic growth and helped the state increase its power. While the Yuan rulers definitely venerated artisanal talents (when campaigning, they recruited them as booty and let them live), the overall impression is still that the Yuan exploited, rather than promoted, crafts, for their contribution to the state war machine. The Yuan introduced a hereditary system, recruiting craftsmen under both civil and military administration.[8] After a century every sector of production including agriculture had been sucked dry. Acknowledging the advantages the Mongolian emphasis on crafts gave to imperial rule, the Ming incorporated craft production into administrative structures. Having seen the devastating effects of a military orientation, the first emperor put the control of state manufacture in the hands of his civil servants.

The Ming engaged with important technologies such as silk and porcelain production to an unprecedented extent. By entrusting the organization and control of his foray into craft production to the scholars rather than the craftsmen, the first emperor tied scholars to the practical arts and craftsmen to the state. Scholars recruited on the basis of their literary skills suddenly found themselves organizing craft production. They had to enter a new field of knowing, while skilled master potters continued to pedal their wheels and highly specialized weavers bent their backs over their looms. The imperial tie that bonded officials and craftsmen faced extreme tensile tests during the three hundred years of Ming rule. Tugged painfully at both ends, it tightened and loosened, tangled and unraveled several times. While the craftsman often cut these ties by fleeing or through passive resistance, the scholar-official was bound by his ideals. To serve the state, the scholar had to bridge his knowledge gap and make the system work. By the fifteenth century, the scholar-officials were committed to protecting the great Ming ancestor's master plan against misuse and sometimes against urgently needed reform. Yet, on a daily basis they were faced with challenges that often forced them to contrive ways to achieve the required results and make ends meet. This tension between the poles of maintaining tradition and dealing with economic necessity characterizes the sixteenth and seventeenth-century elite view of the role and function of craft works in the state and society. We can follow its major shifts by analyzing the relation between state and private engagement.

The first state-owned workshops were established in traditional centers of production where officials could recruit local specialists.[9] In the case of silk these were mostly located in the regions of Zhili, Jiangnan, and Sichuan; porcelain manufacture mainly took place in Jiangxi Province. The smooth operation of the imperial workshops in the first years of the Ming rule indicates that state and private sectors enjoyed a healthy symbiosis with both profiting from the activities of the other. The state filled quotas by using private workshops that welcomed the steady income they earned by supplementing the annual state production. The state also actively supported the infrastructure provided by private industry to keep their manufacture going and growing. Throughout this process many scholar-officials involved in the administration of craftsmen work realized that they relied more on the craftsmen than the craftsmen relied on them, the representatives of the Ming state and court. They learnt to meet this challenge in both their roles; the official answered with firm organizational control and the scholar with intellectual sophistication. Drawing the arts and crafts into their domain, the Ming literate elite left the craftsman standing outside in the cold.

It was the third emperor Chengzu (1360–1424; reign Yongle from 1402), who attempted to expand the imperial silk workshop structure into regions without any expertise in silk manufacturing. The coming decades would all too vividly demonstrate that the spread of innovative technologies was a subtle business, one doomed to failure if it relied on administrative enforcement alone. Throughout the subsequent century, the officials rose to the challenge to keep these artificial nodes of the silk network on life support, often at great financial expense and with a huge personal investment. Silk manufacture required the cooperation of craftsmen groups, appropriate climate conditions, and a fine balance between the sectors of raw material production and finished processing. Lacking the necessary networks and expertise in the handling of materials, most silk workshops beyond the traditional centers never really took up production. Meanwhile state production sites established in the traditional centers of Ming silk production continued to prosper.

The official documentation on craft manufacture that supplements Song's idiosyncratic descriptions in *Works of Heaven* was shaped by varying concerns, many of which differ essentially from Song's. State-controlled sectors such as silk and porcelain are the best documented fields of craft production and thus they have essentially shaped our view of it. In general, this documentation is formed by its administrative purpose. Officials noted down facts and figures. They carefully filed the orders, their fulfillment, and the details of deliveries. Although officials did not directly reflect on the increasing professionalization of the craft sector, the skills of the artisan, or his social role, their documents give the accountant's view of such issues. Occasionally these officials complained of late deliveries or a lack of manpower. And then they included basic, yet precise, technical explanations or quantitative information about raw materials, labor force, and machinery in a managerial context that demonstrates the local official's competence and expertise. As the visible blots on a spotless sheet, these complaints distort our view of an otherwise pristine whiteness. The whiteness is evidence of a system operating as smoothly as a Swiss watch. It is, however, also due to the fact that local officials avoided writing memoranda to the central government to avert prosecution for violating their duty. Or if they did commit their complaints to paper, they idealized and manipulated facts and figures to support their argument.[10]

The memoranda speak of local officials' efforts to safeguard their position after interference from the central government, in particular from the eunuchs dispatched by the inner court. Complaints of this kind increased toward the end of the sixteenth century, to the degree that historians speak of an evident decay.[11] Yet this cannot be substantiated in terms of technological advance, quality, or output by looking at the materials in detail or at the remaining artifacts.[12] These reports mainly court sympathy for being unable to satisfy demands for higher outputs or special requests (*jiapai*, or *qianpai*) at very short notice. In some ways these reports verify the strength of this industry and the usual ability of local officials to produce the output demanded by the court and the state even under difficult conditions. Special requests for higher quality and quantity were always met (though sometimes at the expense of standard production). Indeed the workshops throughout the area of Jiangnan and Zhili, the prosperous centers of Ming China, seem to have met their quotas and solved most difficulties (more or less) on the spot.

The fact that this was not the case for all state-controlled workshops outside the traditional centers of silk production indicates cooperation between state-owned manufacture and local industries. The paucity of historical reflections on this cooperation can again be explained by the local officials' fear of prosecution. Their duty was to control the situation. In their official documentation, therefore, the civil servants constructed an ideal of two independent sectors, the smoothly operating state manufacture and the minor contributions of private enterprises or individual master craftsmen working for the state. The increase in the volume of the documentation compiled by these civil servants in their private lives or by their colleagues shows that by the mid-sixteenth century, large and small work units, some under state control and some run by private entrepreneurs, coexisted in all sectors and places. This is exemplified by the state-controlled Weaving and Dyeing Bureau of Suzhou (*Suzhou zhiran zaozhi ju*), which is well documented by both official and private sources. Suzhou had been a center of silk manufacture since the tenth century, partly because of its location. Situated in the middle of a prosperous region boasting tea plantations, rice, cotton, and factories for luxury goods, the state-owned workshops' many demands for raw materials, services, and manpower could easily be met all year round.[13]

The Suzhou Weaving and Dyeing Bureau is also an excellent example of how manifold factors shaped historical documentation. Officially the Suzhou bureau was subordinate to the regional administration, but its high level of expertise attracted the court eunuchs who regularly seized the entire production for the imperial household. Its close affiliation to the cen-

Fig. 3.1. *Tuopian* rubbing of the original stele of the Suzhou Weaving and Dyeing Bureau (*Suzhou zhi zao ju tu ti ji Shunzhi si nian*) (1647). Originally installed in the compound of the Suzhou bureau, it was later incorporated into the *Suzhou beike bowuguan* as part of *Suzhou Kongmiao*. Rubbing produced in July 2009; reproduced with the kind permission of the Suzhou Museum.

tral state explains why both its structure and various deficiencies are well documented. It also gives us an insight into the kind of information produced by the administration. The ministry of works (*gongbu*) and finance (*hubu*) received a rough, probably idealized, description of the workforce, buildings, and personnel. Thus we know that the institution employed 1705 craftsmen in 25 professions with varying status. The local documentation was more detailed. Whenever the bureau was reorganized or reconstructed because of shifts in demand, damage, or other reasons, local officials compiled detailed reports on the changes. Reports could also be the result of a turnover in supervising personnel, in which case the new official often recapitulated the previous situation, gathered archival documents found in the place, and reported on any new developments. As a state edifice, the bureau had a commemorative stone inscription usually placed at the entrance or other exposed place within the compound of halls as was common for main buildings, bridges, and temples throughout Chinese history. As source material these stelae are the documents providing essential details and presenting a particular contemporary viewpoint. In fact they may be even more useful, as they were literally written in stone. They could not be revised or rewritten, as may have been the case with local recordings. A stele on the Suzhou Weaving and Dyeing Bureau records 173 pattern draw looms, placed in six halls in the year 1647. This particular stele also features a schematic layout of the arrangement of the halls and offices, conveying the overall plan of organization, which was almost never included in official reports transmitted to a higher level. The diagram seems to have served a representational and documentary function as well as allowing visiting officials a spatial orientation (fig. 3.1).

Local administrative records and stelae are hence rich in the details that cannot be found in individual endeavors such as that of Song Yingxing. But neither the still extant stelae nor administrative reports reveal details of the working organization within the bureaus or how problems of the working environment were solved. Producing high-quality silk, the weavers worked on draw looms with a pattern tower (*ti hua ji*), complicated machines that had to be built well to bring forth excellent results. No report, however, ever mentions where the bureau obtained the many draw looms they required or who made them. In addition, the local supervisor responsible for organizing 173 draw looms must have taken into account that the weaver and the so-called pattern tower boy (*ji hua zi*) who sat on the tower to pull the pattern strings needed to hear each other. While working on complicated patterns, the weaver had to communicate with the draw loom boy through a highly sophisticated singsong call and response. This had to be synchronized in order to provide good results. If the state put all 173 draw looms in one hall the weavers would have been forced to drown each other out against the background of the continuously clanging shuttles. How such problems were solved is unfortunately not clear.

From the perspective of work organization the porcelain industry is better documented than silk production. Imperial porcelain production during the Ming took place in Jingdezhen city. Both private and official documents generally emphasize the highly segmented production process. Some imply a manufacturing process resembling Josiah Wedgewood's eighteenth century assembly line.[14] Segmented production required more organization as each step had to be tuned in with the functioning of the whole. It allowed sophisticated expertise and complex action to be fragmented into simple components that could be performed by anyone with a minimum of training. This reduced the official's dependence on any individual craftsman's skills. Conversely, small units of experts could react much more flexibly to demand and produce a more varied assortment, working side by side in distinct styles or on diverse projects. If managed competently both methods of operation could produce high quantity and quality.

Cultural prerogatives suggest that neither operational method aimed at mass production in the modern industrial sense of the word. The Ming emperor and the commercialized society at the end of the Ming valued specialized style and sophistication over uniformity. Officials used module production and a unitization of production steps in all workshops, from the biggest to the smallest, to exercise control over the practitioner's skills and knowledge.[15] Segmented production and small teams required an operative ethos of management once production and distribution had reached a certain level. Thus the official's managerial skills placed him at the center of the production process regardless of his actual knowledge of the craft itself.

We can assume that, in the huge Ming urban industrial centers for silk and porcelain manufacture along China's coastal regions and in Sichuan Province, all local officials were affected by the state's involvement in crafts one way or the other. Taxes had to be collected and delivered. Materials passed through their area of authority and waves of craftsmen working on set-term levy filled the streets and marketplaces. In the midst of this, social order had to be maintained and morals had to be safeguarded from too many material distractions.

With regard to the craftsman's incorporation into state practice, Song stood at the end of nearly three hundred years in which scholarly attitudes to crafts, formed by the authority of the first Ming emperor, adjusted to the stipulations of each subsequent ruler. Song, with his aspirations for an official

Resident Registration in Early Ming-Dynasty (1390-1393)

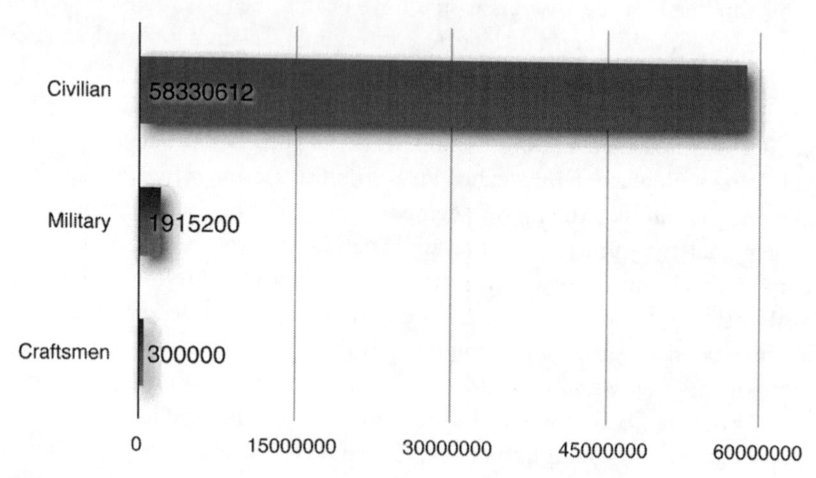

Reference:
《明會典》卷十九、卷一八九
黃啓臣：《明清商品經濟的發展與資本主義萌芽》

Fig. 3.2. Only a small percentage of households were registered in the category of craftsmen by the central government. Artisans, however, were also registered under the category of military, producing weapons or cloth, or doing construction work. Local officials could also recruit artisans. (Source: *Da Ming huidian*, chap. 19 and 189)

career, must have been aware of these potential managerial responsibilities and the resulting social obligations. The ancestor's master plan had indeed been beautiful in its scope, with provisions for almost every conceivable problem. Its keystone was a fixed quota of supply and demand. The plan secured raw material and workforce through taxation and transportation along fixed routes. A constant flow of skill into the state-controlled manufacture was ensured by the inherited household profession system (fig. 3.2.). And yet, was the first Ming ruler ever naive enough to believe that subsequent generations would stay within the fixed limits? And did he really think forced labor would produce the same quality as craftsmen willingly lining their own pockets? In fact, the third Ming emperor, Yongle, began tampering with the quotas in 1403, redefining them ad hoc both with regard to quality and quantity. From 1573 on, during the reign of the Wanli emperor, the court almost annually announced supplementary quotas to meet its growing lust for luxury, an increase in demand the officials had to struggle to meet.

In the beginning the civil servants mastered the challenge of recruitment posed by these changes by combining state pressure with fair reimbursement. As production pressure increased the officials put the onus on the craftsmen, who learnt to curse the compulsory hereditary attachment as their obligations grew. The annual journey to the imperial workshop for levy service had always been a financial burden even with secured fair payment. But it turned into an enterprise causing personal bankruptcy, not to be offset by a year's hard work. Those craftsmen unfortunate enough to have talentless sons had to buy in workers to meet their obligations or to pay fines. With increasing frequency, enrolled craftsmen had to sell all their goods and chattels. Some lost their homes and became outlaws and refugees.

State-owned manufacture entered a vicious cycle in which quality was enforced by coercion that in turn negatively impinged on quality. It was left to the officials to strike a balance between the ancestor's plan, the requirements of subsequent emperors, people's needs, and the interests of their own class. They found the answer in one of the few institutional loopholes the emperor had built in to cushion customized production or to deal with unforeseen events in sectors such as silk and porcelain production, the so-called "workshops of short term requirements" (*gongying jifang*). The officials inflated these bureaus, once little more than filing cabinets for craftsmen's household registration, into an elaborate agency system they used to recruit the best craftsmen from all over the country to manufacture superior artifacts. Silks found in the imperial tombs of the Ming and other items such as lacquer are marked as being produced on demand for these offices.[16] Agencies worked outside fixed quotas or financial restrictions. They also obtained the highest quality because they could choose the best from all sectors, those who used the finest technologies and created the most innovative products.

In 1531, during the reign of Emperor Shizong, the Ming state and court institutionally were managing to stay abreast of their assorted difficulties. Various reforms were made to the former quota production, and tax payments in kind, which had usually been paid in bolts of silk, were changed to payments in silver within all the sectors and institutions that had never produced well or had always worked inefficiently.[17] The imperial workshops and the local outposts established by order of the first ancestor remained unaffected by this move. Private industry in many locations such as Suzhou and Hangzhou profited from it, as workshop owners and free weavers now could be formally approached to produce high quality wares and were no longer bled white by non-official gray requests.

As the state slackened its hold on the reins, craftsmen groups began organizing themselves into various associative forms to seek shelter from social and financial insecurity.[18] Regional workshops in centers such as Suzhou, Jiaxing, and Hangzhou as well as the imperial bureaus in Nanjing had always almost exclusively relied on local workers (*zhuzuo jiang, cunliu jiang*), and hired labor to produce their high-quality silks (see fig. 3.3). Their demand had helped create a market for specializations within the silk trade, and we can assume that the same is true for trades such as porcelain, carpentry, and lacquer production. The agency system was another important factor that promoted the development of open labor markets and fostered private initiatives to organize access to raw materials, specialized labor, and the marketing of finished products, giving much more freedom to entrepreneurs. Unlike European guilds, Chinese craftsman affiliations of the time were based on locality, not occupation. This concurs with a general trend in this period toward localism.[19] But in this case the associative and communal form may have been a result of the craftsmen's hereditary status, which bound each artisan through household registry and tax obligations to his familial roots and thus to local origin.

Studies have demonstrated that in the Qing period craftsmen organizing such groupings often became professional trade agents in fields such as cotton production. This development may have its origin in the state agency system, whose recognition of expertise reinforced the craftsman's professional identity, which led to the emergence of markets for laborers in urban centers. From about the 1550s onward, officials and the state became less and less visible in the craft sector, exerting their control over the private economy, workshops, craftsmen, and traders' various activities through social or religious means rather than institutional ties. In many cases either the artisanal groups or the state left or handed over distribution and marketing to merchants.[20] In the silk sector, for example, the burden of maintaining quotas was transferred from the officials to the merchants. In the salt sector, officials began to promote religious-ethical cults that worshipped local deities of the trade. Management was also left to the merchants, who now organized production and traveled the country with their wares.[21] The mobility of traded goods is another characteristic of this period of the Ming. Song mentioned in *Works of Heaven* that merchants connected North and South and considered regions far in the west part of their territory of trade. This concurs with what we know about consumers such as Wang Zhen (1424–95), who delighted in the variety of wares delivered to the markets of the capital from all over China and beyond.[22]

Ming scholars also experienced mobility at first hand. From their early youth those who had chosen the path to an official career packed their bags and traveled in a triennial rhythm to their various exams; once appointed the less fortunate among them were continuously transferred to new

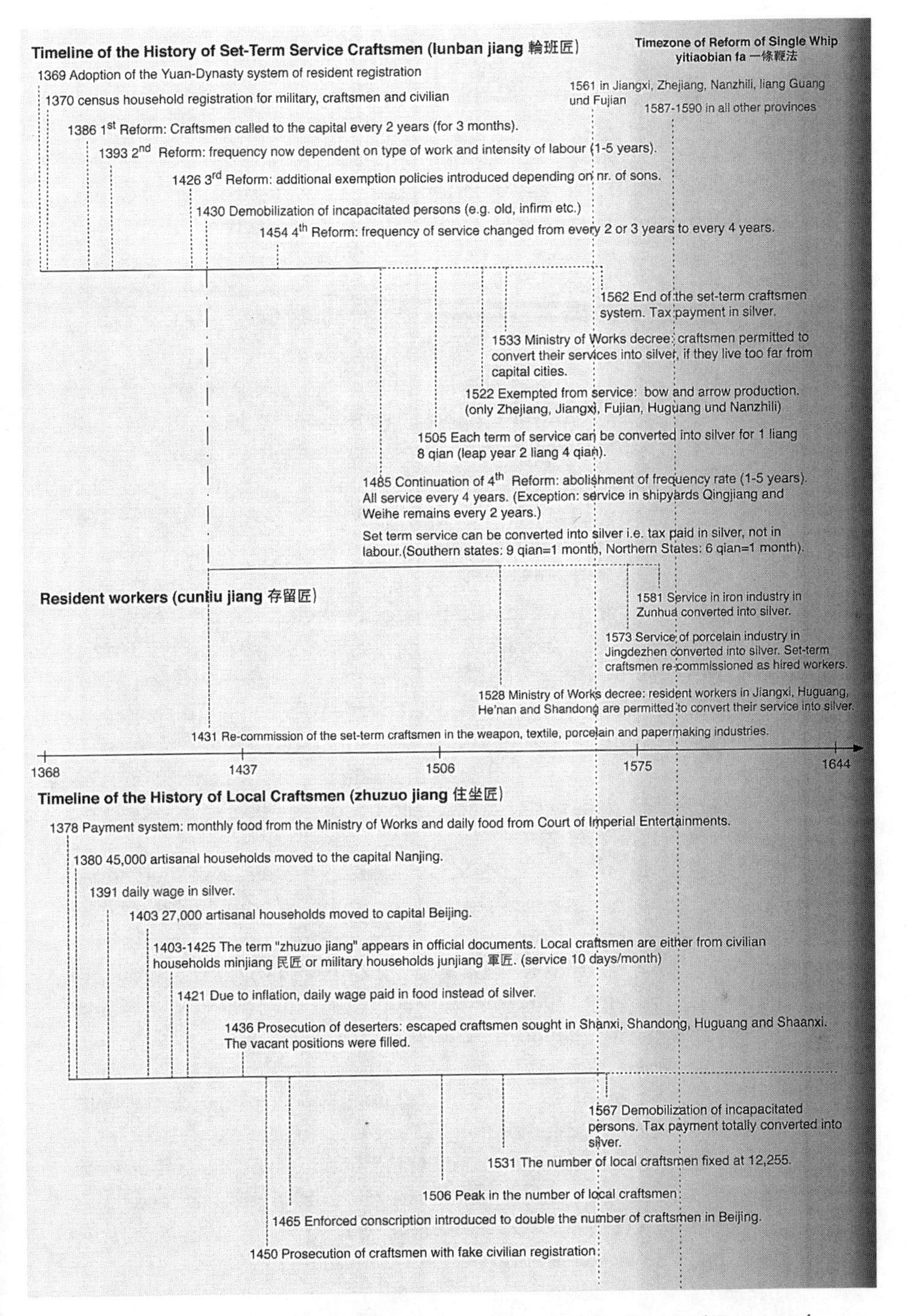

Fig. 3.3. Timeline depiction of the history of set-term service labor (Statistical Overview of Resident Worker Registration in Early Ming Dynasty, 1390–93). Shen Shixing 申時行 (1535–1614) and Li Dongyang 李東陽 (1447–1516), *Da Ming huidian* 大明會典 [Collected Statutes of the Great Ming], chap. 189, chap. 201, 3a.

positions. On the road these scholars came in contact with traders and craftsmen. Levy service had made the latter standard components of Chinese street scenes. While this mobility differed from that of early modern Europe, where political fragmentation stimulated competition between states for expert craftsmen and engineers, it was nevertheless apparent throughout the vast territory of the Ming. All over the country, those who were registered according to their trade had to take to the road every two or three years, leaving their homes for a couple of weeks or months (see fig. 3.3). In Song's lifetime wandering craftsmen were a regular phenomenon on Chinese streets and in its urban centers, even after the levy tax had been abolished. Carpenters of Suzhou traveled southward to work in the shipyards on the coast of Fujian Province or went to search for seasonal employment in the growing commercial publishing industry of that region; talented weavers searched for occupation in private workshops or were sent for by state institutions to fulfill special demands; jade carvers traveled to Yangzhou County to refine their skills in famous workshops; oil pressers from Sichuan Province produced lamp oil and candles that rich traders delivered to markets on the coasts. From a geographical region bigger than Europe, migrating artisans from southern Guangzhou, northern Mongolia, and the Central Asian plains regularly flocked to the flourishing urban areas of Jiangnan and Nanzhili. Song may have encountered many of them when he traveled to his exams via Jingdezhen north across Nanjing to Beijing, although he probably took better compartments on the boats and stayed in more comfortable lodging.

Both local and central officials report state initiatives to move craftsmen, either to meet short-term demands or to disseminate and distribute technological knowledge. The sources seldom reveal if such enforced migration lasted or if these initiatives to transplant expertise were successful. But they show that craftsmen in full-time jobs resisted enforced migration just as stubbornly as the former farmer-craftsmen who were bound to their land and for whom craftwork had provided a supplementary income. Craftsmen traveled but their hereditary status and economic or personal reasons prevented them from being socially mobile. Expert craftsmen were reluctant to leave urban centers such as Suzhou and Hangzhou because they offered varied opportunities throughout the year. Even when officials threatened heavy penalties or offered extraordinary payments, southern craftsmen (*nan jiang*) often refused to move to the capital in northern China.[23]

The Ming state and its society demonstrated its highest ambiguity implicitly in the sociopolitical consequences of the enrollment system. Within the state-owned system and regarding taxes, artisans were distinguished and recruited by trade. Yet the trades were kept within families through state enforcement. On the one hand the first emperor's rulings recognized and finely differentiated occupations within fields such as silk or porcelain manufacture. On the other hand the deep ties to locality and household registry crucially hindered the development of craftsmen identity or any group identity beyond familial ties. This ambiguity reflects as much as it explains the subtly underdeveloped contemporary reception of craftsmen trades in written accounts. It is the cause and origin of the close relation that scholars document between native social role and skills, the idea that the achievements of craft expertise were bound to family morals. Craftsmen struggled to find new ways of training or new forms of collective organization, but, bound to rules of inherited trade, they were always drawn back to their social ties and local origin to learn their skills. Even if some did travel to learn from new masters, they could not stay and practice their newly achieved skills. Apprenticeships were bound to the father-son relationship. The enrollment system moved craftsmen here and there, but it prohibited them from changing their status or moving without permission, which was almost impossible to obtain. Those who escaped their obligations were on the road and on the run. Once they had joined the group of roaming people (*liumin*), they were reduced to the level of vagabonds, indistinguishable from impoverished farmers or tenant employees. Their craftsmen participation is then disguised to the researcher, as David M. Robinson's study exemplifies.[24]

State power and managerial structures controlled the connection of trade to place, family, and craftsman knowledge. These forms of control offered scholars a way to acknowledge the potential of practical knowledge for human civilization without recognizing it intellectually as more than an epistemic metaphor. In this sense the Ming scholarly world chose to watch over the divide between practical and theoretical knowledge rather than attempt to bridge it. Scholars' indifference in scholarly literature toward individual skills could be seen as a strategy to defend their own position. Seen from this perspective, Song's approach to craftsman work is in full accord with his contemporary world. It is, however, false to assume therefore that these scholars were uninterested in crafts, technology, or artisanal skills. On the contrary, we can see the enormous effect that Zhu Yuanzhang's approach to manufacture within the state had on Ming scholarly minds in this era's philosophical discourse on practical knowledge and pragmatic action and the comprehensive discourses on human skills and talents. In this checkered atmosphere Song manifested his interest in crafts and took his stance on craftsman skills and scholarly talents, the practitioner's role in society and its relation to scholarly learning. Song reflects the scholar's control over artisanal expertise when in his *Works of Heaven* he detailed the handling of materials or specified which tasks required specialized labor and the official's particular attention. Discussing silk reeling, Song reported "for gauze (*ling*) or muslim (*luo*) twenty cocoons (*mu*, literally, insects) have to be tossed [into the water pot and boiled]; for Baotou silk from Suiyan ten cocoons are enough." Describing the process of silk wadding, Song emphasized, "the effect of blanched, anachromatic whiteness achieved in Huzhou silk threads (*Hu mian*) is caused by the sophistication of its manipulation (*shoufa zhi miao*). They should be swiftly mounted on the frame so that they can be stretched while still drenched with water. If one is too slow, the water will drain off the thread: Then the pieces cannot be untied and the color will not be pure white."[25] It was their knowledge of effective methods for each step of the process, and their managerial skills that defended the scholar's role as civil servant and kept master craftsmen at bay.

MAN'S NATURE (*XING*) AND TALENTS

> The manipulation of the nature (*xing*) of gunpowder and the following of the instructions (*fang*) for such weapons, requires a man of intelligence (*congming*).
>
> *Works of Heaven*, chap. 3, no. 15, 35a

To guard its frontiers and suppress internal revolts, the Ming state required soldiers. These were recruited in a system of military examination parallel to the civil service examination. Candidates had to execute a catalogue of skills to demonstrate their competency in the martial arts and in leadership. For example, they had to pull a heavy arrow from a strong bow and shoot it accurately into the target. Like most bows used in China during that period, the stave of the examination bow was composed of bamboo with ox horn covering the outside surface. To enhance its resilience the bow makers additionally swathed the bow's horn blade with ox sinew. Coated with glue, the piece was finally covered with birch bark. Song noted in *Works of Heaven* that Chinese archers typically used bowstrings made of silk. Silken threads resisted rain and fog more effectively than the ox sinews applied by many northern barbarian tribes.[26] Song seems to have been well informed about weapons, but his generation in the South grew up largely detached from the deeds of actual warfare. Decades of peace had reduced the necessity for trained military personnel. Under the influence of a civil regime many examination practices had deteriorated to a Confucian ritual performance. Candidates pulled their bows without actually shooting an arrow, parading skills rather than mastering them. Training armies to deal with Japanese coastal incursions in the mid-sixteenth century, general Qi Jiguang spurned the antiquated archery syllabus of the military examination system as a mere recital of "flowery" skills. He advocated the use of firearms, pointing to the Ming founder's use of gunnery to defeat the Mongols. Indeed, almost a weapon fanatic, the first emperor of the Ming is said to have possessed every sort of firearm, gunpowder construction, and cannon in his extensive armories. Chinese military used gunpowder as early as the ninth century, but it was not until the fourteenth century that it was developed from an incendiary weapon to improved artillery. This gave the Ming a crucial edge in warfare with internal and external enemies in the subsequent fifteenth and sixteenth centuries.[27]

An astonishing range of knowledge on weaponry and military strategy was widely available by Song's era. The Chinese military treatise *Wujing zongyao* (*Collection of the Most Important Military Techniques*, AD 1044), which was circulated quite openly during the Ming, contains the earliest surviving written recipes for gunpowder. The ingredients for fireworks, saltpeter and sulfur, were also used in textile dyeing, for medicinal purposes, and in other processes. Sulfur came from the geographical margins of the Ming's sphere of political influence.[28] It was mainly processed from pyrite in conjunction with green vitriol. Fire masters heated the pyrites in simple dome-shaped earthen furnaces with a truncated top. Why was sulfur production monopolized by a few people? For Song this had to do with the fact that specialist skill and training was required. Most meticulously, Song explains that after the appropriate temperature was reached—which was judged by the color of the flames—stokers turned a porcelain bowl over the hole to catch the emerging yellow vapor. The vapor condensed on the bottom of the bowl and then drained, through "a rim and through a small aperture via a clay pipe into a small tank, where it formed sulfur crystals."[29] The second ingredient for gunpowder, saltpeter, was a common byproduct of salt production. Song noted in *Works of Heaven* that the topological and geological configuration of the soil surface determined whether salt or saltpeter was formed in the earth. While salt could be produced in a huge variety of soils, only a thick ground surface adjoining a mountain was favorable for the generation of saltpeter. According to Song experienced craftsmen (*qiao jiang*) knew this and could easily locate deposits. Song went on to explain that for gunpowder, saltpeter was first roasted and then ground to a fine powder with a stone roller. He explicitly admonished against the use of iron roller, "as its spark could accidentally trigger an explosion."[30]

Mixed together, he theorized, saltpeter acted as an upward projecting agent, whereas sulfur mainly behaved laterally. He informed his reader that, regardless of the ratio, forcing these two substances to fuse (*he*) would inevitably "create noise (*sheng*) and transformation (*bian*),"[31] Warning that making and dealing with gunpowder called for more than a knowledge of its ingredients Song, in *Works of Heaven,* stated that the handling of gunpowder required intelligence. Yet, he did not explicitly state whether the required intelligence was one of a steady hand or an intimate knowledge about the reactive forces of materials. To whom did Song assign this intelligence, to the maker of gunpowder, the craftsman, or the scholar in charge of control or men in general? Was this intelligence an innate talent or a trained skill or a combination of both?

As a seventeenth-century scholar with a thorough training in the official curriculum, Song discussed talents and skills, education and schooling and thus also intelligence within the ongoing intellectual debate about the right ways to gain knowledge. As state interference changed the craftsman's life and work and people of all social strands began to test social boundaries, the growing mass of literati began to discuss fervently the issue of human disposition, its innateness, and moral direction. If man's knowledge was innate, was he ultimately good or bad? Which talents were inherited and how much could be learned? Much of this discourse was based on the Yongle emperor's authorization of the Cheng-Zhu interpretation of the classical canon of texts on the *Four Books* and *Five Classics* within the civil service examination curriculum. The opponents of the set curriculum suggested that those who wished to enter the highest political arena no longer studied these texts for their essence; they merely memorized them to further their worldly interests.

By Song's time the debate had reached a point of no return, which is generally identified by historians in the thought of Wang Yangming. Song can be localized at the margins of the academic controversy and political debate on Wang Yangming's ideas on innate knowledge (*liangzhi*) and his assertion of the "unity of knowledge and action" (*zhi xing he yi*). These terms became a catalyst for complex debates and raging controversies around the issue of what knowledge was, how skills and talents were to be judged, and which role classical study, training, personal experience, or social origin played in the making of knowledge.[32] Although not everyone subscribed to Wang's theories, which some saw as a false start, a short survey of his ideas helps to place Song's view of knowledge and action, talents and skills, practice and theory in its contemporary context.

Put very briefly, Wang defined knowledge as "knowledge of how to act in a given situation." This knowing "how to act" was in the mind and thus behavior equaled knowledge. Wang discussed knowledge in the general sense and its relation to morality. Most scholars found the implicit issue of morality and ethical standing the most interesting part of Wang's approach. Wang observed that even men who were trained in polite, proper behavior could not necessarily put their training into practice. He gave the example of a well-educated son who still humiliated his father with spontaneous outbursts or impertinent answers. He concluded that knowledge had to be internalized in the mind or had to be innate in order to be effective. On such grounds Wang and his followers dismissed didactic methods such as memorization and recitation, literary composition, philological studies (*xungu*), and broad learning (*boxue*) as superficial. He argued that they only grasped what was "out there," that is, information and details, but not the actual wisdom it contained. Wang's argumentation on knowledge was mainly directed against philological studies and evidential text research. In accord with many contemporaries, he identified this form of inquiry as a mere gathering of details, labeling it "vulgar learning" (*suxue*).[33] But as his doctrine aimed at putting morality into "action," Wang urged his colleagues not to forget that man's quest for knowledge for the sake of morality called for a deep involvement with all kinds of issues. As the world constantly changed, moral principles were not always applicable. One had to carefully weigh up each situation. If done the right way, the study of facts could be meaningful: "Innate knowledge is to minute details and varying circumstances the same as compasses and measures are to areas and lengths. Details and circumstances cannot be predetermined, just as areas and lengths are infinite in number and cannot be entirely covered."[34]

In the eleventh century Zhu Xi had propagated the "investigation of things" (*gewu*) as a way to discover morality. By the sixteenth century Wang promoted the "investigation of things" as a way to put knowledge into practice. Wang based his ideas on the doctrine of Mencius (third-century BC), who had proclaimed an innately good human nature and considered human disposition to be inherited. Other contemporaries had a negative image of humanity. Regardless of their position on human nature (*xing*), the majority considered the virtues of benevolence, righteousness, purity, wisdom, and filial piety to be the purpose of knowledge. In contrast, the core of Song's discussions located knowledge beyond moral purpose.

Song's ideas about human disposition are most apparent in *An Oppositionist's Deliberations,* where, inspired by the appointment of Chen Qixin, he described the rights and wrongs of human activity. His method of focusing on what were for him "hard facts," historical events and personal observations, did not concur with traditional philosophical procedures of reasoning, as he himself pointed out in the *Works of Heaven*.[35] This concurs with his statement in the preface of his cosmological study *On Qi*, where he noted the difficulty, indeed the futility, in defining human nature.[36] Preferring to concentrate on situations in which moral temperament became relevant, Song used a reference to Zhang Juzheng's interpretation of Wang's notion of innateness.[37] Zhang, Song explained, had argued that morally upright talents emerged at any time and in any place and would comport themselves benevolently and righteously, regardless of the circumstances. If this was true, asked Song, how could the world have fallen into such a distressed state? If the talents were there, why were they not revealed? In Song's view the decay of his era proved that Zhang was wrong and that talents were unrelated to moral behavior: "At present the whole country is destroyed and damaged. Day by day defeats are claimed as victories. The (soldiers) take one prisoner and capture one single ordinary horse, they cut two throats and collect three arrows and impudently [tout this] in their reports to the emperor as an [imposing] success without feeling conscience-stricken. Sometimes they simply compare their fate with that of other cities that are conquered and under enemy control, and claim it a victory that their own city walls and palisades are locked securely and that fortunately no enemy has occupied them thus far."[38] Song detached the question of human disposition, talents, and skills from issues of morality and made clear that heaven-human interconnectedness was at stake. He insisted on a morally independent relationship between heaven and man when it came to political and social life.

Song's ideas embodied the meritocracy of his era. In terms of talents and skills all groups were the same. He placed men in two categories. The naturally intelligent were the leaders of society: scholars, officials, and generals. Common people, farmers, and soldiers were endowed with minor talents, and it was a rule of nature that they would follow the intelligent leaders. In this Song was in congruence with many of his colleagues. Yet, whereas most endowed intelligence with a moral quality, Song saw morality and knowledge as two completely separate issues. Man could be born intelligent and still be bad. His morality was not inborn; it was the result of situational choices, customs, and habits. Therefore a scholar could live an immoral life as a corrupt official but still be able to change his attitudes given the right outside influences. "Among all kinds of people some are intelligent and responsive (*congming yingwu*) and therefore born to be scholars (*shi*). Things like this do exist. But nobody is born to become a soldier (*bing*). Simpleminded and obstinate people (*yuwan zhilü*) are born to be farmers (*nong*)."[39] Song's notion of talents and abilities was thus based on his perception of aptitude as the primary characteristic of a man. Those with the brain were predestined to guide all others and thus should be promoted to official ranks, in either the civil or military sector. Scholarly intelligence and the farmer's simpleminded attitude were basic elements determined at birth. In this sense a basic order of talents and a social order of leader and followers was predetermined. Song, however, also presumed that external factors influenced talents. For example, in his discussion of the human voice, he brought up the factor of human physiognomy. According to Song, the capacity of the human voice with regard to volume and timbre was determined by man's body and shape, just as the drum's loudness and timbre were dependent on the diameter and volume of the chamber's body: "The *qi* reservoir (*qi hai*) of the human body, or the gate of life (*mingmen*), is equipped with paternal semen (*fujing*) and maternal blood (*muxue*). The duration and volume of sound *qi* (*sheng qi*)—be it short or long, loud or quiet—is determined by the embryonic agent (*taiyuan*), and thus is not the result of achievement (*gongli*). The same is true for animals and beasts."[40] Song's ideas were grounded in everyday Chinese medicinal knowledge: The *qi* reservoir *qi hai* is a technical term that describes a locus in the viscera, and the "gate of life" (*mingmen*) is locus of vital functions. Chinese medicine located man's life energy (*jing shen*) in these two body parts. In the male body the gate is also the source of semen and in the female body the linkage between the kidney functions and the womb.[41] Equating life energy with *qi*, Song insisted that if a man lacked the right physiology, singing scales would be a waste of time, even if the singer was musically talented.

Song's notion of the voice also reveals that he did not think of such predispositions as inviolable. On the contrary, training the body and thus the voice was a way to overcome such limitations. Children had thus to be selected young so that their bodies could be adjusted and their talents developed properly. Training was essential, although all the training in the world could not outweigh talent: the dull could not be made intelligent and an intelligent person would have inborn talent. In general, Song acknowledged that external factors, such as actual life circumstances (*yi shi qi*), could combine to hide skills and talents. The descendant of a farmer could be highly intelligent and a potentially gifted scholar. If his talents were not recognized, he would remain a poor farmer. Circumstances enabled a scholar's son, even if dull by nature, to achieve a high position. The strong influence of external forces became most evident in times of war, upheaval, revolt, or any other disruption of harmonic social order, as these were the circumstances that made man act morally right or wrong. Song proposed "nobody is born to become a soldier" and "nobody is born to become a highwayman."[42] He presumed that if circumstances made men immoral, it was the duty of the intelligent man to provide the right environment to encourage moral behavior in his environment by revealing the right order. Wang Yangming, in contrast to Song, considered morality the result of an individual's enlightenment and his predisposition.[43] In line with this argument, Song insisted that he was a bandit because his status was illegal, not because he was per se an immoral villain. Therefore, anybody could and would switch from bandit to soldier or vice versa as the situation required. He substantiated this argument with a historical example: "[Those bandits and highwaymen] who met the generals Zong Ze (1059–1128) and Yue Fei (1103–1142) yesterday, became their soldiers the next day."[44]

Zong Ze and Yue Fei were both high officials commissioned with military tasks to defend the Song dynastic rule against the invading Jurchen. By the late Ming they had become enshrined heroes.[45] Song idolized them because these leaders had judged morals individually in each situation, thus acting according to circumstances rather than blindly following rules. According to the family genealogy, Song even once attempted to put this social idea into practice himself. During his short tenure as a local magistrate in Dingzhou District, Fujian Province, he helped some local bandits return to normal life as farmers. Recognizing their extreme poverty, Song released many of the accused without charge after giving them a moral lesson. His superiors, the governor and vice-governor, then sentenced Song for aiding and abetting the bandits. Consequently Song offered to catch the bandits, alone and with his bare hands. Because of the danger the governor and vice-governor insisted that he be accompanied by military troops. Song refused, abducted the renegades, explained the situation to them, and then let them go again. The genealogy reports that the bandits, honoring Song's benevolence, lit a fire atop a mountain for him and the entire group confessed to their crimes. Song then burned their dwellings in order to disperse the dissidents. When the governor and vice-governor heard of Song's actions, the genealogy reports: "they appointed him subprefectural magistrate of Bozhou (*Bozhou zhizhou*)."[46] This apocryphal story suggests that at least once in his life Song put his social theory into practice. It also confirms that Song opposed his era's firm link between intellect and moral virtue. To sum up, Song thought individuals from all social groups could be born intelligent, but life situations, experience, and training could obscure innate intelligence. Born into a system where a scholarly career was theoretically open to all, Song divided the world into the intelligent and the dullards. He categorized scholars, farmers, generals, soldiers, and merchants as functional groupings rather than unalterable social identities. Throughout his comments on talents, training, and education the artisan is conspicuous by his absence.

ABILITIES AND EDUCATION

> The versed [among the workers] overturn, wrap and cord [the milled and steamed oil cakes] with all speed and thus obtain a higher yield of oil. That is the secret of this art! Few among the pressers grow old without realizing this.
> *Works of Heaven,* chap. 2, no. 12, 66a

According to the huge cooking literature published in the late Ming, Chinese food culture held vegetable oils in high esteem. Considered beneficial to health, some were even believed to prolong life. Sesame oil was among the favorite ingredients and particularly approved for imperial dishes. It enriched fish-paste and crabs and added a delicate flavor to peacocks and sparrows and bean curd. Oils for such elevated purposes were freshly extracted with refined methods and sophisticated handwork in the imperial kitchens and those of high officials. The oil presses and workshops at the gates of the urban areas and in the countryside produced oil for industrial use, for caulking, to make paints and dyes, or to provide lamp light for the ambitious scholars who read texts through the night to prepare for their exams. In these workshops teams of ten men usually worked in shifts to recover the high investment in machinery and tools. Large, affluent workshops owned ox-pulled rolling mills; the smaller, poorer workshops relied on human strength. Labor in such mills was arduous, and only the lowest and most desperate men were willing to break their backs to move the grinding gears or press the hemp, rape, soybean, or tallow tree seeds. Few relished the task of boiling and wrapping the seed meal in parcels to be packed in the press, because men were steamed like lobsters during these processes.

A good deal of brute force was required to drive the heavy ram over and over against the wedge to press out the oil so that it would, as Song put it "spring forth like water."[47] Workers had to perform these tasks quickly and sedulously. Song insisted, "inappropriate handling of the steamed meal will spoil the oil yield, as it allows the *qi* pleasingly raised by water and fire *qi* (*shui huo yu zheng zhi qi*) to evaporate."[48] Noting that not all craftsmen were equally skillful, Song adduced that even though they had no talents, they required training. Through lifelong practice, artisans could achieve better results and higher output. In this regard scholarly pursuit was similar. In order to be a scholar one had to have talent, which had to be trained to be useful. The poet Han Yu (768–824) of the Tang dynasty and the politician and poet Su Shi, better known as Su Dongpo (1037–1101), of the Song dynasty, had both made efforts to speed up the process of official appointment and failed, thus verifying that learning took time.[49] Etiquette could not be learned quickly nor essays compiled hastily, meaning that while heaven gave some people a positive constitution, refinement still lay in the hands of the individual. Thus, Song concluded, education played a dominant role in societal accomplishment.

While Song did not consider the practical arts a form of knowledge, he realized that training improved craftsmen's skills, just as it helped scholars or farmers. In this regard he acknowledged that the same principles applied to all fields and all groups. Nevertheless, he was totally convinced that the minor skills of those less favored by nature could never translate into what he identified as intelligence, namely, the understanding of the principles that structured man's world, the earth, and heavenly realms. In *An Oppositionist's Deliberations* Song reveals that for him knowing (*zhi*) was the highest form of knowledge. This form only suited the scholar. Common people, farmers, and merchants could only rise within inferior categories. They should not be plucked out of the rabble and allowed in the ranks of scholars except in the rare cases when a scholarly mind was born in an unfitting environment, a humble background. The commoner performed within the order of nature, learned by trial and error (not by understanding), and refined his skills by experience. Only the scholar observing how the artisan's work creates the natural transformation gained the true knowledge of the underlying cosmological principles. Song saw no need to exchange his scholarly attire for a workman's black trousers.

Within this framework, Song believed that intelligence and all kinds of talents could be improved by training whether one was a scholar or a farmer. Cunningly he used this to support his argument and insisted that occupations could not be changed like robes; once chosen, an occupation shaped its practitioners to the extent that they became a closed group. Members of this group then refined their capabilities by continuous inbreeding. No one outside the group could ever reach this level of expertise, for he or she (he included housework and cooking) would lack the advantage of suitable training and lifelong experience. "At all times, mighty and important commanders have been recruited from the experienced infantry. The same is true for the civil servants who—surrounded by ink and brush—were always selected from the circle of the chancellor's secretaries (*shixiang*). Straw sandals and wraps should stay where they are and rise from their own ranks."[50] Song would have laughed at any suggestion that the talented scholar should engage in the practice of craftsmanship or farming or vice versa that a craftsman should try scholarly work. In fact, he insisted that any intermingling between groups or even the idea that a person from one group could appropriate the knowledge of the other would not work. In his view lifelong training was essential for all occupations, the farmer, the scholar, the official, and the military. Intermingling would only result in chaos and disorder.

Another important factor for Song in the identification of social groups was mutual recognition. He mentioned that all abilities, mental, moral, and physical, required their matching partner to be valued: "it takes a sage (*sheng*) to recognize another and it takes a worthy man (*xian*) to recognize another."[51] As a scholar interested in the "inception of things and affairs," he stated that in the beginning there must have been at least two intelligent men with innate talents who recognized each other and distinguished themselves from those with other talents. For him only two of the same kind could validate and recognize one another's abilities. Any assessment of expertise by those outside the group would be petty and futile. He accepted that a scholar such as himself would never be able to reach the level of a craftsman's performance; his mode of understanding and his approach was too different from that of a craftsman.[52]

Song's line of argumentation was individual in its pragmatism and in its empirical analysis of the status quo. But, like Wang Yangming, he proceeded from a classical Mencenian view in his identification of the origin and development of knowledge: "What a man is able to do (*neng*) without having to learn is the exercise of his innate ability (*liangneng*); what makes a man know (*zhi*) without having to deliberate is his innate moral (*liangzhi*)."[53] Song expounded on these principles arguing they should be adhered to when appointing talented men for civil as well as military service: "Thus where is the problem in distinguishing men? Whenever civil service officials (*wenguan*) gather at court and start a discussion, one can quickly identify the men of ability (*caineng*), that is, those who react sensitively (*jingmin*) in contrast to those who are dull in response (*mengmei*). And accordingly one identifies those who deserve to be promoted and those who should be demoted in rank."[54] By emphasizing abilities (*neng*) and talents (*cai*) rather than morals, Song rearranged the significant points in Wang's discourse: it was not morality, but rather attentive action that identified a gifted man. This definition provided Song with a coherent theory of human action. Human disposition, when defined in terms of innate ability, trained proficiency, or mental skills, allowed moral behavior to become a rational consequence. It freed man from conformity to a single universally valid ethic.

It is the last point in which Song essentially departs from his contemporaries' emphasis on morals. Song even suggested that leaders could make use of base instincts that led normal people to immoral behavior. Leaders stayed calm, keeping the aims of the society and state in mind, whereas soldiers strove for their personal benefit. Song advised military leaders to make use of this to achieve desired results. If soldiers were offered spoils of war, then they would not ransack and destroy valuable property. Allowing them to loot in reasonable ways, Song argued, would strengthen their ambitions to fight the enemy. "If a military leader can submit the severed heads of enemies, he will receive court rewards and high recommendations. And the army soldiers who can plunder will not collaborate [with the enemy] nor steal army supplies for their personal use. If the clandestine looting of the enemy's funds and weapons is neither investigated nor prosecuted by the supreme commander, then the idea of desertion is completely dispelled and changes into the strong desire to continue fighting and completely take the enemy line by force."[55] Intelligent leaders, he suggested, channeled soldiers' base instincts for their purposes and did not moralize them. This idea may have truly shocked his peers and friends.

Such passages show that Song's worldview envisaged a clear opposition between scholarly knowledge at the conceptual level and practical knowledge as a performative activity. He argued, for example, that commoners could "do or process (*zuo*)," but could not recognize the inner logic of what they were doing or understand (*zhi*) what was happening during these processes. People who received the right training were effective soldiers, farmers, or craftsmen though they had no understanding of what they did. On this basis in *Works of Heaven* Song emphasized how competently the farmers of Shandong worked with different kinds of grains to yield excellent results, and at the same time he was convinced they could not tell sorghum from small-grained millet. Not only was this irrelevant to their work, it was beyond their level of comprehension.[56] The identification of which things were important, the highly theoretical appreciation of the making and existence of things and affairs, was a subject for the scholarly mind. Similarly, Song in *On Qi* praised the common woman who ironed cloth skillfully, yet, he was convinced that she did not have the slightest understanding of the intrinsic logic of the task, that is, why the wrinkles in the cloth vanished or why the water evaporated: "The common people practice these things daily and they do not understand (*zhi*)."

Manifold examples indicate Song's appreciation of craft experience in

Works of Heaven and his concern about appropriate training. This includes his great respect for the kiln master who could judge temperature simply by observing the color of the products inside the kiln, and the dyer who knew how to remove red color from cloths by adding an aqueous solution of caustic soda or rice stalk ashes. Each person performing craft work had to be proficient and experienced in his art to work effectively and obtain good results. A veteran oil presser drawing on lifelong familiarity with his craft produced the highest yield, just as a deft ink-maker was able to tend more than two hundred lamps at a time. Daily routine and practice enabled potters to convert the bulky lumps of clay on the potter's wheel into immaculately formed fragile china bowls without the use of fixed molds. Beginners had not yet acquired this routine. They had to be trained and allowed to "discard their spoiled pieces and renew their effort with new clay" until they were sufficiently qualified.[57] Practiced coal miners may not have grasped the underlying principles of mining, but they could find underground coal by the color of the surface earth because experience had taught them to recognize the various shades. Song remarked that people in Jiaxing and Huzhou had learned to control heat and ventilation to create the appropriate climate for the spinning of the silk cocoons. He realized in this context that the transplanting of new technologies into less experienced regions or those without indigenous traditions was tedious and time consuming.[58] Identifying lifelong practice as the postulate for its successful implementation, Song provided evidence that the difficulties of managing crafts within state manufacture had deeply intruded on the scholarly mind and that scholars had internalized the notion of inherited craft professions.

Song promoted a strict hierarchical assessment of knowledge forms in which the scholar knew the theoretical basis and the craftsman had only experience. His approach to craft abilities, skills, and training aimed at an improvement of the yield and quality of the product, while he ignored skills as a feature that could possibly enhance the craftsmen's social status. An artisan had no "knowledge" nor was he a working mind or a mindful hand, as a few saw him in European premodern culture. When Song explicitly urged the princes in the palaces as well as his colleagues to show interest in and pay attention to the techniques of silk weaving and farming implements, he wanted them to acknowledge artisanal performance as a mirror of cosmological order, not to appreciate skills and abilities. Song's viewpoint was that of the natural philosopher with an epistemic interest. He made craft performance a means rather than the end. The information given in his work, each step and each process, was based on the meticulous empirical studies of an intellectual observing deeper truths and presenting them to his equals. Scholars who acknowledged that crafts and technology were important were on the right path to gain knowledge in general and order the state and society. Song's approach paid tribute to the fact that crafts such as salt, silk, and porcelain had become public affairs and at the same time reinforced the scholar in his socially and intellectually leading role. Fighting a traditional case, Song, however, also rearranged the troops within the social battlefield: craftsmen, soldiers, and farmers were all equally commoners. This challenged his era's high valuation of agriculture and the civil-oriented elite's distaste of military methods. His view of the merchant was that of his colleagues: Song emphasized the benefits this occupation brought to the society and state, but he fervently opposed their intrusion into scholarly realms.

SOCIAL PERMEABILITY AND THE COMMERCIALIZATION OF SOCIETY: THE MERCHANT

> Master Song says: The ten ranks among men range from [nobles such as] kings and dukes down to the ranks of carriers and menials. The great earth creates the five metals for the benefit and use of all under heaven and for later generations. The sense of this [existence of five metals] is the same [as that of the ten ranks], that is, the five metals are an accomplished set in the same way as the ranks of man.
> *Works of Heaven*, chap. 3, no. 14, 1a

In the introduction to the section on boats and carts in *Works of Heaven* Song assigned merchants and traders in human civilization a central role, arguing that after "men, divided into [various] groups, have produced manifold goods, these are traded in order to create the [social] world."[59] Throughout his work Song was quite concerned about state economics of his time observing, "the merchants grow poorer due to the political chaos." He felt that putting financial pressure on merchants through higher taxes would injure the state as a whole. Song warned that merchants had to fulfill their function of ensuring the exchange of goods within and beyond the country. His thoughts on the salt trade, which by Song's time had almost completely been handed over to private entrepreneurs, exemplified this idea: "The small as well as the big commercial traders should deal with the buying and selling of salt as commission agents, in the same way as the five grains. Once this law is put into effect, people from all directions will rush in like a flood of water. And not even the time span of half a year will pass before they have accumulated high amounts of money piled up like mountains."[60]

In direct contrast to the Confucian dismissal of the merchant class, Song suggested the state should ask merchants for advice or even find an institutional way to integrate them into the state economy for the benefit of all: "There are some people in the world who are good at business and commercial enterprise. I would like to debate with them and [we should] advise each other."[61] Song's concern with merchants accords with a trend that, according to Frederick W. Mote, started in the sixteenth century and during which a changing merchant ethos became particularly obvious.[62] A deeper look into this general appreciation of merchants at the end of the Ming, shows that the merchant ethos is less self-affirming than generally assumed. In fact the social trend was one-dimensional. It allowed the merchant to afford the trappings of a rich lifestyle, but in most cases the social position of merchant was not itself respectable or desirable. A merchant could still only gain status by becoming a scholar.[63] Intellectuals such as Wang Daokun (1525–93) or the aforementioned Wang Yuan, who challenged such social distinctions, were exceptions (both equated the functions of merchant's roles with those of the scholar-officials). Merchants themselves attempted to imitate the pattern of the scholar-officials' lifestyle as closely as possible, investing family talent in official careers and family wealth in land to have a claim, at least in name, to the highest intellectual class and gentrified society. Doing this, the merchants held on to a value system that honored intellectual supremacy of a scholarly kind. Their attitudes actually reaffirmed the social order.[64]

Much research has focused on the number of private writings that complain about vanishing social restrictions and the Ming contemporaries who bewailed the crumbling of their society. From the point of view of an unsuccessful scholar like Song the penetration of wealthy merchants' sons into scholarly fields greatly intensified the competition for the few vacant official positions. Even after they passed the exams scholars were no longer nominated, they had to buy themselves an appointment. Scholars facing redundancy from government service, felt that capital was stronger than capability and developed a strong sense of personal and familial uncertainty. Like Song they were disgruntled men, fighting a losing battle on the lowest tiers of their own class. These men feared and fought against any change in the social hierarchy. *An Oppositionist's Deliberations* reveals Song as a near-perfect example of this development in which scholars acknowledged the benefits of using merchants and at the same time advocated traditional values rather than social change. Song considered himself the descendant of a famous high official, although he lacked the credentials of a long-lasting intellectual lineage. He felt victimized by a society led primarily by unworthy and undeserving officials whom he characterized through their ignorance of literature, corrupt administration, and unethical behavior. He could appreciate that merchants were an important element in society and at the same time blame them for several shifts in social behavior. Commoners were giving up farming and losing their appreciation of the simple life; people were losing their respect for intellectual authority and seeking ways to earn quick money; official ranks were illegally purchased; and some commoners lived in luxurious residences and wore splendid clothing. Song accused the merchants in particular of trying to imitate the behavior of the gentry, disobeying the rules that determined their social status.[65] In this context Song repudiated the flamboyant world of consumerism, cultural brilliance, and innovative ideas which recent research has identified in the writings of Li Yu (1611–79) and others during the period of transition from the Ming to the Qing dynasty.[66] Song belonged to the group threatened by liberation from social hierarchy, and he blamed it on the lackadaisical behavior of his colleagues. Instead of serving society, they let society serve their needs.

Socially Song promoted a traditional viewpoint: scholars provided the concept, the commoners performed their daily duties, and the merchant kept the economy going. Scholars arranged this order hierarchically in ten ranks. According to the *Commentaries of Master Zuo Traditions*, at the top was the king (1), then the duke (2), followed by the minister (3), and at the bottom, the slave (9), was followed by the menial (10), (no specification for ranks 4–8). For Song, and here he differs from his colleagues, this order was not simply justified by tradition and references to the classics. It manifested a relation between heavenly and human affairs, defined by a quotient of ten: "For Heaven there exist the ten days [cycle], for man there are the ten ranks or categories. The affairs down [on earth] accord with those in heaven."[67] The proportional presence of all kinds of metal ores in nature displayed the universal order in natural phenomena. Within this framework Song valued the lower classes as vital and accomplishing elements of a functioning system. Nevertheless, he thought of intelligence and stupidity along the same quantitative lines as for metals. Thus, noble men were as rare as gold while stupid people abounded like zinc and lead: "Of the noble ones [of man and metals] one is born every 1,000 miles, or at the most every 500 to 600 miles; the minor kinds are [found] in locations which are accessible, that is, not very hard to reach by boat and wagon. These [minor kinds] are produced by the soil arbitrarily in big swathes. [The relative proportion between] the value of gold, the beauty [among the metals] to black iron (pig iron) (*heitie*) is a ratio of 16,000 to 1. But, if the cauldron, the pot, hatchets, and axes were not available for daily life, just because one found gold, [this would be like a society consisting] only of high officials and there would be no commoners any more."[68]

In this passage order is displayed in nature and material inventiveness alike. In this framework, Song, the scholar, assumed his traditional role as the intellectual interpreter of the relationship between heaven and man. The world in which man lived was a universal and a rational system beyond human influence, while society, state, and human civilization became the result of human behavior, traditions, customs, and habits, a product of man's situational choice to obey or disobey the universal rules of *qi*.

CUSTOMS AND HABITS

> It is the nature of boats to follow the current of the water, just as it is the nature of grass to bend to the wind. Therefore the rudder should be used to control the water flow and regulate the direction of the boat.
> *Works of Heaven*, chap. 2, no. 9, p. 33a

The builders of the large sea-going vessels in which Zheng He (1371–1433/5) had traveled the south Asian seas, found their model in nature: bamboo. Inspired by the septa in the bamboo stem, they built elongated ships with compartmentalized sections that could be closed in case of leakage. The material qualities of bamboo could be applied in various ways in shipbuilding. Sea-going vessels, for example, attached posts made of split bamboo to both sides of the bulwark, like a fence to fend off the ocean waves. Bamboo could stand against the power of the waves and not break. When boiled, green bamboo fibers could be twisted to form ropes for the anchor that could sustain 1000 *jin* (about 600 kilograms per strand). According to Song, this was because the nature of bamboo was "genuine" (*zhen*),[69] bending like grass to the wind without breaking or losing its true qualities. This was also how a scholar should perform his work, bending to custom while preserving his true nature.

Song used long parts of *An Oppositionist's Deliberations* to argue that customs and habits (*feng su*) were basic structural elements in the construction of beliefs and the functioning of society. Once established, he argued, they shaped the human world and made man hold on to certain features while dismissing others. Bad customs and habits were initiated by man's search for well-being and society's approval and then aggravated by materialistic trends in social taste, such as when youngsters "wore the splendid official hat and put on their high horn-shaped hairstyle."[70] For Song man was thus a calculating rather than a rational or moral person. Man was, furthermore, not steadfast (*gu*) enough to move on the straight and narrow when moving through the murky waters of bad customs and entrenched habits. A rudder was necessary to keep him on track: that rudder was the scholar.

Song thought of customs and habits as an involuntary product of the human heart-mind (*renxin*), not generated by active considerations or intentions. In fact, "a single tendency of the human mind [toward a certain direction] can generate customs and morals," which made them arbitrary and artificial. In times of crisis even efficacious long-standing customs could be overthrown in a second: "How long are thirty years [in which customs were effective] with regard to the current condition (*guangjing*) [of despair]?" People's transformation of customs was only restricted by those customs that were already in use. Convention became part of tradition, and soon people forgot that the custom was not by nature or heaven given but made by man. Song saw pragmatism as a factor when he suggested that man would only maintain customs such as that of adequate clothing during times in which this required no huge effort. Those who thus tried to establish proper behavior by maintaining customs would fail. Moral behavior could only prevail when man did not have to fear for his and his family's life: "in generations of peace, the human mind is rather modest and easy to satisfy—[people] just look after what they have. They do not want to live in debauchery or indulge in luxury. It seems as if in generations of peace people are frugal and humble and thus try to avoid trouble. They first look toward protecting that gold which is already in their hands, and do not think about future wealth or do not busy themselves trying to obtain a high official position. In generations of peace it is possible to store the surplus money in deep vaults and there is no need to be anxious about relatives strolling around and leading a dissolute life." In times of increasing social and political insecurity, people, however, started to "squander money and indulge in excess, . . . the poor people, who are in debt, long for clothing of bright colors and for luxurious, wasteful celebrations for themselves."[71] If one wanted to employ customs to initiate moral behavior or establish order one had to understand how customs were created, namely, by man's aspirations for a good life. Understanding this, one could manipulate customs and habits to initiate new social formations and moral attitudes. The scholar hence had to make use of the universal goal for most people in life "to be rich and feel valued. . . . All people yearn for prosperity. So like a herd of wild ducks (*wu*) they hurry after the money, move to other districts and die there. As soon as they realize that the salt enterprise (*yanhang*) provides profit, who would not run after it with great eagerness?"[72]

Within this context Song paid respect to the fact that desires were different for the members of the two groups: the commoner was mostly interested in a good life, the scholar ultimately desired to perform his duties and serve the state. A scholar did not want to enrich himself. He wanted to be valued by an appointment to an official position—to then exemplify "untarnished conduct" (*qingcao*).[73] The leaders of society, the emperor and those in power, had to respect that and act accordingly. Their disregard of the commoners' and scholars' essential goals and desires was the root of chaos. As a result they did a miserable job of running the state.[74] Song further emphasized that even people who had a basically good disposition (*xingshan*) could hardly resist wrongdoing when their life was threatened and others set a bad example: "At first it seems as if their spirit can hardly bear it. But when they are together with such people for a long period they are infected. They are exposed to wrong behavior (*bu shan*) and then become used to these usages (*su ye*), committing the same crimes over and over again. All but maybe twenty in a group of one hundred are able to resist [when confronted with such attitudes]. And eventually even those who hold out have no alternative: they turn around and become followers [behaving badly]."[75]

In such passages Song reflects the insecurities of his era. Identifying man's morality as volatile and subject to the fulfillment of his essential desires made craft look like a stable issue: artisanal production was one of the few domains that still worked in his time, producing the required output in an orderly fashion. Fashions came and went, but porcelain was always made of white clay, molded on the potter's wheel, dried, glazed, and fired in a process with an inherent technical nexus. This provided the kind of cohesion Song was looking for: stability and reliable principles. Crafts displayed universality because they were rooted in antiquity. This proved they were basic to society and human life. The artisan was a cog in the wheel of things and affairs produced by and representative of universal order. Song, passing by the hundreds of stylized vessels, fragile cups, and huge cauldrons stacked in the gateway to Jingdezhen, may not have seen the laborers struggling to complete their daily tasks, but he worshipped the universal principles that artisanal work expressed and entailed.

In sum, Song's approach is ingenuous in many ways, in particular because it subtly rather than rudely defies the standards of premodern Chinese thought on crafts. It does not fit into a simple narrative about knowledge-making in historical perspective. Rather it indicates the multiplicity of historical development and the variety of optional choices that human history has taken throughout cultures and generations. Song's interest in practical things develops from a uniquely Chinese incentive to engage in making them. Observing craftsman skills, he saw the knowledge inherent in its performance and claimed that only the scholar was aware of this particular nexus. In the vibrant world of the Ming, in which the world of crafts diversified and things abounded, Song thus approached crafts as a way to qualify himself and his social status as a scholar. Locating the origin of knowledge in things and affairs, he identified scholarly, theoretical skills as superior and considered the craftsman's work an epistemic object rather than the working hand. Observing the world, Song thus came attired in a scholarly gown with a traditional dark blue aura. Blue as it was, it carried the iridescence of cosmo-political interpretation. Both the blue and the iridescence contributed to the way in which Song meticulously researched his environment. How he gained his knowledge about technology, crafts, and natural phenomena and presented its facts and facets in his writings is the topic of the next chapter.

CHAPTER 4

Written Affairs

> As the consciousness (*ling*) of the myriad things, man has just the right [combination of] the five senses (*wu guan*) and the hundred constituents of the body (*bai ti*).
>
> *Works of Heaven*, chap. 1, no. 2, 23a

Year in, year out, late spring saw tenant farmers swarming over the farmland in southern China, toiling side by side with oxen to prepare the desiccated fields for the second harvest of rice. After the ploughs had imprinted the land with an orderly network of furrows, families took to the fields, flooding them and planting the small green stalks in orderly rows. A week later the old leaves were replaced with a fresh pleasing green that reminded man of nature's power of creation. In *Works of Heaven*, Song highlighted a process called "banking up the roots (*zi*)." He explained it in detail in the text and had it illustrated (fig. 4.1). The illustration shows two stylized men leaning on a stick and pushing the earth around the roots with their feet to fix the plant firmly in the ground. Living in a region with rich soil, Song must have observed such scenes many times before he described this essential but mundane procedure with painstaking exactitude. He may have anticipated the clanging of the treadle and the moaning of the ox-driven wooden waterwheels accompanying him throughout the summer. And then turning around, he would have looked forward to the silence of his chamber, where he could write down all that he had seen and heard.[1]

In the previous chapters, Song is shown to be a Chinese scholar who saw beyond the craftsman and discerned theoretical principles in the performance of their work. How Song crafted knowledge-making and how he shaped the details of production described in my short vignettes into facts in his writings are the subjects of this and the following chapter: they unravel

Fig. 4.1. "Banking up the roots" (*zi*): Song illustrated processes because he considered them central to transformative process, which is in this case "growth." Everyday practices and knowledge such as hoeing are thus given as much attention as hydraulic engineering. *Works of Heaven*, chap. 1, no. 1 *Farming*, 7b.

the subtle interlocking of erudition and empiricism in seventeenth-century written approaches to material inventiveness and natural phenomena, the role played by personal experience, experiment (hypothetical or physical), and quantification. This chapter begins with Song's approach to practical observation in his visual and verbal rhetoric and then examines how he uses practice and theory when making his knowledge. Switching back and forth between the two levels of analysis, practices of inquiry and their rhetorical usage in Song's writing, this chapter reveals how Chinese scholars observed knowledge and wanted it to be observed in their writing and illustrations.

A burgeoning literature on early modern European history has been drawing attention to practices and forms of inquiry and their function in the written discourse of emerging communities. Many works based on a sociology-of-science approach employ this as an indicator of change and continuity in the making of scientific or technological knowledge. These studies show that as hypothesis, calculation, ways of reasoning, and factual evidence received and lost validation, disciplinary fields were formed or abandoned, and domains of expertise emerged or vanished. Such practices can be found in the whole gamut of Chinese scholarly culture. Scholars usually discussed them in terms of sociopolitical bearing and moral agency. The literature of the sixteenth and seventeenth centuries shows crucial shifts in literati views on forms of inquiry, knowledge assessment, and its presentation. Notables and minor officials, intellectuals and politicians reacted to the growing presence of material things and practical matters in their life and work by wrapping these issues up in fine words and, increasingly, expressing them in images. It is the similarities and differences in his conceptual bearing when compared with those of his Western and Chinese contemporaries that makes an investigation of Song's efforts so interesting.

By juxtaposing his individual strategies of gaining, holding, and assessing knowledge with contemporary British culture the subtle implications of the "cultural factor," that is, the impact of contemporary environment and conditions, come to the fore. When Song pointed to the importance of man's senses and his body within the investigation of things and affairs, as exemplified in the quote at the beginning of this chapter, or when he gave items and issues meaning and sanctioned their inclusion in the discourse of natural philosophy by "writing down" facts and events, he did not differ greatly from his British contemporaries. As Steven Shapin and Simon Schaffer have argued, natural philosophers of the seventeenth century also employed "matter of fact" to substantiate their theories. Robert Boyle (1627–91), for example, cautiously focused on experimental and observational "facts," criticizing theories derived solely from thought-experiments.[2]

Boyle described what he saw and then offered theoretical explanations, as did Song. Both presumed that man's suggestions were subject to constant revision and expansion. But as their cultural leitmotifs were not the same, their approach was different. For Boyle facts mirrored the eternal truth of a "nature" in a world that contrasted man with God.[3] Yingxing assumed that facts about his surroundings revealed the universal validity of cosmic principles. A comparison of the methodologies and rhetoric of seventeenth-century scholars in England and Ming China shows subtle variations rather than a clear contrast.

When Yingxing upheld direct observation and personal experience by referring to antiquity (e.g., the sage-kings as discussed in chap. 2) and criticized his era's views, he reminds us not only of his Chinese contemporaries, the philosopher Wang Fuzhi and the political theorist Huang Zongxi, but also English natural philosophers, such as the physician William Gilbert, one of many European thinkers who also rejected the authority of the classic schools of his time and questioned the pedigree that seemed to link these doctrines to their alleged pure, potent, and ancient sources.[4] Despite their differing cultures, traditions and ideals, Gilbert in England, together with Huang and Song in Ming China, all proposed that the errors cumulatively introduced into ancient texts by contemporary scholars (whom they identified as "mere copyists") could only be corrected by personal observation of nature and man. Affected by sociopolitical insecurity, these men from two distant cultures lamented in a similar manner the loss of significant values and the ignorance of their contemporaries.

When seventeenth-century Chinese intellectuals such as Song Yingxing or Huang Zongxi promoted an engagement with nature and material objects, when they placed faith in personal experience or played up the role of the senses and intuition, they laid their trust in nature and the individual mind, as did British naturalists such as Gilbert. And yet the paths of these men diverged because of the varying ideals governing their thinking. Seventeenth-century Chinese scholars developed their ideas on nature and culture along ideals of order and unity, as opposed to the religious discourse influencing early modern European culture (see chaps. 2 and 3). Chinese scholars, achieving their status through literary training, also combined a strong obligation toward society and state with their quest for knowing. This crucially affected their approach to knowledge as well as its presentation and assessment in written discourse. Whether this framework had a restrictive effect and if so, to what extent, is open to debate, especially when we broaden our view to the entire scope of knowledge-making and do not confine ourselves to knowledge we assume is necessary for the development of modern science categories. But we can see that by the seventeenth century the British had already begun to consider elements of nature in isolation, often delving deeper and deeper into the details of phenomena such as magnetism and electricity. Going beyond acknowledged realms, Gilbert and his colleagues had to cope with a loss of confidence in their sayings and findings. His generation in Britain established trustworthiness by supporting their ideas through the personal testimony of reputable and trustworthy people as eyewitnesses. And they developed new methods and rhetoric of reliability, such as the replicable nature of proofs, accuracy of measurements and standardized criteria for verification procedures. Chinese scholars of Song's generation, broadening their view to new fields, also developed new practices of inquiry and revived old ones. They changed their rhetoric of knowledge-making and, observing their world and experimenting in it, they opened their minds and reached out their hands to a realm full of things and affairs.

Knowledge contents are only one topic of discussion in the Chinese history of scientific and technological knowledge. A systematic approach addressing the changing rhetoric of knowledge-making in Chinese literature on subjects such as nature and material inventiveness, practices of observation and experiment is still in the nascent stage. Studies on the lives and philosophies of individual thinkers show they had a huge set of issues to choose from and manipulated their choices with surprising flexibility.[5] In principle, Song's propagation of empirical studies followed an eleventh-century tradition based on literary exegesis that still provided the building blocks of academic education in the seventeenth century. By analyzing Song's choices, seeing where he adhered to tradition and which conventions he rejected, we learn the significance within late Ming intellectual discourse of such seemingly random or everyday observations as "catching sight of something" (*jian*), the "knowledge of seeing and hearing" (*jian wen zhi zhi*), "intended inspection" (*guan*), and "experiencing" (*yan*) or "experimenting" (*jing*) with material objects or natural phenomena. Discussions of "knowledge and action" (*zhi xing*) in the late Ming intellectual world turned away from text-based approaches and new ways were found to establish facts and explain evidence. Joanna Handlin remarks that toward the end of the sixteenth century, personal experience "supplemented, if not superseded, received opinions as a source of authority for knowledge."[6] Her study presents intellectuals such as Lü Kun or Lü Weiqi (1587–1641) as shining examples of a generation that proclaimed experiential recognition (*tiren*) of every word in actual fact. The previous chapters have shown that Song positioned himself intellectually within this trend but criticized contemporaries' rhetorical means and methods. In *Works of Heaven* his matter-of-fact style condemned rigorously the province of moral philosophy, social ethics, and aesthetics that his colleagues Lü Kun and Lü Weiqi used to authorize their efforts. Looking for certainty, Song strongly disapproved of their approach to textual authority, evidential studies, and text-based philology. He insisted that understanding lay in empiricism and the correlation of facts gained from careful scrutiny of one's surrounding.

Song transformed contemporary and classic features to foster his argument and promote crafts and technology as the foundation of knowledge. The illustrations and diagrams (*tu*) in *Works of Heaven* and *Talks about Heaven* provide a view of the visual rhetoric of his culture. They show that Song fruitfully combined acknowledged practices of representation with new content to express his ideas and concepts. Song's textual descriptions connected to the plain matter-of-fact style of the twelfth-century historian and encyclopedist Zheng Qiao (1104–62). Zheng mainly theorized. Song methodically reported things and affairs and in that way, throughout his writings, delineated facts and issues of everyday life into a momentous practical and intellectual exercise divulging universal orderly schemes. I pay tribute to Song's method in this chapter by embedding his theoretical statements into my portrayal of his approach to observation.

RHETORIC OF KNOWLEDGE INQUIRY: TEXTS AND EXPERIENCE

> There are a myriad of things and affairs [in the world] which have to be heard of and seen before they can be truly comprehended. To how many of them is this [principle] actually applied?
> *Works of Heaven*, preface, 1a

Approached by boat at night, Jingdezhen, the Ming dynastic center of ceramic production, must have resembled the glowing embers of a furnace with hundreds of small and large kilns irradiating the dark with a hot red light. Dust rising from the water- and labor-driven pestles pulverizing the kaolin in the mountains east of the city smothered the valley. Small boats transported square blocks of clay to the potteries, where they were transformed into the fine porcelain wares for which the city was renowned. This information regarding transportation is mentioned in *Works of Heaven*. Song may have accompanied the boats on part of one of his journeys to Beijing. We can only presume he took the opportunity to pay a visit to the famous pottery town (see map).

If Song did travel to Jingdezhen, he would have come across huge factories and cramped, poky workshops in which sweating workers and highly skilled craftsmen labored side by side to produce the objects so desired by the emperor and his elite. His journey would have passed through a landscape ravaged by surface mining, deforestation, and deeply furrowed roads scoured by the passing of one-wheeled barrows. These details impressed his contemporary, the art collector and official Wang Shimao (1536–88) and later, in the nineteenth century, William C. Milne (1815–63), the son of a missionary.[7] Song's *Works of Heaven*, however, concentrated on the sober facts of production, not on the man-made devastation of the landscape. A careful documenter of crafts, Song gave details on grading raw material, tempering the clay, forming a body, decorating, glazing, and enclosing it in a saggar (a boxlike container made of high-fired clay to protect porcelain wares during firing), and on the final firing in the kiln. He distinguished the assorted end products and he briefly outlined potter's wheels and kilns in word and image. How did he acquire the knowledge his pen so insightfully transcribed as facts into the domain of writing?

The philosophy and history of science has long stressed that data gathered by the observer is influenced by the individual's background, but it must be added that how this data is presented, the rhetoric of addressing sensual experiences, varies significantly throughout cultures. This is particularly relevant in Song's case. On the basis of his interest in craft manufacture, historians of Chinese approaches to scientific and technological thinking have claimed that he is the perfect example of a scholar-practitioner engaging directly with the issues he described. Yet we only know what was exceptional about his approach when we decode his rhetoric within the context of his era. The eligibility of practical engagement and the analysis of material objects by way of systematic inquiry was as bound to disciplines, content, and fashions in premodern China as it was in Europe (or any other culture). For example, during the Song dynasty astronomers at court favored calculation over observation. The polymath and astronomer Shen Gua criticized this prejudice, arguing that in his era astronomical knowledge suffered because his colleagues did not apply empirical modes of inquiry.[8] Daoist masters (of alchemy) fervently debated whether mixing, distilling, and extracting could lead directly to knowledge, or if theoretical models were preferable.[9] Authors of agricultural treatises almost always pondered the necessity of hands-on experience with planting methods, whereas devotees of tea production, flora, and fauna seem to have shifted back and forth between textual sources and empirical inquiries.[10] As the only source material is their own writings, it is hard to judge if these literati actually had any contact with the sweaty side of personal experiment, or if they sat theorizing on elevated platforms under thatched roofs.

The situation is even more complicated when one looks at the sophisticated rhetoric of scholarly and philosophical discourse, especially that of the seventeenth-century Chinese world. In the uncertain times of the late Ming, the strategies of authorizing one's method of gaining knowledge were volatile. They were involved with, and often subservient to, sociopolitical purposes. Actual practice often developed while rhetoric persisted, or vice versa. Scholars of the early seventeenth century showed an increasing interest in enlarging sensual and bodily inquiry as more than a mere supplement to textual survey in their quest for knowledge. These men saw (*jian*), heard (*wen*), personally encountered (*yu*), experimented (*yan*), or tested (*shi*) what they wrote down as their ideas. Some systematically inquired into natural phenomena (*xiang*), things (*wu*), or events (*shi*), tasks and affairs (*wu*), while others claimed to have come across their object of interest by chance. Some thinkers left their study rooms and watched clouds or inspected trees. Others interpreted practical engagement as a call to self-contemplation and the study of one's own mind, as, for example, Wang Yangming. Scholars explained how they "practiced" gaining knowledge but did not provide any verification for it. Therefore, the actual quest for knowledge may have remained an effort of the mind, a mind where, according to the majority of scholars, universal truth lay.

A common methodology in Song's time was to mark out one's intellectual territory and position oneself through terminological distinctions and reference to classical sources and predecessors. Then the scholars interpreted these references individually by linking to other thinkers or texts. Popular maxims were "practical or pragmatic learning" (*shixue*) and the "investigation of things." In the title *Works of Heaven,* Song associated his work with "the inception of things and accomplishment of affairs/tasks" (*kaiwu chengwu*). He was, however, particularly careful never to discuss his modes of inquiry philosophically or to link them to other thinkers. His literati sociopolitical background suggests that Song approached sensory experience within scholarly tradition as a method of academic training. As the quote from *Works of Heaven* at the beginning of this section emphasizes, for him things and affairs had to be "seen and heard before they can be truly comprehended," that is before they became meaningful. Song used in this context (see quote at the beginning of this section) a classical quote that refers to a scholarly practice of memorization rather than observation. The purpose of direct involvement is, as the original text continues, that scholars encountering what they read, "can recite a passage after one perusal" (*guomu chengsong*). Song's reference suggests that, for him, comprehension was an act of making knowledge by seeing it in action, for example, by watching how porcelain pots were formed and fired. It also suggests Song assumed a priori knowledge or action, not its unity as Wang Yangming assumed. A unity of knowledge and action, Song presumed, would mean that knowledge would always be enacted. The chaos of his era proved that this was wrong. Instead, Song concluded, knowledge came by observing actions such as the performance of crafts or natural phenomena. Within this framework Song also gave relevance to texts. Written reflection, he proposed, could guide one through the order displayed in things, elucidating the essentials. Texts could be the incentive and end result of inquiry, as knowledge was formalized within it. Yet, texts could not replace the close scrutiny of action when it came to lending credibility to knowledge. Texts detached from action, that is, from knowing how things and affairs happened, were dubious. Song hence validated Li Shizhen's *Compendium of Materia Medica* as credible by quoting its contents as facts, but this was done only after verification. The only exception Song allowed were some early classics, such as the *Book of Changes,* which he suggested entailed some essential truths. Song explicitly disapproved of "historic-evidential research" (*kaozheng*) from the Song era up until his own. He entirely ignored the purely philological, text-based surveys and the commentary literature that had accumulated over centuries. Song felt this scholarly approach had polluted true, universal knowledge. Dismissing the growing repository of textual knowledge flooding seventeenth-century intellectual life, Song assigned the observation of the mundane a crucial status in understanding the principles of order.

As mentioned previously, Song emulated the writing style of the Song dynasty author Zheng Qiao. First used in the *Tongzhi* (*General Gazetteer,* 1162), this style was later used in a considerable number of private jottings—the genre to which I argue Song's oeuvre belonged (see chap. 1).[11] Using the same trope as many of the authors of these jottings, Song apologized for paying deficient homage to classical precedents. "Alas, how grieved I am and how poor! I wanted to buy distinguished implements and do some historic-evidential research, but I lack the financial resources of [those who work on] the [orthodox] thinkers from Luoxia," (that is, Luoyang).[12] I suggest this remark should not be taken at face value; Song was employing the topos of poverty to damn his colleagues with faint praise.

On the surface level this is a clear statement of ambition frustrated by financial limitations. On the contextual level, when one takes into account Song's anger and frustration, it can be read as a critique that men of talent have no need of resources, for their knowledge is innate and their actions sufficient to reveal their morality. Song was claiming that his aims were comparable with those of great historical figures of the past. Simultaneously, by stating the unbridgeable difference between their situation and his, he was excusing his limitations.

Song's allusion to *Luoxia*, the third-century name of Luoyang, reminded Song's contemporaries of a scholarly tradition of philosophical issues represented by two brothers Cheng Yi and Cheng Hao (i.e., Cheng-Zhu learning), who were born there. Their learning represented the orthodoxy of Ming scholarly training that, as his family genealogy reveals, was important to Song. His remark mocked those who had access to the huge secondary literature on the original philosophers. They could afford the scholarly paraphernalia to help them pass the exams: primers, private tutors, and letters of recommendation. But in fact their studies were superficial, as they neglected the necessity of seeing knowledge in action. Song's remark expressed his suspicion of the lineages of thought defined in terms of commentary literature that lay at the heart of Ming learning. His approach followed Zheng Qiao who claimed that one should never blindly follow texts. The true scholarly method, Zheng Qiao argued, lay in using different channels of perception. As observer, man could see and hear. The ear represented the verbal aspect that was expressed in writing. The eye represented the visual, for which reason images were central to understanding. Only in the combination of both, could one "understand the usages" (*ming yong*) and "verify the facts" (*fushi zhi fa*). On this basis the scholar could then establish the method for the classification of sections (*bu wu zhi fa*), that is, order the knowledge. One should furthermore learn and read extensively before specializing.[13] It was within this framework that Song approached observation or personal experience as an important form of inquiry.

The surface level of the quote certainly also carries a truth, in that Song may well have lacked sufficient funds. But few of his colleagues would have taken Song's remark as a yearning for material wealth, because most scholars deemed such a desire unworthy of their class. Song addressed poverty to signify his moral rectitude (see chap. 1). This made him more reputable in the eyes of noble-minded scholars. This was the effect Song was after in *An Oppositionist's Deliberations*, where he attacked those with money but no personal talent. He complained bitterly that rich households bought their sons high positions, while assiduous, talented, less well-off scholars fell by the wayside and had to restrict their scholarly contribution to the role of a county teacher—just as he had done.[14] Such remarks, in which Song openly addressed his own situation and expressed his anger at his unjust treatment, indicate his emotional involvement. Leading a life of moderate means and equipped with marginal sociopolitical power, he armed himself against his fate with rectitude and scholarly ideals, not with plough and ploughshare.

Song further regretted in *Works of Heaven* that he was not able to summon his peers for scholarly discussions, because he "ran short of the guesthouse (*guan*) like that of Chen Si" (192–232).[15] Once again, the surface level of Song's remark is a statement of fact; he did not have access to a guesthouse to assemble the great minds of his time. The relevance in Song's disingenuous apology for his lack of resources lay in the reference to the prince Chen Si, addressed as king (*wang*). Chen Si is the nickname of the poet Cao Zhi, the third son of an eminent statesman and military leader Cao Cao (155–220), who founded the Wei dynasty of the Three Kingdoms. Mentioning Cao Zhi conjured up the image of a poetically talented son with high aspirations to quell the chaos of his time. Cao Zhi was involved in power struggles after a period of despotic rule, but he failed and was thus neglected by his contemporaries.[16] Song may have seen himself in a similar situation, but he also invoked the case as an example of academic debate and literary sophistication being insufficient to overcome political chaos. By linking himself with Cao Zhi, Song found fault with the statesmen of his era and promoted his efforts as part of a tradition of righteousness and orderly statecraft.

As in the family genealogy, Song's rhetoric carefully avoided any indication that he had personally engaged in any form of manual work, not even agriculture, an honorable occupation for all gentlemen as were spinning, reeling, and weaving for any virtuous lady. This is surprising when one considers that much of *Works of Heaven* could be placed in this category: the growing of grains (section 1), clothing (section 2), grain processing (section 3), dyes (section 4), and related issues such as oil production (section 12) and the fermentation of beverages (section 17). When authors on agricultural themes mentioned their personal experience or observation of farming technologies, they often did so to stress their altruism. This was a form of attack on politicians and elite representatives who had lost touch with reality and had their heads in the clouds. Geng Yinlou (?–1638) exemplifies this trend during the late Ming. Discouraged by his appointment as a minor official, Geng occupied himself with farming techniques. He claimed to "promote agriculture" (*quan nong*) for the benefit of the state and the people.[17] Song abstained from such claims and also avoided any implication that he had actually performed agricultural tasks. This shows that his view of these fields as described within *Works of Heaven* was not driven by altruism.

Song's notion of the value of agriculture becomes particularly apparent when we extend our view to the visual rhetoric in *Works of Heaven*. In his longest book Song extensively utilized illustrations, subtly employing visual presentations of technical contents to substantiate his arguments. Song could draw on a long tradition of using illustrations in woodblock printing. Since the first uses of woodblock printing in the eighth century more and more books were illustrated, both for didactic and decorative purposes. Nonverbal documentation is evident in a wide variety of fields: astronomy, alchemy, pharmacopoeia, mathematics, and tracts with religious or sociopolitical content. Agricultural writings started to tap into the potential of this method after 1313 when Wang Zhen published his extensively illustrated *Book of Agriculture.*

Wang Zhen's combination of image and word attracted a broad readership from landlords to state officials and helped to promote agricultural knowledge as an elementary field equally relevant to the farmer and the scholar, as Francesca Bray has pointed out.[18] With the second expansion and commercialization of the printing industry in the early sixteenth century, illustrations became a competitive factor, attracting new readerships and markets. By Song's time, illustrated household encyclopedias, ritual guides, works on medicine and farming, and treatises on geomancy or divination, as well as lavishly decorated novels, poetry and private jottings flooded the Chinese book market. Experimentation with the new medium coincided with the discourse on how things and affairs revealed knowledge and which forms of inquiry were legitimate. Song was affected by these dynamics when producing *Works of Heaven,* but probably his approach to illustrations was the most significant result. In the tradition of Zheng Qiao, Song considered their active side as well as their reflective purpose and thus effectively deployed visual presentations as a complementary argument to express his philosophical viewpoint. This rhetoric reveals that Song considered "observation," the study of things and affairs via images as well as in real practice, central to comprehension.

IMAGES, TECHNOLOGY, AND ARGUMENT

> Every scholar drills during his childhood education the verbatim sense of words such as "order and chaos" and "good government." Yet, at the end of his life, he has still not seen the very objects that represent (*xingxiang*)[19] these issues, the "ordering of threads in silk reeling" (*zhi*) and "badly entangled fibers" (*luan*), "the warp thread (*jing*) and the woven pattern (*lun*)." This is so disappointing!
> *Works of Heaven,* chap. 1, no. 2, 23a

Since the time of the southern Song dynasty (1127–1279), the lower Yangtze valley had been China's main silk-producing region, particularly the northern prefectures of Zhejiang Province and the Jiangnan area on the shores of Lake Tai. Silk brought prosperity and production reached unprecedented heights with the increase in imperial silk-weaving factories during the Ming dynasty. In some districts of cities such as Suzhou, Nanjing, and Hangzhou, silk looms may have been seen in nearly every household. The draw loom with a figure tower was truly impressive. Five meters long and requiring two operators, it was used to weave exquisite silks with complicated patterns that only the court and the rich elite could afford. Setting up such a loom required time. The warp threads had to be carefully wrapped in parallel order to prevent entanglement of the fibers. Special care was necessary whenever the pattern was warp- and not weft-based, because in such cases the warp threads had to be arranged beforehand in keeping with a given scheme and the driving shafts had to be adjusted accordingly. Then, if the weaver gave clear commands at the right moment to the draw boy, who sat on the figure tower to lift the shafts, the pattern would be flawless. Indeed, the loom was a perfect example of order; it could not tolerate any element of chaos.

The concept of quelling the chaos was Song's major authorization for his approach to crafts (see chap. 2). Substantiating his argument, Song, in the section on weaving in *Works of Heaven,* employed both text and image to promote his major concern. In the text Song offered an etymological analysis, reminding readers that phenomena such as "warp and weft" or "badly entangled fibers" were associated to words such as "good government" or "chaos," respectively. His illustration of a draw loom with a pattern tower (*tihua ji*) (fig. 4.2) does more than simply represent the actual object. Analyzing it in relation to the other images in *Works of Heaven,* we can see that it is by far the most sophisticated, propagating order in its very design, both refined and austere. Snugly embedded in the textual argument, the two-page illustration contained in the original 1637 edition depicts the main details of the draw loom with its pattern tower. Extraneous details and embellishment are kept to a minimum. Major components are labeled, guiding the observer's eye through the image, revealing the methodical internal working of a complicated technical apparatus. To emphasize Song's claim the majority of images in this section showed the ordering of threads and illustrated men wearing puttees and sporting a long fingernail on their little finger, both marks of the scholar-official.

With a pages ratio of 1:4 images to text, *Works of Heaven* can be deemed an extensively illustrated book. Unlike the elaborate illustrations in the eighteenth-century reprints that charm the reader with their appealing, elegant style, the originals are, as E-Tu Zen Sun and Shiou-chuan Sun remark in the preface to their translation, "striking for their simplicity and

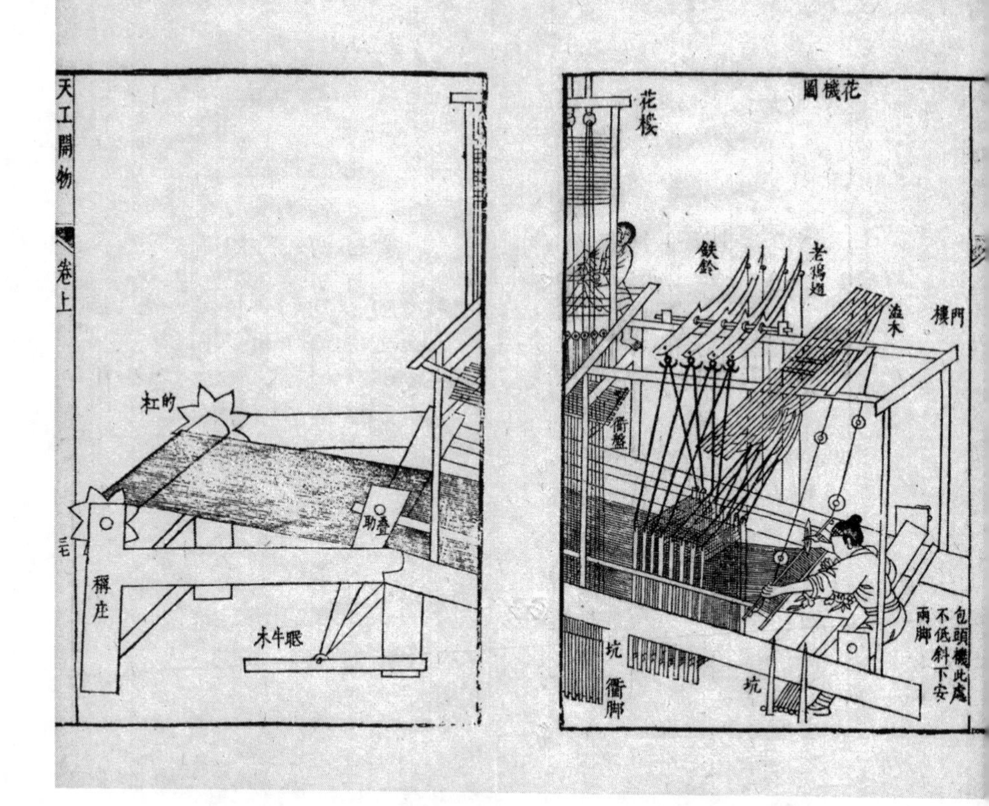

Fig. 4.2*a–b*. Order personified—"Diagram of the draw loom with pattern tower" (*huaji tu*): Accurate in drawing style, the author also meticulously names crucial elements. The note at the left downside warns that at the headband the loom should not be slanting and indicates the place to rest one's legs. *Works of Heaven*, chap. 1, no. 2 *Cloth Manufacturing* (double page image), 37a/b.

clarity."[20] Some of the first edition's illustrations depict complicated machinery, others show everyday issues such as weeding. Images of pearl diving or the calcination of stone depict topics that were hardly ever addressed in Chinese literature, either in writing or image. Illustrations in all three editions up until the nineteenth century have received ample attention from researchers who have used them as a source to examine the history of technology in premodern China, or the history of illustrated book printing.[21] My focus is on Song's perception of the relationship between observational practice and knowledge-making, how he looked at things and wanted the reader to look at them. Thus I consider only the illustrations from the first edition, which I see as integral to Song's argumentative strategy and a constructive and supplementary part of his textual argument. They were not meant to be viewed in isolation. Song himself stressed the importance of illustrations for his approach in his introduction: he argued that those emperors who had never actually seen a plough or loom could "open the illustrations and contemplate them (*pi tu yi guan*), which amounts to having obtained a treasure of great value (*zhong bao*)."[22]

In Chinese written sources on natural philosophy there is a specific category of technical images called *tu*. As Francesca Bray notes, the classification *tu* is functional in its implication rather than stylistic: the images convey skilled, specialist knowledge and complex meanings with more or less succinct methods.[23] New approaches in the history of Chinese book illustrations and visual culture have refocused attention on this important nexus, disclosing that throughout the centuries Chinese intellectuals subtly and diversely combined images with words and words with images to communicate and educate and to transform knowledge. This is the category into which the illustrations of *Works of Heaven* belong, including the later versions as well as the diagrammatic schemata (also called *tu*) that Song employs in *Talks about Heaven* to explain eclipses or sunset and sunrise (fig. 4.3).

Song's visual compositions followed a common generative grammar. He depicted technical features to emphasize their importance and substantiate his argument on the importance of knowledge and activity. His illustrations conveyed a technical nexus and he used the technical features to bring in sociopolitical ideals or theoretical meaning. The illustrations of *Works of Heaven* thus charted an intellectual construct on the basis of technical details. With this approach Song connected to the efforts of men like Lou Chou (1090–1162), who had designed an album on agricultural techniques and silk and rice production for the emperor (i.e., the *Gengzhi tu, Pictures of Tilling and Weaving*) in the 1130s. At the same time he associated with progenitors such as Wang Zhen who in his agricultural study had "intended his visual-verbal depictions of technical devices to serve as the equivalent of blueprints, allowing officials to introduce more advanced technology to backward regions."[24] Wang's work indeed created a complementary symbolic communicative system that employed realistic impressions of the actual object as well as representational, diagrammatic, or conceptual portrayals of complex structures. In this sense Wang Zhen prepared the way for Song's application of text and image. But, by Song's lifetime, the world of *tu* had even more to offer. Complex structures had emerged in which tableaux, diagrams, or charts elucidated theoretical assumptions or illustrated textual

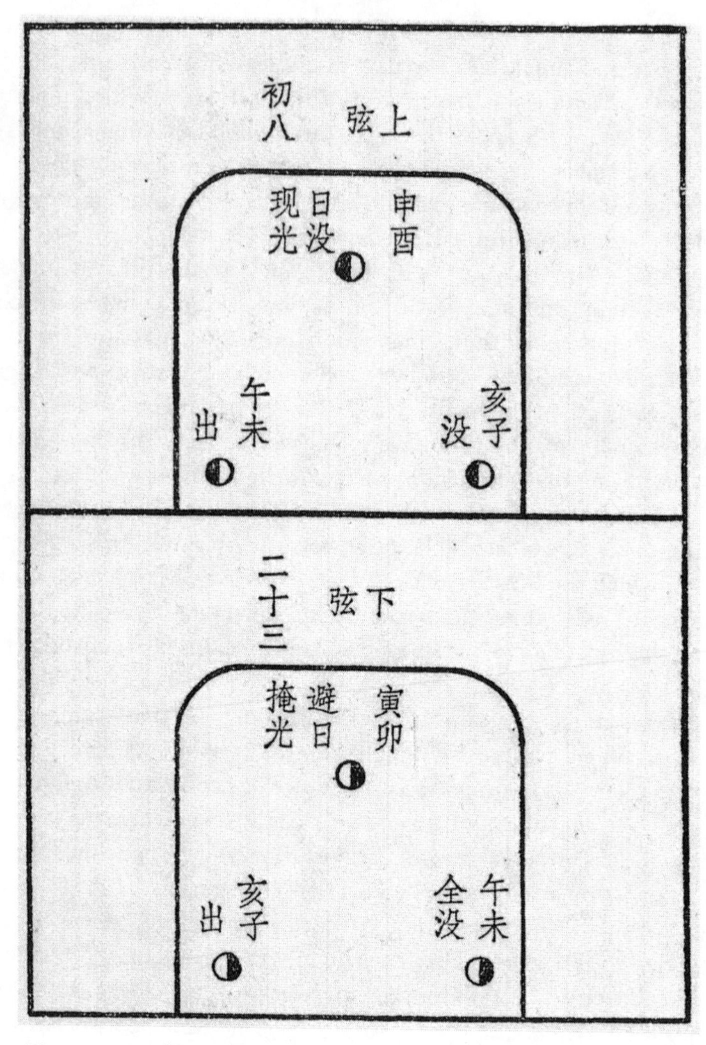

Fig. 4.3. Diagrammatic scheme of various sun and moon constellations. The top is labeled "First crescent" normal on the 8th. The bottom says "last crescent," which normally happens on the 23rd of a lunar month. Song noted the hours during which moon and sun could be seen. *Talks about Heaven*, 17b.

contents and texts delineated metaphoric scenes. Scholars described objects in words and depicted the manifold features of objects in images. Within the discourse on knowledge and action that characterizes the seventeenth century, Song, like his colleagues, increasingly depended on illustrations as templates for action.[25] Employing images, he called for practical action, following Zheng Qiao, who had argued that a scholar when he "sought appearance" in image (*suo xiang*) and "sought principles" (*suo li*) in writing followed the sages' key method for learning. Sages had known that they had to look for the appearance in images while looking for the reason in writings. Therefore, the sages had still known "how to learn things, and it was also easy for them to put things they had learnt into use." It was only later on, as Zheng Qiao suggested, that "scholars left images for writings, and focused on wording and talking. That is why it was hard to learn and hard to put the knowledge learned into real use. Although they had thousands of books in their minds, when it came to practice in reality, they became confused and did not know what to do."[26] Song's usage of images thus entailed a harsh criticism of his colleagues' focus on contextual knowledge. Adding images to his writings, he expected his readers to unfold the knowledge within them and then act. Combining these ideals and attitudes with the presentation of tasks such as casting a bell or drying paper sheets, Song gave technical details a challenging new meaning; they became features integral to the revelation of universal knowledge, an intellectual exercise.

From this perspective, the 122 illustrations of *Works of Heaven* were more than decorative elements employed to make the text attractive to the consumer, as previous research has argued.[27] Neither were they a mere visual replication of the technological nexus. I suggest that Song chose features and selected the contents to clarify and augment his point. This becomes evident when the images are read in correlation with the complex issues that Song raised in his other writings. They were part of his strategy to place both technical and sociopolitical facts and explanations into his argument. Each image dovetails with the text inasmuch as the image itself is a meaningful combination of diagramed features, a symbolic depiction of material and natural objects, people, and charts. In addition, issues such as cultural standards or budget limitations may have influenced the execution of the design. A comparison of recurring features in various contexts such as oxen drawing a plough, or folds of drapery, suggests the woodblock carvers may have used templates to reduce costs. The abilities of the woodcarver may also have affected the overall style in that some features were only roughly executed. Nevertheless, in sum the content of each illustration relates so directly to the textual content and Song's sociopolitical ideals and shows so many idiosyncrasies in detail that the illustrations can only be considered a conscious effort on Song's part.

Working within standard conventions, Song filled his illustrations with details his colleagues could easily decode. He used schematic features that represented cultural constructs. Social rank could be symbolized by a long fingernail, as in the loom illustration, and by facial hair and headgear. Gender distinctions in occupations were sometimes illustrated even when not touched on in the text. He also employed symbolic features such as stones adjacent to banana palms, which were archetypical in southern regions (fig. 4.4).[28] Placing the stones and leaves together in his image of cotton separation, he indicated to the reader that this production process was sensitive to local climate. In the image in which bamboo was pulped in water basins, Song added stone pillars to denote the institutional and scholarly value of papermaking (fig. 4.5), although this production step was in general performed in the open air next to the plantation.

A particularly nice example of this combination of social issues and technical details is the illustration of the production of salt at Lake Jie in Jiezhou County, Shaanxi Province. It depicts an area encircled by a high, brick wall topped with battlements (fig. 4.6). Two scenes are arranged around the small puddle in the center that schematically represents Lake Jie. In the upper part of the picture two men stand sweeping together the crystallized salt. In the lower part of the picture a third man guides an ox and plough, exemplifying the annual practice of salters to allot the land next to the lake into small basins. Heaping up embankments (*qilong*) of mud, the ploughs simultaneously carved a fresh drainage system (*die*) into the crusty surface. The image depicts a surrounding fortified wall for whose existence we have no archaeological evidence and which can actually not have had any productive purpose. The wall fulfilled a symbolic function, showing that the state took measures to prevent people from illegally entering the production area or stealing the precious raw material. Salt is portrayed as an issue under state control.[29]

In some cases the content of Song's illustrations and his textual description deviate from each other. Then the modern reader has difficulty bringing them in line. In the section on weapons, for example, one image depicts two levels of bow and arrow construction. Concentrating on quality control, the upper scene depicts the checking of the arrowheads. The lower scene illustrates the weighing of bows, titled "examination of the bow (*shi gong*) and the determination of the pulling weight (*ding li*)." The scene depicts a man holding a scale. The bow's cord is attached to the scale and a heavy bundle is attached to the bow's center (fig. 4.7).[30] The text explains in relation to this image that "the force of the bow is generally tested by stepping on the bowstring [and pressing it] to the ground. To measure [the pulling force] (*gong li*) attach the hook [of the scale] to the flexed bow (*gongyao*) with the string fully extended. The pulling weight can be known by moving the scale's weight until it is flattened (*ya*)."[31] The text hence de-

Fig. 4.4. A southern technology—"Ginning the cotton [seeds]" (*ganmian*) depicts a cotton gin for separating fibers from seeds. Banana leaves frame the production of cotton, placing it geographically in the South. The caption below explains that the seeds were heated by fire (*huo hong*) beforehand. *Works of Heaven*, chap. 1, no. 2 *Cloth Manufacturing*, 43a.

Fig. 4.5. The pillars of civilization—"Chopping the bamboo and steeping it in a pool" (*zhanzhu piaotang*). The insinuation of a building in the background underlines the central role of paper for the state to which Song points in the text. *Works of Heaven*, chap. 2, no. 13 *Paper Making*, 75a.

Fig. 4.6. Sowing and reaping salt—"Lake salt" (*chiyan*): The lower scene shows how "lake brine is channeled into plots" (*yin shui ru ye*). The "southern winds [then] accomplish the ripening" (*nan feng jie shu*, caption upper middle). Men then sweep together the dried salt, which the upper scene depicts. The terraced wall symbolizes that salt production was under state power. *Works of Heaven*, chap. 1, no. 5 *Salt*, 73b.

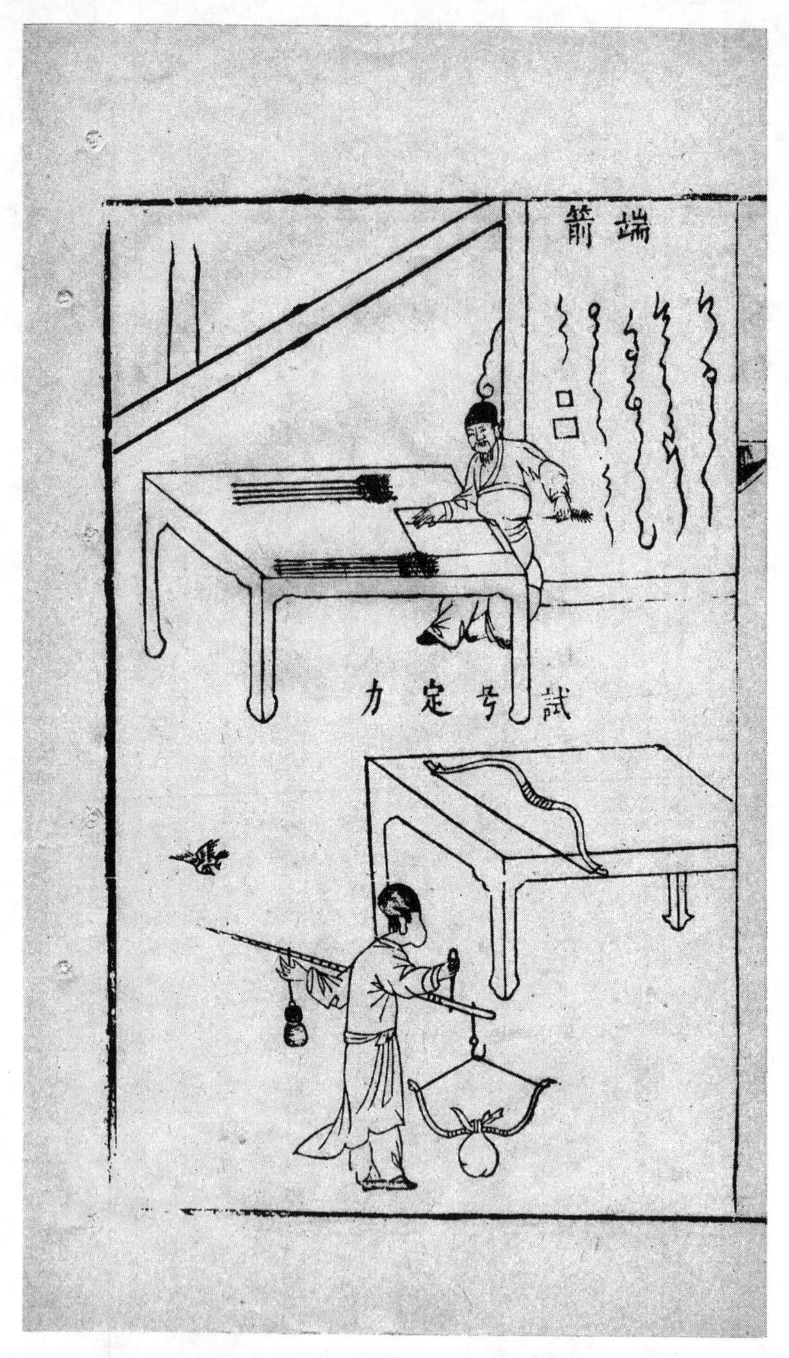

Fig. 4.7. "Targeting the goal, while drawing the bow" (*duanjian*): Song's title of this image refers to a quotation by Zhang Zhuo (ca. 660–740), *Chao ye qian cai* [Records of the Official and Unofficial Matters], which Ma Zuchang (1279–1338) then used to entitle a poem (*Minghuang duanjian tu* [Illustration of Emperor Minghuang while drawing a bow]). Both texts issue the obligation of the ruling class to act considerately and be knowledgeable about warfare. The upper scene illustrates the straightening of the shaft of an arrow. In the bottom half a man "tests the bow and determines its pull" (*shigong dingli*). *Works of Heaven*, chap. 3, no. 15 *Weapons*, 31a.

scribes two steps of which the illustration only depicts the last. Stepping on the bowstring allowed one to generally test the stability and expand the bow. For the exact pulling weight according to which bows were classified, a weight had to be attached. The text only vaguely describes the actual process of measurement, whereas the image shows that a bundle was attached to the bow's body. Presumably this allowed weight to be added in the form of sand, thus establishing the pulling weight. The tension we identify between the contents of the text and the images may result from functional deficiencies, such as a lack of ability or a misunderstanding on the part of the block carver. We could, however, also argue that these tensions only result from the fact that our perspective is different to that of Song's world. In fact, text and image may also have addressed two different issues. Song's text gave a description of the construction of bows and arrows in the seventeenth century. Meaningfully presenting additional features, the image effectively communicated the importance of exact measurements in the quality control of weapon production.

Analyzing the illustrations, we also see that Song kept silent about some issues that interest the historian today. In the section on ceramics he arranged metaphorical scenes of workers doing individual tasks to demonstrate the manifold steps of production. Song's organization does not specify a chronology or which form of cooperative work was involved. The text only states that for completion each piece of porcelain went through 72 hands (*guo shou qishi'er*). Neither medium specifies how the work was actually organized; was it 72 people or 72 steps of production? Equally the image does not specify if three steps in the work process are being depicted or a tripartite division of labor. Yet if the work process is somewhat vaguely explained, the varying emblems of social status are crystal clear. Only one of the potters forming a jar wears a cap and has a straggly beard. The two other figures performing easier tasks such as the addition of handles or the production of bowls are bareheaded and look much younger in their schematic depiction. Thus Song exemplified difficult tasks as an issue requiring experience and probably also performed by craftsmen higher in the occupational hierarchy. Using signifiers of senior status, especially whenever groups of people are carefully arranged, Song assigned a relation between rank and tasks within the conventions of his era. In other cases the positioning of the figures within the composition additionally emphasizes the hierarchy. In section 14 dealing with the five metals a man dressed as a scholar dominates the group of people assembled beside a molten puddling pit (see fig. 4.8d).[32]

Although the illustrations in *Works of Heaven* are unique in content their stylistic elements comply with contemporary conventions. When Song described specific tools in the text and then showed their application in images, he followed didactic methods similar to those in Wang Zhen's (fl. 1333) *Book of Agriculture.*[33] This allows us to decode the initial purpose of each illustration. Yet it is only when we look at the overall arrangement of the illustrations that we can appreciate the particularities of Song's methodological approach. In the section on weaponry Song fruitfully combined the visualization of the weapons in action with a strongly diagrammatic view of the whole canon, leading to a close-up depiction of the various components. The series of six images (one is double page) depicting fire/gunpowder weaponry systematically guided the reader into the topic, starting with generalities and then providing graphic detail. Thus, the first image shows how Western muskets rout the enemy, and then what cannons look like in detail. A caption informed the reader that the depicted gunpowder bomb was called "ten thousand people killer" (*wan ren di*): Its design suggests that this weapon could be efficiently employed to end a siege, destroying (city) walls and defeating cavalry, which the text further emphasizes.[34] The sequence ends with schematic depictions of various kinds of land mines (*dilei*). Clarity also seems to have been a goal, as Song systematically titled the illustrations and added explanations whenever illustrations were too abstract. Details and longer descriptions are included to indicate purpose or explain abstract functional aspects. Song also used this style in *Talks about Heaven,* where his rough sketch of the movement of the sun and moon to illustrate his concept of eclipses was carefully named and annotated (fig. 4.3).

Following the lead of his literary descriptions, Song's diagrams and illustrations employed an acknowledged set of issues, rhetorical strategies, and modes that inspired the reader's confidence in the presented contents and lent credence to the information delivered. The seemingly realistic depiction of technological features induces a sense of familiarity, of the accepted. This was a purposeful persuasive strategy employed to foster a central argument. In the case of the depiction of the loom, the argument was the concept of order. His images did not simply record what could be seen when observing technical procedures; their purpose was to make knowledge visible, that is, comprehensible through observational inquiry. In this sense Song's images require a trained reader as does the analysis of modern scientific images.[35] When Song discredited scholarly negligence of issues such as the loom, he was pointing out the right way to observe things. He urged his colleagues to acknowledge the important implications of things and affairs. Seeing the illustrations in *Works of Heaven* as strategic tools places the particular selection of contents of the images in a meaningful relation to the textual explanations. A closer examination of those illustrations whose contents seem odd to a modern reader with technical interests reveals strategic suggestions behind the illustrative narrative of *Works of Heaven*—a reading backed by Song's theoretical approach in the treatise *On Qi*. The section on the five metals (14), in which Song explained in detail where to find the five metals—gold, silver, copper, iron, and zinc—contains four series of illustrations, delineating the production of these ores from source to the smelting process (figs. 4.8*a–d*, 4.9*a–h*). The first image of each series depicts the geographical conditions in which the ore was normally found and thus promoted Song's theoretical assumption that the production of ores depended on the creative forces of *qi*. In Song's view gold and silver ores required a special rare blend of *qi* conditions that only took place deep inside the earth and then only surfaced in very few remote mountainous regions. By contrast the text stressed also that "less valuable metal was found in large quantities in places easy to reach"[36] and that they were "repeatedly produced (*fansheng*) on plain grounds and hilly regions (*gangbu*), and not in high mountain ranges."[37]

In accord with this assumption the first illustration of the processing of iron ores renders men on the foothills next to a river. One man is collecting metals from the ground while another one tills the ground with a plough (fig. 4.8*a*).[38] The visual image underlined the concept that ores were offspring of the *qi* phase of earth (*tu*), a theoretical concept of Chinese thought on nature that the common mind translated quite directly into physical reality. The text says what the depiction of the plough implies; iron grew (*zhong*) from the earth like plants: "It is also possible after a rain to turn the soil with an ox-drawn plough and gather [the pieces] which lie a few inches (*cun*) under the soil. After being ploughed, the pieces will continue to grow, and they can be harvested continuously without ever being exhausted."[39] Song used the term "planting" (*zhong*) in various contexts. It was not unusual in his culture. The terminology "planting salt (*zhongyan*)" was, for example, quite common. Nevertheless it is worth noting here that he took it literally. His concept of the creation of metals is an analogy to plants that grow and multiply. Both are equally a transformation of *qi* into *form* (*xing*). It indicates a particular model of explanation and original idea. Song may have seen or heard that in some regions Chinese farmers collected pig iron from their fields. This information fit his theory. He explained to his readers that ores, as products of the earth *qi* phase, grew in the soil as did plants. How both theory and practice came together in Song's knowledge-making process is the issue of the following section. Based on an overview

Fig. 4.8*a–d*. This series illustrates in four images the processing of iron ore. Figures *a* and *b* show two possible sources, namely, by collecting it from the fields or by panning, or "Cultivating the land and collecting the ore" (*kentu shiding*) and "Washing iron sand" (*taoxi tiesha*). Figures *c* and *d* are entitled "Furnace for the creation and refinement of iron" (*shen shu lian tielu*). The left part shows a scholar surrounded by workers assembled by a puddling pit. The caption terms the substance the scholar drops into the puddle "wet loam and ashes" (*chao nihui*). The molten iron is then "flowing into a rectangular pool" (*liuru fangtang*). "Pig-iron" (*ban shengtie*, caption right) is gained. *Works of Heaven,* chap. 3, no. 14 *Metals,* 19a/b, 20a/b.

Fig. 4.8b

Fig. 4.8c

Fig. 4.8d

of his theoretical ideas, it shows how Song combined theoretical assumptions, observational experience, experimentation, and deductive reasoning with the scholarly tradition of literary exegesis to make sense of the world around him.

OBSERVING THE NATURE OF *QI:* THEORY AND PRACTICE IN KNOWLEDGE CONSTRUCTION

> Well, it appears as if *kan,* the water, depends on *li,* the fire, and together they manifest the changing modes of the five phases.
> *Works of Heaven,* chap. 3, no. 16, 40a

Reigning over a state with large geographical dimensions, Chinese dynastic rulers had a vested interest in keeping bureaucracy intact as a way of maintaining control over the regions and of communicating the demands of central government effectively to the people. The proper functioning of the state, Song remarked, was not solely dependent on the attitudes of its officials.[40] Rules and orders, ideals and notions had to be put down in writing to guarantee they were transmitted accurately to every corner of the state. Craftsmen supplied the implements of administration, paper and ink, as Song noted in the sixteenth part of *Works of Heaven.* Vermilion, the colored ink that scholars used to punctuate and comment on texts, and the emperor exclusively applied to endorse an official's memorandum, allowing or denying word to be put into action, caught Song's attention. Honored for its color, which did not fade over time, it was considered to be as unchanging as gold. It also promised perpetual youth and long life. Chinese craftsmen in general manufactured vermilion by combining one hundred parts by weight of mercury with twenty parts of molten sulfur that was usually extracted from iron pyrites. Carefully mixed, these ingredients were heated in an earthenware pot to the point of sublimation. At this point, the pot broke and the red mercuric sulfide could be scraped from its inside. Finally, the workers added an alkali solution to remove all traces of free sulfur. Indeed, the making of vermilion was not an easy task, and Song remarked that it was only with the help of divine forces (*shenjing*) that this treasure of the study room came into being.

Song described vermilion and all the other processes, phenomena, and events in *Works of Heaven* to give an objective, real view of universal order, the sum of which corroborated his understanding of *qi.* His descriptions exemplified how the course of nature could be gradually constituted through inferential steps rooted in "matters of fact." His *On Qi* applied the same

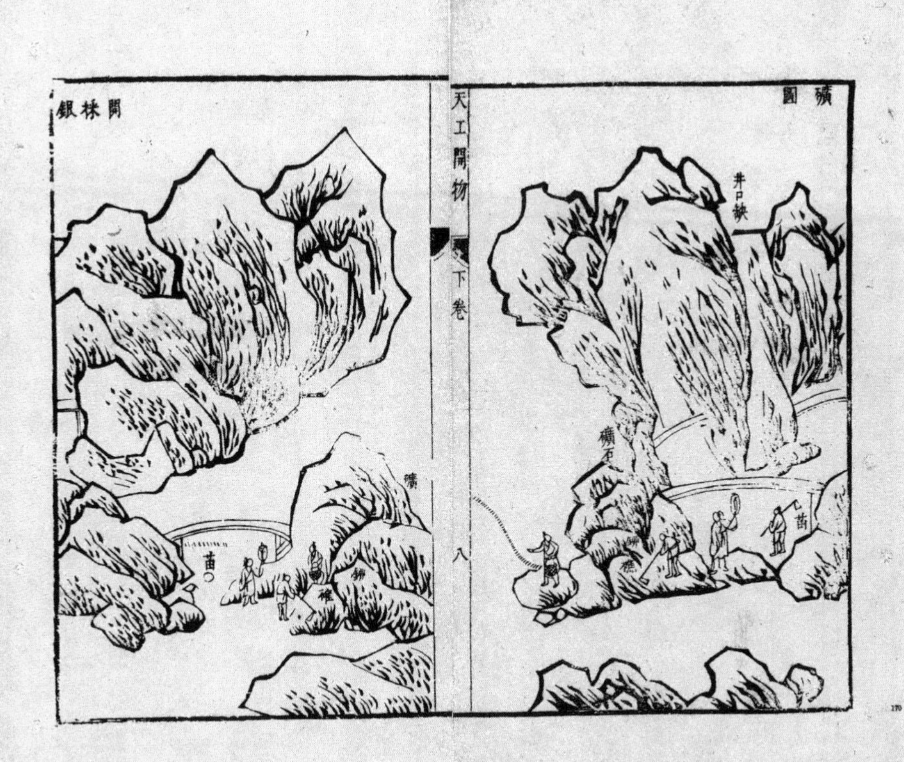

Fig. 4.9*a–h*. Origin of metals series consists of four double images. Figures *a* and *b* are "Illustrations of the collection and exploitation of silver ore" (*kaicai yinkuang tu*). Men are digging and collecting silver in a river valley surrounded by high mountains. The "sprout" of silver (*miao*) is flagged in both cases as lying in the river bed. Figures *c* and *d* are an "Illustration of the smelting of the rock that binds the silver to the lead" (*rong jiao jie yin yu qian tu*). Figures *e* and *f* show separating lead from silver; *g* and *h* depict the purification of silver. *Works of Heaven*, chap. 3, no. 14 *Metals*, 8a/b, 9a/b.

methodology to natural phenomena from weather to sound production and presented the subtleties of the theory in more detail. These elaborations provide the key to Song's approach to knowledge. They show the models of explanation that informed Song's observations and thus also explain what he thought he had observed and in which way it made sense to him. Thus his particular approach to *qi* deserves a moment of our time.

In chapter 2, I discussed the way Song approached the world in terms of *yin* and *yang* and the five phases and the way this structured his worldview. It has been demonstrated that all of Song's notions about his surroundings

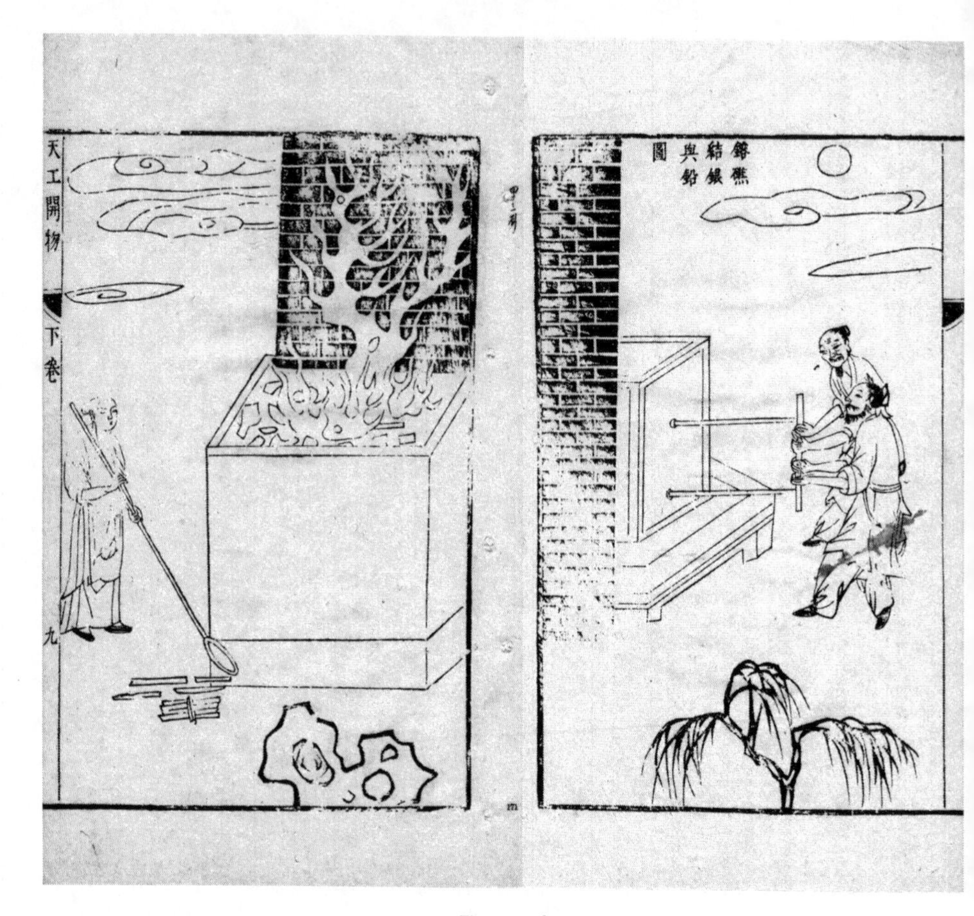

Fig. 4.9c–d

were based on the theory that the unifying, eternal agent of the world was *qi*, "saturating heaven and earth."[41] Ontologically this is a staple in Chinese physical conceptions. In the twelfth century, Zhang Zai gave it new recognition and philosophical vigor. It was later redefined by a number of seventeenth-century scholars in their approach to explaining the world. Drawing on this multifaceted basis, Song found his own way and identified the world as an interchange of the varied appearances and phases of the universal *qi*. In this section, considering the margins Song drew between theory and practice, we delve further into the details of Song's *qi* theory, scrutinizing how he explained things and affairs in detail, and how he saw *qi* acting in natural phenomena and material inventiveness.

For Song, *qi* was omnipresent and could always be ascertained. In its

Fig. 4.9e–f

subtle variations it could, however, easily escape human senses (which include seeing, hearing, tasting and feeling. Song never refers to smell). *Qi* in its original condition was invisible. This invisible *qi* represented the great void (*taixu*) that constituted the entire universe. In man's world it saturated the transparent space; air, however, was *qi* containing mote-like particles (*chen*), a concept I will discuss in detail in chapter 5. The invisible *qi* consisted of the two phases *yin* and *yang* that, matching one another, resulted in harmony. Invisible *qi* opposed and complemented *qi* manifesting a "form" (*xing*), which designated any appearance of *qi* that was (more or less) tangible. Form was constituted by the fact that one of the *qi* phases, *yin* or *yang* had accumulated and could not find its matching partner to dissolve to the void and thus it took on a shape. Song did not, however, think

Fig. 4.9g–h

of form as matter; it was a transitional stage of *yin* and *yang* in interaction. *Yin* and *yang* Song equated with *yin* water and *yang* fire, arguing that in attitude and behavior *yin* and *yang* resembled their materialized representatives. Terminologically water and fire could represent the abstract concepts *yin* and *yang,* but also the phenomena of water and fire. I distinguish in the following between the compounds *yin* water, and *yang* fire and use "water" and "fire" when Song means only the natural phenomena. Song occasionally also equated the agents *yin* water and *yang* fire with *kan* and *li,* the corresponding hexagrams in the *Book of Changes.* These cases I translate into *kan* water and *li* fire, respectively.

Song thus adhered to the "two *qi*" (*er qi*) theorem that defined *qi* as the substratum of *yin* and *yang.* Traditionally the axiom of the "two *qi*" was integrative with the "five phases axiom" (*wu xing*), which means that the *yin* was equated with the phases of water (*shui*) and *yang* with the phase of fire (*huo*), and both stood in relation to the three phases: earth (*tu*), metals (*jin*) and wood (*mu*). This so-called "*two qi*—five phases" axiom gave water and fire a role superior to that of the other three phases. This was contrary to the traditional viewpoint of the various cycles of the five phases. Song signposted this viewpoint in various ways, for example, in one of the headings in *On Qi* he openly opposed traditional interpretations which presumed that "water brings forth fire." He suggested that the *yang* fire and *yin* water phases equaled each other as the major phases of *qi* transformation and that they were not fighting but fusing on equal terms. In different proportional relationships the two participated in the formation of the phases of earth, metal, and wood, which, however, were different from *yin* and *yang*, as each embodied an idiosyncratic phase of transformation. They were less forceful than the *yin* water and *yang* fire yet still essential to the making of all other forms, that is, the myriad things and affairs of the world. Section four of *On Qi* was highly eccentric, because in it Song amalgamated several ancient concepts of *qi*, including elements of alchemy as well as the Buddhist concept of dust motes (*chen*) or the idea of ashes (*hui*), a condition of *qi* in a very particular environment. Dust motes distinguished atmospheric air from the pure *qi* of the void. Explanations in this section of Song's work often misleads modern readers looking for an atomic worldview or occasionally inspire an interpretation of *qi* as what modern chemistry names oxygen. But Song used particles to verify that the basic unit of cosmological order was *qi*. Functionally both particles and ashes fixed a bug in Song's theory; as we will see in the course of discussion he used them to explain the resistance of stable forms toward transformation.

Song's observations and his knowledge about the natural world were motivated by his interests in and concern about a universal comprehension of *qi*, in which he addressed shared, public viewpoints. Discussing everyday events, nature, and the making of material objects, he wanted to make them rational and explicable issues. He was certain all pertained to one logic: *qi*. He admitted things or events did not necessarily look consistent at first sight but argued that the anomalies or oddities his colleagues discerned were only caused by their deficient understanding.[42] Song claimed that the singular could always stand for the universal, and vice versa. Exceptions to the rule or the replicability of events were not an issue for Song as they were for his European contemporaries. From this perspective Song employed the observation of things and affairs as a way to illustrate his point and as proof of his theory. At the same time he presumed that observation was suspect in its essentials if it contradicted principles of general validity. Observation was only reliable when the mind guided the view. Contradictions between observation and theory indicated that the observation was deficient. Either the analyst had overlooked some essential factors, or focused on the wrong issue.

How both theory and observation come together is best exemplified in Song's view of the void. Accepting Zhang Zai's notion, Song considered the void a stage of *qi*. At this stage *qi* formed "a completed primordial (undifferentiated) whole (*hunlun*). Nothing can be in between."[43] The two phases *yin* water and *yang* fire were unified. The realm of the void, removed from immediate human consciousness, was invisible. It was not empty or a "nothingness," as the Buddhist conception of the void (*xu*, also called the *taixu*, superior void, or *daxu*, great void) presumed. It was full of *qi*, made up of *yin-yang* phases in full congruence with each other, completed by their matching partner. In this condition of the void, *qi* could have a spatial dimension, as we will see in Song's ideas about sound (see chap. 6). Each time the *yin-yang* phases of *qi* separated from each other or fused, that is, interacted, *qi* became graspable by the senses, as materials or phenomena. When water extinguished an open fire and nothing was left, or steam rising from a pot disappeared, the *yin* water and *yang* fire agents fused back to the invisible *qi* of the void. In such processes *qi* was "not visible (*bu jian*), but also not fully veiled (*yin*)."[44] Materialization meant conversely that *yin* water and *yang* fire had been separated and were out of balance. Either one or the other phase had accumulated in a way that it could not find matching partners. *Qi*, generating a graspable form, became apparent in all kinds of technical processes. In the production of the oil so important for color, lighting, and nutrition, scholars could see how material was generated seemingly out of nothing. Song drew attention to the fact that although the seeds used to extract oil were themselves dry they were, nevertheless, a form of *yin-yang qi*. In fact, they consisted of the *yin* water phase that had accumulated and taken shape. When heated, the intensified *yin* water *qi* took the opportunity to fuse with the *yang qi* floating in the void. Part of the *yin* water phases achieved unity and dissolved together with the *yang* fire to the void. But not all could find matching partners and the remaining composition liquefied, taking shape as oil.

In the case of oil production one could also observe that the *yin* and *yang* phases were volatile stages of *qi*, for the phase of *yin* water followed its "inherent tendency" to match with *yang* fire and revert back to void. Oil pressers thus had to handle the oil cakes quickly and carefully to prevent them from cooling before they were pressed. If the oil presser was too slow, he allowed too much *yin* water phases to find a matching partner and vanish into the void and the yield would be diminished. Oil, Song assumed, "originally exists because of *qi* created from the void."[45] Revealing this nexus of *qi* conversion was the purpose of Song's detailed descriptions on oil pressing. In chapter 3 I have already shown that Song recognized the experience of the oil presser as an important production factor. But understanding why the correlation existed required more than experience or empirical observation. It required a mind capable of abstraction.

In accordance with Zhang Zai, Song explained processes of transformation by the fact that "the origin is the void (*xu*) and *qi* exists in relation to it."[46] Contained in this concise statement is a significant presupposition for Song's theory of *qi*, namely, that *qi* always longed to revert to a stage when *yin* and *yang* were in balance as the great void, and when this great void was at rest. In this stage *qi* was invisible and inaudible. The existence of the world systemically relied on this tendency because it triggered (*ji*) change (*yi*). Change was for Song a form of unrest, in which *qi* mutated from one stage to the next and its *yin* and *yang* phases defined their momentary relationship in some way other than balance. In principle, all *yin* and *yang* could find their matching partner, but in general there was an imbalance, in which satiation countered hunger, which means either all or only a few could find matching partners. This took shape as a material or phenomenon. In satiation or hunger, *yin* or *yang* polarized: The heavenly sun was an intensified sphere of *yang qi* accumulation that, together with the earth, an intense *yin qi* accumulation, engendered the world of things and affairs. This complementary opposition of *yin-yang* and its opposing polarity characterizes Song's cosmology. It explained the very existence of the world as all materialization, all objects, and every phenomenon that could be observed and sensed. All were the product of a complex interaction of *yin-yang qi*. In Song's view this embraced both the production of oils and ores and the entire cosmological system. The void *qi* in its imperceptible condition was for Song the pinnacle of harmony to which all visible and graspable things eventually dissolved. However, the system could never achieve this imperative because the accumulated *yang qi* in the heavens (the sun) and the accumulated *yin qi* on the earth were too intense and maintained a state of oppositional polarity. In principle, only Song's model allowed the possibility of a full dissolution.[47]

Song's approach was within conventional lines when he identified the void within terms of *qi*. He drew on the thought of Zhang Zai, who also equated *yin* with water and *yang* with fire, giving them a superior position in the five agents theory, but he abstained from explicitly referring to Zhang

Zai to authorize his theory of *yin* and *yang*. This may have been due to his prominent progenitor's occasional emphasis on morality. But there are other discernable differences in their understanding; Zhang discussed and classified *yin* water and *yang* fire primarily as bipolar aspects of *qi*, whereas Song intimately linked them to all kinds of creative (or destructive) processes.[48] For Zhang Zai *yin* water and *yang* fire were attributes of transformational process; for Song, they were themselves the *qi* that drove change. The two thinkers also differ in their terminology: Song only rarely applied the terminology of *yin* and *yang*. He preferred to address them as water (*shui*) and fire (*huo*). Li Shuzeng as well as Pan Jixing valued Song's emphasis on the materialized representatives of *yin* and *yang*, water and fire, as a decisive addition to, or indeed advancement on, Zhang Zai's approach.[49] In fact, such chronological assessments, as mentioned in chapter 2, must be looked at with care. The apparent qualitative difference between the thinkers is the result of different aims. Investigating metaphysical contents, Zhang Zai aimed at perfecting the self and individual knowledge for moral reasons.[50] Song explicitly aimed at ordering the world by unraveling the universal principles of a world into which human speculations had brought disorder and uncertainty. For Song, morals were no more than the result or a side effect of an order embodied in physical and metaphysical appearances.

Song's emphasis on the completeness of *qi* (nothing can be in between) in its primordial stage of *hundun* varied considerably from standard *li* cosmologic and cosmogonist ideas that *qi* in this stage was indeed "unstructured" or "confused" (*luan*) and thus had to be structured and was given order and form by *li*.[51] By ignoring *li*, Song had to show *qi* provided the structure alone. Song was radical in this, but he was not alone in rejecting the primacy of *li* as the structuring principle. As Michael Leibold has shown, Wang Tingxiang, a *qi* thinker of Song's generation, also speculated on such issues, emphasizing that the *qi* void was more than the principle of differentiation in a world of *li*. Wang Tingxiang suggests that *li* thinkers were utterly mistaken about the idea of the primordial *qi*.[52] Particular in Song's ideas on the *qi* of the void is his consideration of it, like all other conditions of *qi*, as a provisional activity or status of *qi*. His conceptual approach solved the problem of impetus or stimulus (*ji*) by arguing that the system had an inherent tendency toward change, hovering between the opposition and harmonious fusion of *qi*.[53] The myriad things and affairs were all equally subject to *yin-yang* interaction.

Song rooted his notion of the *yin* water and *yang* fire agencies in the *Book of Changes* by identifying them as *kan* and *li*: "*Kan*, the water, is female and paves the way. It ploughs the irrigated fields and weeds the plants;

no matter whether people of high or low social status observe it or perform their compulsory labor, it is done in front of people's eyes and ears. *Li,* the fire is male, its proper place is in the inner palace, hidden secretly in the inner rooms (*aoshi*). As soon as it becomes aware of people, it withdraws. Such is the character (*qing*) of *yin* water and *yang* fire."[54]

Song explained that the interaction of *yin* water and *yang* fire happened in front of man's ears and eyes: the workings of *qi* could be observed in all kinds of natural phenomena. To claim general validity for his notion of *yin-yang* fusion, Song usually provided samples of observational evidence. He explained that whenever "a droplet of water spirit (*shui shen*) is added to the charcoal, [the water] will take the opportunity and entice the fire spirit (*huo shen*) to fly up and dissolve."[55] The subtle process of the conversion to the void could be observed with human senses, by looking at everyday processes:

> when one strains an ear at the outer surface (*jie*) of a vessel (*fu*) or pot (*ying*) and listens carefully, [it sounds] like someone weeping, or like longing for something or like singing or even like moaning: this is the prelude to the fusion [of the two spirits of *yin*-water and *yang*-fire] to the void and their ensuing conversion. As soon as the water [in the pot] is no longer consistently replenished, or somebody comes and removes the brushwood so that the fire stops and there is no more condensation, [even] Shi Kuang [i.e., the Chinese embodiment of excellent hearing skills; 572–32 BC] will no longer be able to hear a sound. That is the reason why the great void is the origin and where the two *qi* obtain their name. The origin is the twofold *qi* of *yin* water, *yang* fire and form. *Yin* water and *yang* fire go together (*can*). There are various ways in which this is applied by the people. *Yin* water and *yang* fire *qi* fuse and the great void (*taixu*) appears (*xian*).[56]

In the passages following this quote, Song used *yin* water and *yang* fire *qi*-agency on par with *yin qi* and *yang qi* and, more provocatively, even interchanged *yin* and *yang* with water and fire. After giving this or similar examples, Song often noted that processes of *yin-yang* fusion happened in many technical procedures, but the influencing factors varied. These factors could be the manifestation of *yin* and *yang* in a special condition. Whenever *yin* water or *yang* fire phases enfolded (*yunjie*) its primeval spirit/body (*yuanshen*), it was "incapable of borrowing a form (*yi xing qiu*)."[57] In this stage the *yin*-water *qi* was in extreme opposition to the *yang*-fire *qi*. Song laid out the notion of primeval spirit (*yuanshen*) as a characteristic of the five *qi* in the tradition of medical physical conceptions, based on permutation studies.[58] They meant that the *yin* or *yang qi* was without its matching partner: it was pure. Objects of this kind originated from special places: The earth was already quite pure. *Yang*-fire took a pure form in the body of the sun. The place of the purest *yang* was in the heavens; *yin*-water in it purest form could only be found in the deep caverns down in the deepest earth. Both places were beyond human reach, but as objects and substances were created and issued forth, man could come in contact with them. Embodying the "primeval spirit" of *yang*, it resisted any change unless man forced these equals to mingle and exchange *yang* with *yin qi*, as was done in the case of gunpowder production.

In his explanations on gunpowder Song suggested that the hunger of pure *yin* and *yang* of the primeval spirit for each other was extreme. For this reason they were "not able to meet (*jian*)" without dissolving immediately to the void. Polarizing one another, the high tension erupted forcefully in reunification. Song suggested that saltpeter, embodying pure (*chun*) *yin*, repelled sulfur, which was pure *yang*. In this way the concept of the primeval spirits helped explain the huge effect of a gunpowder explosion.

Many objects in man's world existed because *yin* was hindered from fusing with *yang* or their fusion was only possible in a particular way: charcoal was caused by hindering the *yang* fire *qi* from fusing with the *yin* water *qi* of the air. Since the *yin* water *qi* in the wood could not find enough matching partners, the charcoal remained behind:

> If the fire burns in the open (*wai*) every single, independent water essence (*ziyou shuiyi*) in the air (*kongzhong*) will join (*hui*) with [the fire]. The fire joins the void and the wood (containing *yin*-water agents) is exhausted, too. If a fire is sealed up (*bi*) in a pit in the earth, and is not able to reach the void by fusing (*he*) with the water essence (*shuiyi*), the quality (*zhi*) of fire *qi* returns to the mother skeleton and takes shape as charcoal (*tan*): it is the changed body (*bianti*) of fire.[59]

This example also reveals that *yin* water and *yang* fire *qi* transformed even when they were not able to meet and fuse. Indeed this kind of transformation was observable throughout the world of the mundane. During cooking both phases were separated from each other through materials, such as metal or earthenware. In this case the transformation took place because the phases evaporating from the pot and the fire rising from below (as heat) dissolved into the void, leaving a transformed remainder. In such descriptions of the transformative power of *yin-yang* fusion Song draws attention to the fact that despite varying spatial arrangements between *yin* and *yang*, water and fire, there was always a general principle: "The water is above and the fire beneath, metal and earth (*tu*) are in between them: That is a cooking pot (*dingfu*)." Estranged from each other, water and fire also helped iron clothes (*yundou*). According to Song, in the case of ironing "the fire is above and the water beneath." When heating wine in an amphora (*yingfou*) "the *yin* water is left and the fire right" (the amphora was inserted into the glowing coals). Song assumed that "in all of these three [examples], two spirits (*shen*) encounter another (*xianghui*), but their two forms do not contact each other (*xiangqin*)."[60]

The interaction of *yin* and *yang* not only created things as such. It also helped transform things, explaining to a large extent the attitudes of the varied material world around Song. The attitudes and characteristics of *yin* and *yang* were according to Song similar to those of water and fire. For example, the *yin*-water *qi*, representing the process of liquifying, explained why sugar syrup was liquid. Solubility conversely meant that the *yin* water phase was contained in sugar. Song acknowledged that these were simple qualifications that only explained one or two attitudes of things or one part of a process. Most phenomena did not fit into such simple schemes, because the processes of *yin* and *yang* creating things and affairs were complex and many processes were interwoven with each other. For example, why did the sealed earthen pot that was used to extract vermilion not transform into the invisible *qi*, even though the two *qi* interacted when the pot was fired from below and continuously brushed with water? For Song, the answer to such subtleties lay in understanding that the entire texture of *qi* phases consisted of the three other *qi* phases of the five *qi* cycle: earth (*tu*), metal (*jin*), and wood (*mu*) and their relation to water and fire. Song conceptualized these three phases as conditions of an interaction of *yin* and *yang*, which had a characteristic agency in transformative processes: they could generate by themselves (*ziwei sheng*). Similar to water and fire, their attitudes could be largely explained by observing their materialized representatives.

THE COMPLEXITY OF *QI* TRANSFORMATIONS—COMPOSITES AND COMPOSITIONS OF *QI*

> Master Song says: Among the five phases (*wu xing*), earth (*tu*) is the mother of the myriad things.
> *Works of Heaven*, chap. 2, no. 11, 53a

Elegant pictorial designs, colored glazes, broad washes of enamel, and manifold polychrome patterns cover the bodies of the porcelain and pottery of the Ming and Qing dynasties. Elite tables provided a feast for the eyes, seductively splendid and lustrous. Table and ritual wares were decorated in turquoise blue, aubergine, dark violet, deep green, and yellow. The glazing material for these wares was obtained from a huge variety of minerals that were mined in China, imported, or obtained as tribute from its neighbors. All these glazes were added to varying quantities of lime mixed with ashes of ferns, alkali-lead-silicates, soda, lead and silica, to control their fusibility, perishability, luster, and permeability. In Song's classificatory scheme, together with coal, alum, vitriol, sulfur, and arsenic, lime belonged to the category of stones or minerals (*shi*). Lime was a valuable "son" of the earth phase of *qi*, similar to metal—except that this most beneficial, yet disregarded substance was "obtainable without having to seek it in distant places."[61]

Lime had numerous applications in Chinese culture. It was used in construction work, to caulk ships and plaster walls, in paper production, for medicinal purposes and in articles used for personal hygiene. It was calcinated by heat and placed in the open air to be dried by the wind. Song explained that this treatment exposed lime first to *yin* water and then to *yang* fire *qi*. He noted that lime persisted in its visible quality and did not vanish to the void. From his observations, Song concluded that lime was a product of earth *qi*, which provided it with the capability to persist as a stabilized form and resist an instant conversion into the void through the fusion of *yang* fire and *yin* water *qi*.

As shown in the previous section, Song gave *yin* water and *yang* fire a special role, but he still adhered to the five phases concept, considering *yin* and *yang* two of the five basic qualifications of *qi*. The other three, metal, wood, and earth, also had a distinct roles in transformative processes. The variety of things and affairs that surrounded man were in fact composites and compositions of water, fire, and earth, metal and wood *qi*. In the next section I discuss how Song related theory to practice in the context of earth and metal and their relation to *yin* and *yang:* Song's ideas on the phase of wood *qi* are dealt with in chapter 5 as they nicely illustrate his view of growth and decay.

Song suggested *yin* water and *yang* fire were superior phases, because earth, metal, and wood acted only in relation to these two. In this interpretation Song connected to an important classic, the *Hong Fan* [*Great Plan*] chapter of the *Book of Documents*. The *Great Plan*, he asserted, disclosed wood and metal as a mixture of *yin* water and *yang* fire, "for which reason these materials can change and become crooked or straight, liquid or solid." The text also gave a distinct role to earth because of its "original disposition (*benxing*) to create things (*sheng wu*)."[62] Song went beyond the classical source in his definition of the five phases in relation to attitudes of the three phases wood, metal, and earth. He suggested that metals and woods were subordinate to earth, as both were created from it. Furthermore the phases had the characteristics of their material representatives, which made the observation of the material world vitally important.[63]

The strong relation Song saw between material objects, their behavior, and theoretical imperatives distances him from contemporaries who approached the five agents as issues of moral values or for political authorization. Yuan Huang (1553–1606), who compiled a moral guidebook is a prime example.[64] The pragmatic politician and chancellor Zhang Juzheng discussed the five phases within a political perspective as a theoretical concept of legitimate dynastic succession.[65] The aim and achievement of Song's approach was to restore a theory his colleagues had almost completely detached from practical issues to what Song considered solid empirical ground, namely, facts and events that everyone could observe—the natural and the mundane. Looking for order, he systematically dissected natural processes and material inventiveness in which the attitudes of the five phases and their theoretical implications became apparent.

The most common five phases cycles were based either on conquest or on creation. In his theoretical model Song constructed a hierarchical order based on the phases's active participation in change. The earth *qi* became the creative component of metal and wood. Earth represented in Song's view a stage of extreme imbalance in which *yin qi* was so highly accumulated that it could not find enough *yang qi* to return to the *qi* of the void. Because of its enormous intensification of *yin qi* the earth *qi* had a particular transformational potential. But it was still governed by the interaction of the *yin* water and *yang* fire *qi* and also created by their enactment. How this creation happened is not explicit in Song's descriptions. Placing the earth as a potent phase within his *yin-yang* and five phases theorem, Song aspired to coherent reasoning and vindication of his *qi* theory. Comprehensive coverage of all relevant details was not his concern.

Man could observe that earth, subject to *yin* water and *yang* fire fusion, could transform to invisible *qi* and vice versa. During the firing of an earthen pot some *yin*-water phases left the clay and would fuse with the *yang* fire *qi* to dissolve to the void. To understand how the remainder, the fired earthenware, transformed back to invisible *qi*, Song explained, men had to look at a pot after it was crushed into small parts. The shards would entirely disintegrate to the void, step by step, piece by piece, when *yin* found its matching *yang*. As *yin* was highly accumulated within them, matching partners were not easily found. For this reason such processes took place over long periods of time. Most people, concluded Song, either did not have enough patience or live long enough to learn this. For this reason they misinterpreted the logic inherent in the making of things and falsely thought that some things, such as soil and stones, endured forever. In fact, however, all things were constantly dissolving to the invisible void only to then materialize again.

Song's world, its things and affairs, was continuously changing. Thus sands and stones were in Song's view not just crushed fragments of soil, although they may appear that way to a general observer. They were materials that came into being as another appearance of earth *qi*. "Sand (*sha*) and stone (*shi*) are originally brought forth (*sheng*) from the earth (*tu*). As they are born, they are also transformed." Men of his era, Song bemoaned, often disregarded that stones or minerals originated from the earth *qi* and fully exploited a mine without realizing that this would disable earth *qi*'s capacity to create new lime, sandstone, gold, or even jade. The lode would hence disappear. Those who understood the principles of *qi* knew that a mineral lode should never be exploited down to its very origin (*quan*) as the remnants of stones (*shiyu*) were indispensable as part of a regeneration process.[66] Song explained that coal miners, for example, refilled depleted mines with the remnants of stones and sand not only to prevent soil subsidence or nearby shafts from breaking, but also to allow earth *qi* to restore the lode. Societies who disregarded this nexus were doomed to lose their material wealth. Those who handled lodes as sustainable resources would, after twenty or thirty years, be able to mine coal again: "a lode dealt with in this way is inexhaustible, indeed!"[67] This argumentation fully complies with Song's view on the creation of metals. He argued they contained *yin*, as they were able to liquefy.[68] Observing processes of transformation, it was obvious to him that metal was dependent on earth *qi*: it grew out of it. "Well, earth is the mother who creates metals. Given form, metals can be efficiently used by many generations. The mother is the mold for the filial son, who fills it in close resemblance."[69]

Metal's dependence made it inferior to earth *qi* and also more viable to change and transformation. The participation of earth in metal, however, explained why metal did not lose weight: it was earth that prevented its *yin* from dissolution. Earth also explained how metals liquefied and hardened again. Song's view of the world of ores and metals was quite comprehensive, influenced by the fact that the processing of metals was integral to Chinese culture. Ritual devices were made of bronze and weapons and working tools were made of cast iron or forged by smiths. Silver, tin, and copper were used as the raw material for monetary purposes. Gold was mostly reserved for decorative treasures. Ore deposits were found directly under the surface as well as buried deep in the soil, so mining technologies included deep drilling and tunneling similar to the methods used in sixteenth-century Europe. Often small irregular deposits found near the surface led to underground mining through vertical or near-vertical shafts as Song illustrated in the context of coal mining (fig. 4.10). By the late Ming, horizontal tunnels, adits, and drifts posed major problems because they required wood for bracing. Wood had become extremely scarce because of its many uses, in particular its function as an energy source for the industries supplying the growing market for consumer goods. Vertical shafts were prone to flooding with poisonous gas, so miners tried to channel these poisonous smoky gases (*du yan qi*) off with pipelines made of bamboo canes. Other images in *Works of Heaven* indicate the dangers of underground extraction by illustrating unconscious workers.[70] In the hot season, the mines were often closed because then the gas could not escape. During these periods the experienced miners went to search for new coal or ore fields. According to Song the color of the surface soil indicated the kind of ore hidden underneath. It seems as if he had questioned a miner when he noted that a brownish shade in the surrounding gangue raised hopes for a vein of gold, whereas heaps of surface stones that are "slightly brown in color and scattered in such a way as to present the appearance of forked paths" promised the presence of silver ore in rocks and caves. Song lent theoretical credence to his observation of the relation between ore and surface structure, arguing that iron had "a light quality and thus floated on the surface of the soil."[71] Gold, in contrast was according to Song created "in a process of amalgamation with the true fires (*zhenhuo*) of the heaven and the earth. This takes place deep in the earth."[72] Stones and metals were thus emergents of a creation process and related to each other, not solid, unchangeable materials. This tradition is prevalent throughout Chinese culture and is demonstrated in the twelfth-century treatise *Yunlin shipu* (*Stone Catalogue of Cloudy Forest*) by Du Wan (fl. 1126). Kong Chuan, who compiled the preface to this work, put it poetically, saying "the purest energy of the heaven-earth world coalesces into rock. [The ore] emerges, bearing the earth (*tu*). Its formations are wonderful and fantastic."[73]

For Song metals were "created by a loss of the capacity of the mother body."[74] But once formed, metals always resumed their original quantity and quality, never losing weight or volume, and therefore no metal's *qi* detached and returned to the void with its complement. In its balanced stage it thus contained a specific proportion of *yin* and *yang* that could not es-

Fig. 4.10. "Coal Mining in the South" (*nan fang wamei*). The bamboo, as the caption explains, channels "poisonous smoky gases" (*du yanqi*) out of the mine. *Works of Heaven,* chap. 2, no. 11 *Calcination of Stones,* 56a.

cape from its materialized body. The *yin qi* in metals, however, could become activated whenever metals were fired. Man could observe the principle when ore was melted and hardened through the exposure of *yang* fire and *yin* water agents.

Song may have based his theory on his personal observation of the smelting process in his native hamlet. In this case, his assumptions were amply supported by traditional theory and thus not unusual in themselves. In the five phases cycle, metal was traditionally perceived as the product of an intermingling of *yin* water and *yang* fire *qi.* The eleventh-century thinker Zhang Zai had also emphasized the properties of metal in transformation through the axiom of *yin-yang.* Zhang suggested that *yin* water and *yang* fire were latent in metals, yet imperceptible.[75] *Yang* fire phases of the coal fire would attract the *yin* water phases inside the metals. The *yin qi* would become volatile as it tried to dissolve with the *yang* fire *qi* but the earth component prevented this. The metal would consequently liquefy, but not lose *yin* water phases, which were not able to detach. When the molten metal was quenched in water for hardening, they would return to their original coagulated relationship. Finding the correct balance between *yin* water and *yang* fire, the metal could be configured. Wrought iron and steel showed that the same material could attain various degrees of hardness. But the composite of *yin* and *yang qi* remained the same. As metals were created from the true fires, Song identified liquefaction as their original condition. Song delineated the details of his theory on metals within the *yin-yang qi* theorem in *On Qi,* but his explanations in the section on casting in *Works of Heaven* are consistent with this view: "Tools are manufactured by using moulds [in which the] metals [are poured]. Inside the furnace the metal is liquefied returning to its true condition (*zhenyi*). The essences (*jing*) of the *yin* water and *yang* fire *qi* then complete the form of metal." The expression "true condition" was a standard reference to illustrate the idea that deep in the earth there existed a place of intensely accumulated *yang* fire *qi,* in which metals first appeared in a liquid form before coagulating. The origin of this concept may have been the observation of volcanic activity; for Song it was an epistemological standard. He explained in detail that hammer-forged articles were first exposed to fire and then immediately quenched in clear water until all "soft properties of the iron and steel are removed." Song thought of this process as an interaction of *qi,* a procedure in which "the divine *qi* (*shen qi*) supported the amalgamation (*meihe*)."[76]

Observing metal production and coal mining, Song concluded that the relationship of *yin* water and *yang* fire determined the attitude of materials and how they behaved in transformation. Metals liquefied because they contained *yin* water phases. The *yin* water *qi*, however, only became apparent with the help of fire, for which it longed so it could dissolve into the void. Encapsulated in the *qi* of metal, *yin* water was unable to escape and thus returned to its coagulated form as soon as the fire was extinguished. This was the theory behind Song's idea that metals such as gold never endured a loss of capacity, even after transformation by the *yin-yang qi* of water and fire.

Song's elaborations on metals in relation to earth confirm that he considered the world in terms of conditions, not substances. For him all things and affairs resulted from the interaction of *yin* water and *yang* fire, and their inherent tendency to fuse in harmony. The bipolarity of *yin* and *yang* established earth (*kun*) and heaven (*qian*) and was evident in all kinds of complementary and bi-polar couplings, from female to male, hard to soft, and stable to volatile. *Yin* water and *yang* fire components in the surrounding *qi* could be unbalanced, creating heat and cold. Heat and cold were thus not caused by the sun, but by the unbalanced presence of *yin* water and *yang* fire agents. Song's insistence on his *yin-yang qi* paradigm is the reason why quantification played such a crucial role in his descriptions. Depending on how much *yin* water or *yang* fire agent was present at a particular time, the variety of transformative procedures could be explained. If one of the phases was extremely intense, no change would take place; if both were floating around in imbalance, a transformation was volatile. If the quantities or volumes did not match, one *qi*-phase would remain. This leads us to another unique aspect of Song's approach to the world in the realm of *qi*. He claimed that factors such as quantity, motion, speed, and power had a crucial influence on the way things were generated or vanished. He assumed these factors could be deduced by more systematic experimentation and a refinement of correlative practice. The following chapter examines how Song established a relationship between a natural event and experiment. An event had to accord with the findings of an experiment. Otherwise the theory needed to be adjusted.

CHAPTER 5

Formulating the Transformation

> To convert copper into brass, people use "self-fanning coal" (*zifeng meitan*). Place 100 catties of these coal briquettes in the furnace, then put ten catties of copper, and six catties of smithsonite (*kuang*) into an earthen crock. Place the crock in the furnace to melt the metal.
> *Works of Heaven* 1637, chap. 3, no. 14, 11b/12a

Upon entering the market, copper had to meet two competing demands: the manufacture of goods for everyday and ritual use, and, alloyed with zinc, tin, and lead, providing the feedstock for coins. Controlling its supply to stabilize finances, the state had to cope with considerable fluctuations in the comparative value of coin and copper. Throughout Chinese history people traded on these fluctuations, either melting down copper coins to mold marketable brass goods or reclaiming the alloyed copper to coin money. In the twenty-four standard histories that constitute the official historiographic canon of Chinese civilization, the chapters on food and money (*shi huo*) are crammed with complaints about counterfeiting, often holding it responsible for monetary crises. From a technical viewpoint, coining was indeed a trivial matter, only requiring a fire, a crucible, and some experience in melting. The difficulty lay in preparing the mold and determining the appropriate ratio of the various metals. Forgers spent considerable time ascertaining the maximum amount of cheap lead they could add to the coins, while officials had to be up to date in the latest techniques to be able to detect the deceit. Proportional composition in the production of copper was one of the many examples Song used to show that the creation of things and affairs underlay a proportional logic of *yin-yang qi* transformation—even if man could not fully observe these processes.[1]

The previous chapter showed how Song observed the world and tested its phenomena within the framework of his theoretical assumptions. This chapter examines how Song integrated issues unverifiable by personal observation and experience; how he sometimes expanded the phenomenological framework of investigation and sometimes adjusted his theory to maintain his claim to universality. By carefully aligning his theoretical paradigm with what he considered facts, Song essentially reinterpreted the traditional framework of *yin-yang qi* and the five phases *qi* theorem, enlarging it with concepts such as particles (*chen, ai*) and ashes (*hui*). This chapter shows how Song experimented with the world. It discusses the role quantity and calculation played in his worldview and delineates how he established credibility by pointing to the repeatability of events in nature and material production. Song used experiment and quantification, as did his European scientific and technological contemporaries, to convince readers that the information he obtained was objective and reliable. In *On Qi*, Song, for example, suggested that the relational behavior of *yin-yang qi* could be verified by measuring systematically the weights and volumes of water and fire in transformation. Contextualized within his culture, Song's proportional approach shows correspondences with ritual and cosmological speculations often employed in alchemical or medicinal writing. As we delve into Song's approach to numbers, we can also see that while he was very careful throughout his work to maintain a coherent approach and argument, he purposefully varied his didactic and rhetorical strategies within each text. In the realm of experiments Song was, like seventeenth-century British and French writers on natural philosophy, content with proportional relationships and estimates. In *Works of Heaven*, Song was more meticulous. When he noted down what in his view were the exact quantities of raw materials and ratios of ingredients or measured widths and lengths of beams, Song's accuracy inspired confidence in his technological descriptions. This distinguishes Song's approach from that of local officials who, in the process of organizing production and collecting taxes, measured, calculated, and recorded their procedures as a way to fix the world in numbers. In sum we can see that in Song's view the world of matter and phenomena emerges as a varied set of *qi* conditions whose transformation relied on factors such as time, motion, or environmental circumstances and whose behavioral principles and material qualities were the outcome of their creation. In this complex world, Song argued, correlation helped to explain events man could not verify by sensory experience, such as the decomposition of corpses or urination.

READING THE SIGNATURE OF *YIN-YANG QI* IN GAS, SALT, WIND, AND RAIN

> To use the essence of fire (*huo ji*) and not to discern the form (*xing*) of *yang* fire—this is indeed one of the strangest things in the world.
> *Works of Heaven*, chap. 1, no. 5, 73a

Yunnan Province was situated on the very margins of the great Ming Empire. Scholars knew it as a region that provided valuable resources to the state. Luxury goods made of precious woods from Yunnan's jungles and the rare herbs gathered there were in great demand among the elite living in the prosperous urban centers. Yunnan also offered huge resources of copper, tin, and coal that, feeding into the growing minting industry of the Ming empire ensured the central government's interest in this area. Bureaucrats were regularly sent there to supervise and investigate the various mines and enterprises under state control, to collect taxes, or to accompany goods and raw materials being transported to the coastal regions and the court in Beijing. Traveling back and forth to this remote region, scholar officials contributed to the spread of stories about the oddities at the margins of the Ming empire, such as the gas fire wells (*huo jing*) found on the border of the wealthy Sichuan Province. Song portrayed them in *Works of Heaven* as the "essence of fire" (*huo ji*): a *yang* fire phenomenon that produced heat without flame.

It is impossible to estimate how many Chinese literati were intrigued enough to embark on the strenuous journey to the uncivilized regions of Ming imperial power, to endure being cooped up in small boats on swampy rivers or trundled over rough roads only to suffer in verminous cots at seedy inns or sleep under the stars at the end of the day. Song, at least, had a reason to make an excursion through the regions in which gas had been used to boil brine until the salt eventually crystallized since the second century BC: it was on route if he visited his intimate peer Tu Shaokui in Sichuan in the early 1630s.[2] In fact we have no historical proof that Song made this journey, except that he describes in *Works of Heaven* in vivid detail how the salt miners hollowed out the bamboo stalks, patched them together, and carefully wrapped them with cloth to channel the gas safely from the wells to the huge boiling pans. Song concluded this passage by expressing his fascination that no charring or burning substance was visible at the end of the open bamboo stalk. This raises the question, had he seen this process with his own eyes?

We do not know how Song gained his intimate knowledge about methane, but we do know he elaborately discussed heat and combustion, gas (which he consistently calls "fire essence") and its attendant burning flame, in *On Qi*. Song's approach to gas reveals how he dealt with issues and phenomena that could not be observed by the senses and the role that conventional models of explanations or theoretical assumptions played in this situation. In this context he drew on the universality of *qi,* arguing that its schemes were always comprehensible. We can say Song thought the interaction of *yin-yang qi* and five-phases *qi* always left a "signature," a pattern or effect that a knowledgeable scholar would have no problem analyzing, if he just implemented the right methods. *On Qi* unveils some details of the schemes relevant to the phenomena Song meticulously described in *Works of Heaven.* For example, "the highly amazing phenomenon," gas wells, were depicted in *Works of Heaven* as containing water at ambient temperature in their inactive state. There was not "the slightest evidence of the *yang* fire *qi,*" until they suddenly displayed "a fiery intention" (*huo yi*) in an eruption of steam, boiling and gushing through the water without achieving a tangible or visible *yang* fire form. Song added that the gases coming from a well or bubbling from a hot spring were "a surplus of *yang* fire *qi*" (*huo qi yu*).[3] It is only in *On Qi,* however, that Song provided the essential backdrop to his observations. He explained that in the case of gas wells it was important to be aware that one was observing a direct interaction of *yin-yang qi* dissolving into the void. For him the fugacity of fire, heat, and gas, were caused by the fact that the *yang* fire *qi,* like its matching counterpart, the *yin* water *qi,* was volatile and its conversion was prompt (*jie*) and pronounced (*zhu*). It was for this reason that man was unable to follow the changing states of the gas wells with his senses.

Reading the descriptions in *Works of Heaven* together with Song's exemplifications on *yang* fire *qi* in *On Qi* reveals that Song grounded his distinction of the two appearances of the *yang* fire agent on man's sensory abilities: the flames, which could be seen, were "the nature of fire" (*huo xing*); heat (exposed in the steam) and gas, which could only be smelled and felt, he identified as the "fiery intention." Pointing to the deceptiveness of human senses, Song suggested that an inquiry into nature and material inventiveness had to be systematic and purposeful to reveal true knowledge: one had to know what to observe and when. In *On Qi* he thus discussed at great length the fact that, although transformative processes were always due to the interchange of *yin-yang qi,* a distinction had to be drawn between observing the transformation from *qi* to form (and vice versa) and *yin-yang* interchange. In the first case one looked at a transformation "from *qi* into form (and vice versa)" with the help of *yin-yang* interchange: "the origin is *qi* and it transforms into form. Form repeatedly reverts (*fu*) to *qi*."[4] In the second case, the interchange of *yin-yang* itself became the target of inquiry. To the human mind, Song continued, form and *qi* looked stable and steady, because the transformation happened so subtly (*shu*) and slowly (*xu*) that man's senses were not sharp enough and his lifetime not long enough to grasp it.[5] In the case of the invisible *qi*, the slowness was caused by the fact that *yin* water and *yang* fire *qi* were almost completely saturated; in the case of form, one of the *qi*-phases had accumulated so intensely that only few matching partners were left. When one looked directly at the interchange of *yang* fire and *yin* water, as was the case with gas wells, "[one] may think one sees *qi* but it is form, [one] may take it to be form, but in fact it is *qi*." It is important to recall here what I explained in the last chapter: according to Song, *yin* and *yang* were "not able to meet (*jian*)" and preserve their form without the help of man (*ren li*). In general, they would always immediately dissolve into the void. "If they are separated," he further suggested, "they always long for each other (*yi*), like a wife longing for her husband and a mother looking for (*wang*) her son."[6] *Yang* and *yin* "transform mutually into and out of *qi*. [All this takes place] within a millisecond (i.e., *skana*); within the twinkle of an eye it exists and disappears (*you wu*)."[7] On another occasion Song spoke of "a short instant" (*qing ke*).[8] The notably erotic imagery of longing and the many metaphors of familial relationships Song employed in such contexts, in contrast to his generally austere style, are part of the traditional terminology of permutation studies. Referring to the *Book of Changes*, Song also suggested that *yin* water and *yang* fire could "embrace each other in love (*aibao*), wrapping up each other closely," they could interchange (*jiao*), look for each other (*wang*) or intervene (*gan*).[9] All these metaphors stress that *yin* and *yang* felt insufficient on their own. They were "always thinking of (*si*) conversion (*zhuan*)."[10] These descriptions illustrate that for Song the intimacy of *yin-yang* and their interaction gave the system its dynamic. He also saw, however, that it was this aspect of *yin-yang* attitude gave the system its obscurity: the speed and variation of interaction meant man was unable to perceive it with his senses and thus overlooked or denied its existence.

In *On Qi*, Song, tried to verify universal formula for *yin-yang qi* conversion in cases of natural phenomena or technical production processes whose logic was hidden from man because they happened too quickly or too slowly or because they were too complex or occurred in obscure locations. He argued that meteorological phenomena such as wind or rain, and man's bodily processes, could be explained by the fact that (a) everything was *qi* and (b) *yin-yang* conversion, its detachment or dissolution to the void, always took place in a proportional relationship. To verify (b) Song included descriptions of experiments with water and fire in *On Qi*, documenting which volume of water was necessary to extinguish specific weights of burning firewood. If each phase of *qi* was present in an equal amount, he suggested both water and fire would fully dissolve to the void. Only in cases where the volumes did not match, would there be a visible remainder: "If one pours water onto a fire, the fire in the bundle of firewood disappears (*wàng*). Subsequently, the water vanishes (*wu you*) too. The fire in the bundle of firewood is exhausted (*xi*), then there is also something left in the basin of water." Song did not take into account that water might soak into the soil or that fire might continue to smolder inside the wood. Instead he insisted, "when ten liters of the water in a cooking pot have been exhausted, one *dou* of fire has been expended (*jian fei*). The volume (*zhuliang*) of a wagon full of firewood and a basin full of water correspond."[11]

These principles were universal. Hence quantitative suggestions could be made about all kinds of issues. The weather would be cold or hot, dry or humid depending on the quantity of *yin* water or *yang* fire in the void in relation to each other. The variations throughout the seasons and at different times of day could be described accordingly: "Within that 100 *ren* thick particle-atmosphere [that fills the space between heaven and earth], the two, *yin* and *yang*, are joined (*can he*). When the winter comes, then the *yin* water obtains seven parts and the fire's portion amounts to three parts of ten [in the particle *qi* sphere]. On a summer day, the particle *qi*-atmosphere consists of three parts of *yin* water *qi* and seven parts of *yang* fire *qi*. During the equinox, they are balanced. On a beautiful day at noontime *yin* water *qi* interacts with the *yang* fire *qi* in the same way as an imperial concubine chasing after a married man: it [first] departs from the *yang* fire *qi*. After having detached from the *yang* fire completely, it snaps up (*danwang*) again, longing for the *yang* fire. The *yang* fire [detached from the *yin* water] recollects itself. [The day] warms up and as the moisture vanishes, the *yang* fire burns down at a constant low level."[12]

With regard to meteorological phenomena, Song concluded that once man understood that the proportional relationship between *yin-yang* was defined by their ability to find matching partners, it was possible for him to control water and fire in a way that "the fire in the furnace will not be extinguished before the water in the amphora is completely depleted." Performing experiments to verify the relationship between *yin* and *yang*, Song suggested, however, that man should first consider how the interaction took place and which factors affected the process. In order to interact and dissolve, water and fire had to meet directly. This one could observe when looking at steam or water extinguishing fire: "If water is poured on the fire in order to put it out in an amount equal [to the burning fire], the form (*xing*) and the spirit (*shen*) are identical (*yizhi*)." When the two spirits of water and fire meet as rising steam, *yin* and *yang* also approach one another closely and then dissolve. In this case, man just does not see much, because "the proportional measurements [of *yin* water and *yang* fire *qi*] are already well balanced in relation to each other (*junping fencun*)."[13] As few *yin* and *yang* phases were unmatched, the transformative process was subtle and difficult to observe.

Having verified the proportional relationship of *yin* water and *yang* fire in direct transformation by way of his experiment, Song admitted that most forms of the interaction between *yin* water and *yang* fire were obscure and, at first sight, might appear to contradict his theory. A particularly interesting example Song uses to show that the same principles were still at work relates to the seemingly disproportional appearance and disappearance of liquids in the digestive processes of man's body. The fact that man urinated even after hours without drinking verified his view that man absorbed *qi* not only from nutrients but also from his surroundings and then extracted *yin* water from this absorbed *qi*. Similarly one could also see that in the human body *yin* water eventually reverted back to its original stage of invisible *qi*. In fact, man constantly ate *qi* and transformed the *yin* water in it. Sometimes storing and sometimes exuding, the amounts changed in proportional relationship to each other. "Although some people drink several hectoliters (*dan*) per day, their urine does not exceed more than one pint (*sheng*); after the water has entered the intestines (*chang*), it is transformed to *qi*. Sometimes, people do not even drink a centiliter (*shao*). Yet, when they relieve themselves before going to bed, they urinate copiously. It is clear that in these cases *qi* has [inversely] penetrated the intestines and transformed [back] into water."[14] Part of Song's ideas of these bodily processes may have come from personal experience or experimentation; others he may have found in his era's scholarly literature on bodily processes, interpreting them in line with his thoughts on *qi*.

In bodily processes one could see whole sequences of *yin* and *yang qi* conversions in which each and every instance tried to resolve *yin* and *yang* in equal amounts to restore a harmonized *qi*. This was true not only for urine; all water that vanished in front of man's eyes verified in Song's view that the *yin*-water had dissolved to the void with its matching partner, the *yang* fire, in a specific relationship. Here, Song did not draw easy assumptions or blindly follow a paradigm. He suggested that man, trying to explain the obscure, had often leapt to the wrong conclusion because he believed the evidence of his eyes more than the existence of an underlying rational order. Focusing on the "in-between stages (*za*) of *qi* and form," which means *yin* water and *yang* fire *qi* in interaction, one could see the provisional character of all being and would then realize that the whole world was constantly changing *qi*.[15]

The distortion of man's view was a result of his overlooking that the fusion and detachment of *yin* water and *yang* fire could happen in a variety of speeds ranging from "delayed and subtle" (*chi er wei*) to "prompt and pronounced" (*jie'er zhu*).[16] Fire was quickly lit but could not be grasped, as everyone could verify. Its shape changed constantly and vanished rapidly. In contrast to *yang* fire, *yin* water *qi* could form a material body, but this form was flexible and liquid. The example of vanishing dew also showed that many material expressions of *yin* water were not as stable as *qi* or form, tending to dissolve into the void much more quickly. With a reference to the *Book of Changes* Song distinguished the attitude of *yin* water and *yang* fire bodies as "minute" (*miao*) or "diminutive" (*qiao*). According to the commentary of Zheng Xuan (127–200), minuteness addressed the fact that "the myriad things in transformation have a phenomenon (*you bianxiang*) that can be investigated (*kexun shen*)." *Yang* and *yin* were also "diminutive" in transformation, which means, according to Zheng Xuan's commentary that "the myriad things in transformation have a phenomenon that cannot be sought" (*bu ke xun qiu*).[17] *Miao* was, for example, the conversion of *yang* fire *qi* from a discernible shape (*you*) into a nondiscernible condition of existence (*wu*), and vice versa: "The phenomenon into which it transforms is diffuse and never obtains a searchable or tangible form. It can be observed with the *yang* body of fire." *Yin* water dissolving with *yang* fire *qi* to the void or detaching from it was diminutive because it established a transformational appearance "perceptible even in fog, although *yin* water *qi* also quickly disperses."[18] In Song's scenario, the dissolution of *yin* water and its reversion to *qi* differed in quality from that of *yang* fire because "fire is swift (*ji*) and water placid and tardy (*xu*). Water congregates (*ning*) and fire scatters apart (*san*). That which scatters first, goes first; that which congregates, follows later."[19]

Yang fire, Song explained, always "waited" (*dai*) for water to dissolve into the void. Knowing this was essential in the analysis of when and how *yin-yang qi* transformed. Observing a burning fire, Song fixed the crucial moment when *yang* fire transformed into the void as the moment when the fire sparked. This was when the *yang* was "attempting to obtain a form, and then loses it while the spark (*dian*) ignites (*guang*)." For the analysis of the conversion of *yang* fire to *qi*, observation had to be focused on the light. The conversion was "not related to burning firewood. [*Yang* fire] converts after burning at full blaze (*chuanxin zhi hou*)."[20] This passage is not only revealing regarding the way Song considered that the mind leads the eye. It also provides the opportunity to remind oneself of the various layers of concerns pervading Song's descriptions or conclusions. Describing the right way to observe fire, Song alludes to the Song dynastic chancellor (*canzhi zhengshi, zaixiang,* rank 1a) and economic reformer Wang Anshi (1021–1086), who suggested that one master teaching his pupils would guarantee the transmission of knowledge from one generation to the next.[21] In this case Song's interpretation implied that neither the teacher nor the pupils, neither the blaze nor the brushwood, were significant in this process: the important element was the knowledge, the spark that ignited both.

Let us come back to Song's delineations of how to do systematic and "knowledge-revealing" observations. With regard to their behavior in interaction, Song suggested that *yin* water *qi* was the dominant part of most materialized issues in man's world. Pure *yin*, as mentioned in chapter 4, only existed in the deep caverns of the earth. All phenomena of *yin* water *qi* in man's world, Song suggested, were a mixture of *yin-yang:* in water, *yang* fire was still present. Even when man could not see the *yang* fire he could see the effect of its presence; as the *yang* fire in the water decreased it turned from a liquid to a stable form. It froze. The changing temperature of water was another indicator. Cold water contained a little *yang* fire. Warm water contained *yang* in abundance. The local texture of *yin-yang qi* changed because things and affairs moved, attracting each other's *yin* or *yang*. A detachment of *yang*-fire *qi*, Song suggested, could be initiated by or through a fierce wind: "when the soil is soaked by heavy rain and a fierce wind blows over it, the *yang* fire is detached and ascends some feet (*xunzhang*) above the ground. Consequently water cannot remain as it is (*zi huo*), it aggregates and converts into ice (*bing*). This is a water catastrophe (*qiongzai*)."[22] Song makes no remark on *yang* fire, but we can conclude from his explanations that for him the *yang* fire *qi* had no equivalent on earth to the stabilized body of *yin* water as ice, as he presumed that *yang*-accumulations could only happen in the heavens. The sun was a *yang* coagulation man could watch. But the variance of transformations was concealed from man's view because he lived on earth and could not approach the heavens.

Song's interest in meteorological issues is in line with traditional *qi* thought, but his assumption that *qi* alone provided the structure and his singling out of influencing factors, in particular movement and distance, was highly idiosyncratic. Compare, for example, Song's ideas on the cause and effect of rainfall and thunder with those of the Song dynastic iconoclast Zhang Zai, who suggested: "whenever *yin* is condensed, *yang* must disperse. Their positional relation (*shi*) is balanced, they are equally dispersed (*san*) [throughout]. If *yang* acts as a consummation (*lei*) of *yin*, then they grasp another and rain falls (*xia*). Whenever *yin* acts and achieves (*de*) *yang*, then they are blown into the air as clouds and rise (*sheng*), and the *qi* of the *yin* [kind] entirely condenses. Whenever there is *yang*, which cannot come out, there is a strenuous attack that performs like thunder (*leiting*). Whenever the *yang* is outside and not allowed to come in, it revolves in its surroundings without ceasing and appears as wind."[23] In Zhang Zai's view the continuous interaction of *yin* and *yang qi* and its relational positioning (*shi*) initiated natural phenomena: *yang* consummated with *yin* produced rain, *yang* confined caused thunder. For Song rain, snow and hail were caused by *yin* water *qi* passing different levels or spheres in the atmosphere. Song defined the atmosphere in this context as the great *qi* of the void filled with particles (*chen, ai*), a concept I will come back to. As the length of the fall and the speed varied, more or less *yang* fire *qi* detached from the *yin* water *qi* or conversely, an increase in distance or speed caused more *yin* water *qi* to accumulate, thus creating a range of increasingly concrete phenomena, from rain to meteorites: "*Qi* swoops down several ten thousand *li*. Crossing the sphere of particles (*ai*), it fuses (*rong jie*) to take on form (*xing*); one is the form of a meteorite (*xing yun*). The *qi* that swoops down more than several hundred *ren* transforming into form (*xing*) without being able to solidify becomes rain or hail."[24] Song further explained that "in the heavenly pool (*tian ze*) they change (*xiang jiao*) when the sky is covered, but it does not start to rain. It moves quickly upwards (*feishen*), and reverberates with angry sounds, and the clouds start raining. Then the clouds evaporate and the rain scatters and silent flashes (*shenguang*) [light up]. Then there is thunder (*lei*).[25]

We see several differences here: For Zhang Zai the thunderstorm with wind, rain, thunder, and lightening were the several effects of one momentum of *yin-yang* interaction. Song, however, concentrated in his analysis on carefully distinguishing a variety of interactions creating a multitude of phenomena that then affected each other. A thunderstorm hence became a complex process of *yin-yang*, void *qi* interaction, in which thunder was a reaction of an uprising movement, an aspect whose stimulus Song does not further elucidate. Song also objected to Zhang Zai's idea that wind was produced through an interaction of *yin* and *yang qi*. Wind was a moving body of void *qi* that affected and influenced the interaction of *yin* and *yang qi*: "The wind initiates a compression (*ya*) of the *qi* [of the void]. The *qi* of the void possesses the original spirits (*yuanshen*) of *yin* water and *yang* fire harmoniously in a stage of equilibrium (*junping canhe*). This *qi* then experiences (*shou*) a compression or impact, and moves towards one or the other direction [and thus is unevenly distributed]."[26]

As a moving body of void *qi*, wind detached *yin* water and *yang* fire *qi* or brought it together. This was also how Song delineated the wind in the context of the evaporation of dew or surface water: "Whether one [hits] a drum (*gu*), [waves] a fan (*shan*), or [blows] (*chui*) a wind instrument, if one does it with some consistency, [the movement] dries up the water [that was formerly there]. We must assume the water [phases] are stored inside the *qi* [of the void]. This is so obvious when one investigates phenomena such as water drops that vanish when only moments before they were pearls of dew on the surface of varnished wooden furniture or oily stones!"[27]

It is the following passage that shows Song did not consider wind an ephemeral phenomenon. For him it was a body of *qi* that could retreat but was always there: "The water is in the hundred springs, the rivers and the sea on a constant average level. Wind cannot be calm for a long term in the depth. If its *qi* is wet, the wet [i.e., *yin* water] rises upward, longing for the unification [with *yang* fire] in heavy clouds and dense fog. Then it transforms again to water (i.e., rain) and simply returns back into the caverns (*xu si*) and that is it. If the wind is in the territory of the clear void (*qing xu*), it is stored there in full satisfaction, though it cannot completely coagulate (*ning*) above in the heavens. Its *qi* is latent (*yin*) and thus sinks downwards. [During this movement] the *qi* [of water *yin* and fire *yang*] accumulate and encounter each other. They move around tiny and numerous (*wei'ai*). As they transform, wind comes up, either blowing to the north or the south, spreading and circling around itself. After that the wind returns to its territory (*fu*) and rests (*xi*)." In a way it was the movement, its ability to separate and unite *yin* and *yang qi*, that enabled water to be liquid. Thus it mattered "whether a spring or stream is calm or in flow. This [principle] prevents the coagulation (*bu ning*) [of the *yin* water *qi*]." In his delineations on wind, weather, and water Song divulges complex, but in fact rational chronologies of *yin-yang qi* interaction. By analyzing a phenomenon step by step, one could see that the same principles were at work. Wind was not an effect of *yin-yang* action or their relational positioning in the void *qi*, as Zhang Zai presumed. It was compressed *qi* that returned to rest in its territory by resuming its originally harmonious spatial distribution at a specific place. The various meteorological phenomena were all temporary manifestations of *yin-yang qi* in interaction: specific stages in the chronology of *yin* water *qi* accumulation or its conversion by a reunification with *yang qi*. Dew was just an interim stage in the dissolution and appearance of *yin* water, as was fog. As fog, *yin* water could persist because of a coincidence of slowdowns at a point in time, "when it manages to obtain a vague form that then vanishes." Dense clouds that "covered the heavens without providing rain" (*miyun bu yu*) were another manifestation of such an interim stage. These were not the same as the volatile stage of hot water "simmering in a *fu*-pot" (*fei fu zhijian*) and producing steam, because in the case of clouds the *yin*-water slowed down whereas steam was "just a precursory stage of the *yin* water *qi* fusing with the *yang* fire to the great void" that then dissolved in a millisecond. The persistent appearance of clouds, fog, and mist consequently meant that conversion had halted at an interim stage. The *yang* fire, "which always waits for water to return to the void," was the reason that movement was important to interaction. In an almost balanced stage *yin* became slow and *yang* had to approach it to dissolve to the void. As such man had to be careful what to observe, considering that conversion often happened in a way that man could not follow it with his eyes. In this way the moment of observation and the velocity of conversion combined to form the meteorological phenomena of the world.[28]

Song's detailed explanations of meteorological phenomena show that he saw various stages of *qi* as affecting each other in transformation and that he considered special conditions and environmental factors as effecting the transformative processes of *qi* and *yin-yang* interaction in weather phenomena. He argued that *yin* and *yang* had to be taken into account to explain the variety of things and affairs in man's world: first, the moment in time at which transformation started or paused; second, whether *yin-yang qi* met in movement or tranquility, directly or indirectly. In fact, the movement of things was essential to Song's ideas about the creation of things and initiation of phenomena which we can see best using the example of his view of the effects of wind and weather on salt production. Song assumed wind, that is, the void *qi* that detached *yin* from *yang* by movement or brought it together to dissolve, produced a different form of salt than that produced by the sun, that is, an accumulated *yang* phenomenon detaching *yin* water *qi* because the local presence of *yang* fire *qi* had temporarily increased. In both cases dehydration was the result, but still one and the same substance could obtain a multitude of appearances: "When the southern winds arise in autumn, [*yin* water and *yang* fire] associate overnight. This is what [brings forth] the so-called [coarse] grain salt (*ke yan*)." In contrast, "salt which is [produced] through the evaporation of sea water is finely granulated, producing fine crystalline structures."[29] The essential factors in Song's formula of the physical processes of salt crystallization were the interaction of *yin* water and *yang* fire *qi*. The environmental factors were daytime and season, the place in which the interaction happened and the velocity with which a saturation of *yin-yang* or their detachment took place.

Yin-yang interaction thus had to be observed in all circumstances and stages. This was particularly relevant as *yin-yang qi* was volatile and thus transformed quickly, and form only appeared stable because it transformed slowly. Multiplicity resulted from the various interacting conditions of *qi*-change, and the accumulation or dissolution of *yin-yang qi*. Apparently in those cases when the transformation happened very subtly and slowly, man's lifetime was too short to recognize it, but that did not mean change did not take place. The transformations of mountains and rocks, earthenware or stone buildings could, however, only be deduced by comparing them with cases whose pace was faster and hence observable: the birth and death of plants and trees or the human corpse. Growth and decay emerge as relational, bound to the moment of observation and the target of investigation. The knowledgeable man realized that the ever-changing world of *qi* was a self-contained regenerating system. Nothing could be lost.

GROWTH AND DECAY: WOOD, CORPSES, AND THE PROPORTIONAL RELATION OF *YIN* AND *YANG*

> When the soil on the hills is withered and scorched, then the heads of all kinds of rice will droop. A diligent farmer fertilizes the field, thus taking several remedial actions.
>
> *Works of Heaven*, chap. 1, no. 1, 4a

Burning weeds and brush on fallow soil was the first step when preparing a field. In sparsely populated regions the farmers planted the soil enriched by slash-and-burn for two or three years and then let it lie fallow for four or five. In densely populated regions, such as Jiangnan, the soil was intensively deployed with two or three annual harvests, and peasants and landlords eagerly searched for ways to stabilize or improve the soil's quality through fertilization. Irrigating or flooding their cropland seasonally allowed alluvial silt to enrich the soil. Moreover, farmers also collected pond silt, canal mud, and spread it on the fields. In most regions of China, farmers also regularly applied human excrement and experimented with a broad range of animal and vegetable waste, composting grass, rice plants, and tree leaves. Oil pressers and brewers earned extra income by selling their residue to the neighboring peasants as fertilizers. In good times, soybeans were even directly cast onto the field for higher yields of more valuable crops, as Song noted in *Works of Heaven*.[30] Soybean, milk vetch, trumpet creeper, radish and mung beans were used in rotation with gramineae crops. Materials such as limestone and sulfur, shells and bone ash were known to rejuvenate acid-sulphate soils.[31] Although they knew the value of these remnants, hardly any farmers, in Song's view, ever inquired into the underlying principle of these processes of enrichment and decomposition. They only saw the utility. None of them realized that decay required as much time as growth, not only for grains, plants, or animals but also in the case of metal ore and stone. For this reason all resources required careful handling to avoid damage, erosion, or depletion. Depicting the world as an organism of *qi* conversion, Song thus called for sustainable development and urged in *On Qi* that even mines needed to be fertilized by filling them with the remnants of stones to trigger the creation of new ores or coal.

Song was very concerned that men often failed to think issues through to the end, in particular when they looked at growth and decay. Why did no one ask why the level of soil on a field was never higher even though new things were added to it? Or why was it, he asked in *On Qi*, that although the potter (*taojia*) blended new clay for his daily use and produced new vessels day by day, the supply of clay never vanished? Even "after a thousand or ten thousand years have passed there is no lack of earthen pots, nor is there ever a lack of soil."[32] Song admitted that as clay was a composite of *yin* and *yang* *qi* with earth *qi*, it grew and decayed at a leisurely pace. But it changed, as all things changed. If one carefully watched ceramics and porcelain and followed how an earthen pot "explodes (*liebao*) and transforms to soil, when exposed to high firing," one could reason that the soil vanished and accumulated again.[33] In fact, one could see that clay, like all things and phenomena of the *qi* world, needed as much time for decay as it did for growth.

A cyclical understanding of life and the processes in the cosmos and on earth were widespread in Chinese philosophy. Song further deduced an intimate parallel between growth and decay in general. Exemplifying by way of plants, animals, stones, minerals, and, in particular, man's birth and the fate of his corpse, Song claimed that the period of growth always equaled that of decay, a proposition which he delineated with regard to wood. Each thing and affair of the world and its specific speed of transformation from *qi* to form and back were proportional. Benignly, he sympathized with doubts about this principle, assuming that it was hard to believe something that could not be witnessed.[34] Observing the growth of plants, Song affirmed his theory that the creation of concrete matter depended on *yin-yang qi* conversion. Some plants like ramie (*zhuma*) seemed to materialize their form from earth *qi*, coagulating one concrete *yin* water *qi* shape in interaction with *yang* fire into the form of wood and stems. Other plants, growing directly out of bare rock, directly translated the surrounding *qi* into leaves and wood. If it was not by the accumulation of *yin* water and interacting *yang* fire *qi* in its surroundings, how could a seed, as petty as a nit, become a tree that "could hide a whole cow?" To illustrate his point, Song meticulously deconstructed the decay of wood in the following passage to a retraceable chronological event.

> Man arranges [his tools] and reaches for a hatchet in order to cut [the wood] down. He wants to use [the wood] to produce implements and [as construction materials] for the palaces. [The people moreover] use an excessively high amount as firewood for cooking. That is what humans can grasp by observation (*ren de er jian zhi*). Catching fire, wood burns. It leaves ashes and particles. If one calculates their weight proportionally [before and after burning], less than a seventieth [of the original weight] remains. Remnants of bare branches, leafless trees, dropped leaves, dried twigs, bushes, and withered vines lose their vitality, drop off and become mud (*tuni*). They lose no less than nine tenths of their form (*xing*) of blooming vitality (*sheng mao*). Man still seems to be able (*you jian*) to recall the [original] form of grass and wood. He thinks that once it has changed into particles and mud, [transformation] stops and [the remnants] do not again transform any further (*bu fu hua*). [That is wrong].[35]

Almost every Chinese agricultural treatise deals with the natural putrefaction of plants and both animal and human excrement, more or less in detail, and a few medical texts cover putrefaction and human digestion. A quantitative analysis as Song pursued it in *On Qi* is, however, is rare.

Song's logic of the world as a closed system of *qi* interchange aimed at demystifying transformational processes and philosophical speculations about it. He contrasted this with his colleagues who sought refuge in supernatural paradigms. He complained that they theorized a spiritual level operating behind the perceptible instead of realizing that nature displayed its principles literally in front of their eyes.[36] Song in particular attacked Buddhist conceptions of decay. He accused Buddhism of superficiality for suggesting that "the skin, hair, bones, and flesh revert back to the soil (*tu*) and the body fluids such as sperm, blood, mucus, tears, and sweat to water (*shui*)."[37] Astonished, almost infuriated by these theories, he urged people to open their eyes to the truth and see that all things were eventually and ultimately a product of *qi*. Song accused his scholarly colleagues of quoting from one another's theories on concrete things without questioning them. Often, he believed, they pondered only philological questions and surmised wildly on the things and affairs described. Why did they not simply go out and look and then deduce the unknown from the known facts?

Arguing for the importance of verifying facts, Song also criticized acknowledged idols, such as the second-century physician Chunyu Yi) (205–150 BC), who had argued that putrefaction and plant growth resulted from a combination of factors: temperature, nutrition, and the available amount of water.[38] When Song rejected such theories as too sophisticated, we must take into account that he was dealing with ideas that generations of text-exegesis had stripped to vague abstraction. Song, by contrast, kept his interpretations about the cycle of a plant's life closely linked to concrete examples. The growth of a certain gourd, he explained, required 180 days. It needed the same time for its complete putrefaction. Song was adamant in the application of his theory, which led to remarkable conclusions such as his explanation that the different cooking times for a chick and a hen relied on the correlation between growth and decay. Boiling together in a pot, he observed, "the chick is already well done, whereas the skin and meat of the hen are still tough (*jianren*)."[39] For Song, it was not the hen's greater size or the increased expenditure of energy that caused the prolonged cooking period. The hen took longer to cook because its period of growth had been much longer than that of the chick.

With this in mind he drew the reader's attention to man, asking why, if growth and decay were actually related to one another, did an old person not decompose more slowly than a youngster who had died early? Song pointed to Peng Ziu, whose longevity was proverbial. Comparing him with a seventeen-year-old exam candidate, he argued that the decomposition of a corpse was in some ways similar to the cooking process. In both cases the decay was caused by fire and water interacting. In the cooking process fire reached the boiling point quickly and maintained a rolling boil when the pot was centered on the fire; dragged to the side, the water would reach boiling point slowly and only maintain a low simmer. When cooking a chick or hen, the actual cause of the conversion was the interaction of *yang* fire and *yin* water *qi*. Song admitted that an intensive fire and more water accelerated the process, but he remarked that the period of cooking and the quantities of fire and water used changed in proportional relation to one another. In the soil *yin* and *yang* were also concentrated in some places, interacting and transforming things, whereas in other places they were almost inactive. The location of a grave hence either prolonged or accelerated the decay of the corpse, simmering it slowly or cooking it to rags until the corpse dissolved as could occur when someone overcooked the chicken in a soup. This was Song's way of pursuing his theory that all relationships of a *qi*-world were universal.

Song further made the correlation that, like the chick, which had to enter the hot water to start the cooking process, all stable phases of *qi*—a human corpse, wood, stone, pottery, or things made of these materials—had to penetrate the *qi* of the soil before they would decay. He proposed that this was a particular kind of transformation (*bian, yi*) that had to be distinguished terminologically from the conversion of materials by *yin-yang qi* in the air. He illustrated his theory with an example from everyday life. The wood used for buildings, bridges, or any other construction, was able to resist decay for "some hundred years"[40] as long as it did not touch the soil. He gave the example of the Jiaowei Qin, a special kind of stringed instrument made of grass and wood, materials that in general decayed quickly. It stood the test of time, "because it does not come in contact with the earthen *qi*." He felt people applied this principle when they preserved or mummified corpses: "Human bones and flesh alloy with soil and mud, requiring one hundred years until they vanish into the void. As man does not wish [his body] to rot away, people treat the corpses with mercury (*shuiyin*) and plug the body [openings] with beads, pearls and jade. Avoiding these items, the vaporizing *qi* does not enter [the corpse] and attack it."[41]

The transformations man recognized as growth and decay depended on the *qi* involved. Furthermore, the researcher had to take into account that not all principles worked together at the same time. The time correlation of growth and decay of wood only applied to the periods when the wood really grew and not when it was used for construction and just persisting, that is, taking a halt in conversion. When wood caught fire, however, the velocity of transformation was not subject to the principle of growth and decay, because the *qi* of *yang* was intensively involved. In accord with Chinese epistemology, Song identified wood within this context in relation to its material existence as one of the five phases. He also suggested that, like metal, the growth of wood was subject to earth *qi* within it. Yet metal drew mainly from earth *qi* and the *yin* and *yang* in metal were deeply enclosed, whereas wood additionally relied on interaction with the *yin-yang qi* of the void for its growth. In wood, the *yang* fire as well as the *yin* water *qi* could easily be made visible and their material existence verified, when one "wrings out a green leaf, a fixed amount of water will emerge. When a withered leaf is burned, a specific amount of fire will result."[42] Consequently wood lost its materialized form much more easily than earth and metal *qi*, vanishing to the void *qi*. Thus it was the last and weakest in the sequence of the five *qi*. When burned, the *yin* phase within the wood could detach, fuse with the *yang qi*, take shape briefly as smoke and vanish to the void by fusing with the rest of *yang* fire present in its surroundings. At the same time the *yang* agents in the wood agent became visible in transformation as flickering flames. Burning wood was a rather drastic form of transforming its inherent *yin* and *yang qi*. It happened in a short time and thus man could easily observe it. The same process occurred when wood or plants were scorched by the sun, but man seldom took the time to observe it: "When one burns wood, smoke (*yan*) rises. This is the *qi*, which emerges while the *yang* fire and *yin* water agents are still fighting (*zheng*) with each other. Exposure to strong winds and intense sunshine exhaust (*jingjin*) the *yin* water *qi* in the wood making it completely return to the void. In this case the *yang qi* are retained in the wood form, while [the water] is completely transformed (*hua*). Consequently, when burned, only a little smoke arises from the blaze."[43]

In *Works of Heaven* wood and its use as a fuel for fire are not important. But in *On Qi* wood becomes a major topic. Within the context of growth and decay Song described woods in almost mechanical terms. Wood could be completely explained as the product of its own growth. It decayed, as it had been created, with the help of the volatile *yin-yang qi*, the earth *qi* and the *qi* of the void. This elegant solution allowed Song to reconcile the properties of wood as a material with his theory of the five phases. Throughout his writing Song posed rhetorical objections to his theory, to substantiate that he had an incorruptible, investigative mind. This is especially apparent when his delineations on experiments with water and fire shift focus to the issue of how wood behaved in transformation. Song realized there was an important gap and significant problem in his explanatory framework; when burned, wood did not fully extinguish to the void. In fact, a considerable quantity of ashes and dust was left behind.

These ashes puzzled Song extremely, as his *On Qi* reveals. Describing several fire experiments, Song initially insisted on the idea of a quantitative relationship. He argued that a specific weight of wood required a defined quantity of *yang qi* (in the form of flames) and *yin qi* (in the form of water) to be dissolved to make the wood fully vanish. After each burning he measured (or estimated) the loss of the material form in relation to the leftover ashes.[44] Song's descriptions, which sound like administered tests, generate trust. The units, however, were vague: bundles or stacks of wood and undefined vessels of water. But Song took the trouble to vary figures and volumes to verify that no matter how he arranged the experiment, he always found a residue of ashes, something that resisted the dissolution into the void *qi*. It is at this point that Song gave priority to actual facts, conceding that the traditional theory of *yin-yang qi* and the five phases was deficient when tested against the mundane world. With many references to his experimental studies, he informed his reader that ashes were a particular kind of *yin-yang qi* that had to undergo another process before fully transforming to the void. Recognizing a discrepancy between his theoretical model and actual experience, Song chose the authority of experience and accepted that there were moments when practice outweighed theory. He thus introduced the concept of "ashes" into his basic framework of *qi*, form, and the five phases epistemology.

GLITCHES IN THE MATRIX OF *QI*: THE CONCEPTS OF ASHES AND PARTICLES

Ashes

> All coke that undergoes fire changes its substance (*zhi*) and vanishes because of the transformation of the fire *qi*. It never leaves ashes. This means that it is a special manifestation of transformational processes (*zaohua*) intermediate between those of metals and soil and stones.
> *Works of Heaven*, chap. 2, no. 11, 55b

Boiling salt, smelting ores, firing porcelain ware, calcinating minerals, dying cloth, or burning resin for lampblack to manufacture ink, all these processes required fuel to produce heat. Song knew from visiting or talking to his friend Tu Shaokui or others, distant regions used gas. But living in Jiangxi Province, Song was more familiar with wood, charcoal, and coke. In *Works of Heaven* he explained that each furnace for smelting iron had a capacity of more than 2000 catties (*jin*), approximately one metric ton. Ore and coke were put together in these furnaces. The coke burned down completely without leaving a residue and the ore flew out of a hole after several hours of burning. In this case Song only approximated the quantitative relation of burning materials required for one firing.[45] Describing the production of tile, he shows his interest in proportional calculations. Song specified that the burning of one hundred tiles required exactly 5000 catties (approximately 2.5 metric tons) of wood.[46] His experiments with water and fire in conversion in *On Qi* convey that Song believed the balanced dissolution of *yin-yang qi* could be verified by measurements of material weights and their losses. Song's descriptions in *On Qi* also show that he may not only have relied on information given by the craftsman responsible for clearing the kilns but that he may himself have measured the residue ashes left behind in order to calculate his estimates. From this, Song concluded there were some things that, though composed of *yin* water and *yang* fire *qi*, would not entirely revert back to *qi*. Challenging his theory, Song turned this glitch in the matrix of his *qi*-world into a concept and introduced the ashes, defining them as a form of *qi*-materialization that required special conditions to revert back to the void. Song suggested that all things that originated, directly or indirectly, from plants or woods left ashes when they were burned. Yet, understanding the ashes was complicated because plants and woods were created by subtle and complex processes that included most aspects of *yin-yang qi* conversion. In order to understand why a remainder of ashes was left whenever such things were burned, one had to retrace every step of growth and decay. Furthermore one also had to consider that plants and woods also participated in the growth of other things, events, and beings:

> Grass and trees contain ashes. The bones and flesh of humans and animals depend on grass and trees for living and growing. Tigers and wolves live without eating grass and trees. But the animals and beasts eaten by them, all grow by availing themselves of grasses and trees. [Accordingly] the body fluids (*jing ye*) of these animals mutually interchange and as a result the grass and trees in their bones and flesh are from the same *qi*-category (*lei*). And in fact the sediments and foam eaten by the fishes and prawns in the waters, are, if we trace them back to their origin, actually grass and trees.[47]

The *yin-yang* condition in ashes could only translate to the void-*qi* with the help of "steaming *qi*" (*zheng qi*) or through a melting with the "primordial *qi*" (*yuan qi*). This only happened when "the ashes are able to rejoin the yellow springs (*huang quan*) and reassemble with the primordial essence (*yuanjing*) of heaven and earth in the cold caves (*huxue*)."[48] The yellow springs are a standard, somewhat poetic way of addressing the place on earth and in heaven where the *yin* and *yang qi* found rest. This place is where all things originated and thus it was also the place where all *qi*-

conditions, no matter how complex, could revert back to the void. "Above, the clear original *qi* (*taiqing*), the two *qi* are in balance and wait to bring forth (*sheng*) the myriad things. Below in the deep abyss of the nine springs (*zhongquan*) the two *qi* are in balance and wait to let the hundred earth spirits (*bai hui*) flow out."[49]

Refined *qi* was an especially intensive kind of *qi* called steaming (*zheng*) *qi*. This *qi* was only present in the yellow springs. When the ashes had fully converted back to the void, they participated in the primeval essence (*yuanjing*) of heaven and earth in the cold caverns. The notion of the cold caverns implies the idea of full reversion to the great void.[50] Special domains where *qi* was created existed both in heaven and earth. In the yellow springs even extremely subtle materials, such as gold, pearls, or jade could revert into the void with the help of steaming *qi*.[51]

Song's concept of ashes verifies that he acknowledged inconsistencies in his theory and adjusted his conceptual approach to explain the world of matter and being. The concept of ashes allowed him to find consistency and logic in things his colleagues considered strange and outside orderly schemes, such as mirabilite, also known as Glauber's salt. While his colleagues assigned each a separate category, Song suggested their behavior was clearly caused by the fact that they did not belong to the world of plants. Thus they could not contain ashes. Neither were they behaving like earth or metal *qi*. "If we now want to see [things] which transform to the void without leaving the slightest quality of ashes (*huizhi*), then [consider the firing of] such categories (*lei*) as cinnabar (*zhusha*), hermaphrodite stone (*xiongci shi*), sulphur (*liuhuang*), coal (*meitan*), kui-stone (*kui*), and sulphate (*puxiao*)."[52]

All items that burned like wood but left no ashes were hybrids; they belonged neither here nor there. Mirabilite, he explained, tended toward metal *qi* despite belonging to the earth phase of *qi*. Because of their in-between status, hybrids could be fully transformed by the *yang* fire *qi* and leave no residue. Hybrids revealed their nature by their attitude in transformation. Their outside appearance was misleading and should not be used for classification, the mistake his colleagues had made. Another substance Song identified as a hybrid was mercury. It "did not have the quality of leaving ashes." Song's enumeration of substances was not unusual. This suggests that he, reacting to ideas current in alchemy, simply wanted to appear all-inclusive.[53] He explained that mercury (*shuiyin*) "flows from soft granules (*nensha*) [of cinnabar when heated]" or that "bright pearls are the fetus of old shells (*laofeng*)."[54]

In sum, Song identified ashes as a vital condition of *qi*, inherent in some things and affairs in nature. This again shows that Song was primarily interested in how natural items supported his theory that the world was unified in *qi*. Another element that fulfilled a function similar to ashes in Song's concept were particles which could be found in the world of invisible *qi*. These particles, like ashes, were an extension of the traditional theoretical model of a world of *qi*, a bug-fix, that helped Song to include all facts and phenomena that would otherwise have challenged his matrix of a universal *qi*-world.

Particles

> Molten vegetable tallow is poured into a paper cylinder to form a candle. This kind of candle resists the particles (*chen*) whirled up by the winds.
> *Works of Heaven*, section 2, no. 12, 67a

A diligent emperor's schedule normally began between five and six o'clock in the morning with court gatherings. By that time servants had already refilled the oil lamps and made sure the wicks were trimmed so that the passageways to the audience hall were illuminated for the emperor and his court officials. Work thus also began uncomfortably early for the civil servant who had to cater to his emperor's needs. Indeed, daylight seldom sufficed to accomplish the manifold tasks of an official's household in the Ming state. And this was even more so for the dedicated scholars who were studying to ensure their future careers. Every household that could afford it was illuminated even in the dead of night by the light of oil lamps and candles to allow their inhabitants to improve their literary acumen and study for higher goals. An excellent lamp oil was produced from the tallow tree (*jiu*, *sapium sebiferum*, also known as popcorn tree, or Florida aspen), as this burned for long periods and gave a calm, clean light, producing almost no smoke to sting the eyes.[55] Like the candles made of the tallow tree, the flame fed by its oil was constant. Put in containers or protected by jars, it could resist fierce gusts of wind that would otherwise extinguish it. However, there was also something in the air—this realm of invisible *qi*—that the flame required. Otherwise why did the flame go out when the container was too tightly sealed?

Being fond of observation, Song recognized that several natural phenomena and events could not be fully explained by the traditional constituents of *qi*, neither its invisible appearance, nor its constitution as form. He realized that recourse to the *qi* of *yin* water, *yang* fire, earth, metal, and wood was not the answer. Neither did ashes offer a feasible solution, as ashes were remarkably visible and material. Whatever fed the flame, whatever it required in the invisible void *qi*-atmosphere, was something not easily perceptible to man. It was not invisible *qi*, because *qi* was everywhere. This component could not penetrate the lantern's structure or the jar. Song thus included another singularity into his theory of *qi*, the originally Buddhist notion of particles.[56]

Before I explain the details of this concept, it must be emphasized that *chen* do not correspond to molecules or atomic theory and should not be confused with such modern models. Like ashes, particles filled a gap in Song's theory of *qi*-change. But whereas ashes constituted a specific condition of *yin-yang qi* that required a combination of factors to be able to transform to the void, particles were defined by their material quality. Song's concept of particles is in fact not easy to understand because it lacks some important specifications. He never clarified, for example, if particles could or would ever convert to *qi*. Neither did he give any explanation as to how their existence related to stable form, the phases earth, metal, wood, or ashes. Particles constituted an autonomous, fundamental constituent of the physical world. Song mentions, however, that he saw particles conceptually on a par with *qi*, form, *yin* water, *yang* fire, earth, metal, and wood. Particles were only in the invisible *qi* of the void and in water. They were a substantial quality in the atmosphere surrounding man: "in general everything in between the particles is entirely saturated (*chongxu*) with *qi*."[57]

Song emphasized that the particles he was referring to were not to be confused with ashes; he signified that this was among the difficulties that arose when explaining the world of *qi*. Song was, in this context, willing to take observation as the gauge of assessment. Talking about particles he also warned: "Particles are invisible items—they are translucent and clear. Actually, all particles which are inherent in the original *qi* have to be distinguished from dust motes." Particles, Song explained, could best be compared to the reflection of the moon on water. They were a mirroring empty material not created from the void and thus also not returning to it. In this regard particles do not actually belong to Song's universal construct of *qi*, although he himself never drew any particular attention to this fact. "In their role as things (*wu*), particles linger in the void (*xu kong*). They have a form without moving (*bu dong*). Their entire body (*ti*) is transparent (*touming*). [Man] can see through it over 1000 miles (*li*)." Song insisted that man, "grasping for independent (*ziyou*) particles, may encounter them although they have no [material] quality. Yet, they are a phenomenon (*xiang*) that completely fills up (*man*) man's sphere (*yanfu jie*). Hence at dawn, when the rays of the sun stream through a clear window and the heavenly obscurities are cleared away and the phenomena (*xiang*) are obvious, man may catch a glimpse of them." Yet, at the same time this very characteristic of invisibility demonstrated how misleading it could be to trust one's eyes alone. Subtle as they were, they were difficult to grasp. "In the human world, [we] recognize and see dust motes (*ai*) at a clear window. This should not be misunderstood as particles, as those are the earthen ashes. Particles are everywhere in daily usage and [man] does not recognize them at all. They are not the only thing of this kind."[58]

Song hence concluded that while men could observe the effects of particles' presence or absence, he could not see or encounter the particles themselves. Together with *qi*, he suggested particles constituted the air, that is, the atmosphere in which man lived. But the particles were not everywhere and observation proved they could not traverse specific materials. In fact, man desperately needed particles. To prove his point Song discussed the requirements of life for man and fish. Song suggested "if a human being does not eat *qi* with each breath, he will die, as the fish will die if it does not eat water." Man ate *qi* and emitted it through the nose, like a fish that ate and emitted water via the gills. The particles in the invisible *qi* atmosphere in which man lived were the equivalent of those in the water in which fish lived. "[Whether] man enters the water or the fish defies the water environment, both die immediately. The other environment opposes the one in which each is born."[59]

The inhabitants of each realm may be aware of the characteristics of the other realm. But they were ignorant of the characteristics of their own atmosphere, because it was overly familiar. "Fish are born in the water, man is born in the *chen* particles. From a bird's-eye view man looks down and perceives the water consciously. Fish are not conscious about their realm. Fish glance up and perceive the particles (i.e., in the void-air) consciously. Man is not conscious of them." If one admitted to this nexus, Song proposed a "simple" experiment would show that man also depended on particles, an experiment which we can only hope was hypothetical: "Just try and fill a fishbowl with water—but not up to the brim. Then seal the top completely. Afterwards wait some hours before lifting the cover and one will find the fish dead. [Then try placing] a person to sit upright in an official examination cell (*shiwuge*). Brick up the walls closely and seal them with plaster (*hu*) and then wait for a period of three meal times before having a look. One will find the man dead." Describing the experiment, Song assumed that *qi* always communicated with each other (as everything was *qi*), even through all kinds of barriers such as walls or membranes. Both deaths could only be explained by the fact that there was something else inside the fishbowl and the examination cells that could not communicate or exchange with the same thing outside: the particles.[60]

Song's experimental arrangement and his stipulation that the fishbowl should not be filled to the brim, suggests he wished to prevent the argument that it could have been the *qi* of the void the fish required. Song himself does, however, not elaborate on this point. He just emphasized that there had to be a special attribute in air and water that was required for species like man and fish. He assumed that "fish are nurtured by water. They must have a *qi* that is penetrated by particles. Only then can they live and breathe. If in a breath of water there is no penetration of particles, the water is said to be dead (*si*). And the same is true of the fish." Similarly man was "nurtured by the *qi* [of the void]. He must move in an atmosphere of *qi*, if he does not want to die. If the *qi* of the breath has no access to the surrounding *qi*, it is said that the *qi* dies and man's fate is death."[61]

In these passages of *On Qi* Song employed a rhetorical strategy of analytic uncertainty to raise doubts and then dismiss them; bringing in counterarguments and oppositional views, he tried to persuade his reader of the inevitability of his conclusions. He built up an argument based on various examples and the description of experiments and experiences. He even suggested that any observer at this point could argue it was not necessary to introduce a concept such as particles to explain the behavior of fish and men in their opposing environments. Why should one not instead propose that *qi* itself was responsible for such attitudes, in that *yin* water was a transformational agent of *qi* and as such was only breathable by fish and not by men? He countered with the remark that this proposal did not take into account that *qi* in its original condition was ultimate and indivisible. *Qi* was a universal that was always there and of which everything consisted. To dismiss any last doubts about his concept of particles and its presence in the environment, he asked why did a man sitting in a sealed up room suffocate? How could his death be explained, given that food and water was provided in sufficient amount and *qi* was everywhere and could penetrate everything? Man obviously required both particles and *qi*, to survive.

It was the experiment with fish and water that demonstrated that in its various manifestations, *qi*, despite being one and the same, could develop a multitude of material qualities that affected its potential to communicate or transform. The fishbowl, despite being *qi*, did not allow the particles to penetrate; the sealed cabinet showed the same qualities. Song explained that the divergent properties of materials were the result of the divergent processes of transformation through which these articles had gone. Thus they were able to ward off other kinds of *qi*; in the case of the fish and man it was the *qi* of the void containing particles. Song made this idea of varying *qi* qualities and penetration his next topic, suggesting when cakes were cooked, they would build up a skin that could no longer be penetrated. This happened whenever *yang* fire and *yin* water *qi*, instead of fusing, exchanging, or melting, amalgamated in a harmonious balance (*huijie*).[62] *Qi,* despite being one and whole and thus always in communication with itself, could nevertheless shape out forms that *yin* or *yang* could no longer penetrate: "Sorghum paste (*daoshu*) is used to produce round cakes. They are steamed while swimming in water and fired by flames. The *qi* of fire and water has braised the outer crust but has not yet reached the inside. If the fire then goes out because the firewood is spent, the interior will remain raw although the outside is finished. Lighting the fire to heat and boil it another time, the inside will never be totally cooked even if one continuously replenishes the logs or refills the water. This is because the *qi* of water and fire has solidified the outer [skin], and the [*qi*] that comes later can no longer penetrate. (The same happens with the chicken. When it is boiled insufficiently and then taken out, one can boil it for a month with no satisfactory result)."[63] This also explained why apparently sturdy walls built with dried bricks were not stable. They had not formed an impenetrable skin so water could again enter the brick and the wall would collapse.[64]

The different cases introduced above exemplify the challenges in Song's systematic application of his *qi* theory to mundane circumstances. In some cases Song met this challenge by adjusting his theory; in other cases he stressed the specifics of certain transformations. He meticulously analyzed processes and materials and identified their various properties and qualities. Focusing on physical issues and their empirical verification, Song identified flaws in his thesis of a universal *qi,* introducing particles and ashes to mend the matrix of *qi* he had identified. His use of such caveats or elaborations proved that Song was willing to expand the traditional epistemological approach to nature using the two-*qi*–five phases theorem *(erqi wuxing)* whenever it did not suffice to explain natural phenomena or material inventiveness. The epistemological system, conversely, proved to be flexible enough to include such new approaches and explain what nature presented to men. Within this framework, Song argued deductively and inductively to give his world what, in his opinion, it required most, an inspiring model of order.

Song used his basic theory to develop this model and account for every case. He did not introduce the structural element of *li,* used by many of his contemporaries. For him all things were one *qi,* and this provided the structure. He then introduced qualifications such as time, speed, and strength to explain the variety of forms and phenomena. In this Song is rather exceptional even among the broad range of *qi* thought in seventeenth-century China. His lack of concern with cosmogenesis is also exceptional: his world was eternal in a stable bipolarity of *yin-yang qi*. Heaven was the place where the *yang* fire phase of "*qi* accumulates and does not again transform (*fuhua*) into form; like the sun and moon." The earth was a place of extremely accumulated *yin qi* in which "form is created (*cheng*) and does not again transform into *qi*, but remains as soil (*tu*) and stones (shi)."[65] The stability was provided by the fact that both changed in a proportional relation to each other, thus keeping their polar balance. This was the reason why the resolution of the world's basic dichotomy was very unlikely to happen. "The earth carries all things. If it would fully transform to the void (*xu*), then the world would cease to exist."[66]

In *Works of Heaven* Song continuously praised the divinity of universal schemes. In *On Qi* and *Talks about Heaven* his rhetoric pointed to the reasonability of his suggestions. After explaining in length the principles of decomposition in *On Qi*, Song, for example, concluded with the strong rhetorical question "why does no one realize the causal nexus of these issues" despite it being so obviously revealed?[67] In other cases a hypothetical inquirer affirms his understanding (*zhi, wen ming*), acquiescing the presented logic before bringing up a new issue. In *On Qi*, Song employed a special didactic rhetoric that, going beyond the descriptive style in *Works of Heaven*, stressed his method of knowledge gaining. This didactic has three components. First, he described a phenomenon and posed a theoretically oriented motivating question, often one that indicated a commonly held general assumption. Sometimes this was presented in a dialogue, starting with a hypothetical inquiry, "someone said (*yue*) or asked (*wen*)" this or that. This is a traditional and widely employed method in premodern Chinese writing. In a second step, Song proposed his theoretical view and then detailed the phenomenon, unraveling its manifold implications. In some cases Song presented a series of experiences or built up systematic observations or (real or hypothetical) experiments. In a third step Song formalized the knowledge, shaping a larger model of explanation and general patterns. To inspire confidence in his theoretical conclusions, within each step Song systematically correlated his notions, in both word and image, to familiar contents and shared experience. Then he went on to a related, new phenomenon to extend his theory. A nice example of this strategy is the way Song introduced the concept of "particles" (*ai*) to his *qi* theorem. In this experiment Song cunningly invited trust by invoking his colleagues' memories of a time when they may have struggled to catch their breath: he confined the man in one of the small brick cells used for official examinations. As such, Song's observations of natural phenomena and material inventiveness within his endeavor were directed by his belief in the universality of *qi*. He used speculative parallelism to correlate the structure of indiscernible phenomena with discernible events. He also allowed anomalies such as minerals and seemingly awkward facts into his outline of transformational processes, testing the coherence and reliability of his theoretical models. Song's intention was to decipher the forces of nature as rational, cogent, and explicable by the characteristics of *qi*. The universality of this *qi* principle, he argued, was not only verifiable in the world that could be seen, touched, and tasted. The same principles also worked in the invisible sphere of void *qi*. Thus Song went on to explain the world of sound and silence.

CHAPTER 6

Acoustics

Would the legendary [officials] Gongshu Ban and Chui have been able to manifest their skills if their generation had not profited from tools? [No!] Among the five weapons and the six musical instruments, the consummate examples of the toolsmith are very intricate. The triggers (*ji*) of life and death [have to be] perfectly calibrated.
Works of Heaven, chap. 2, no. 10, 44a*

During celebrations, such as the Spring and Mid-Autumn festivals, Chinese cities rang with noisy clamor; sedan bearers cursed vilely when the pressing crowds forced them to halt; herds of swine squealed and gaggles of geese honked indignantly on their way to the slaughter; boisterous children exulted in the detonation of firecrackers, and shrewd pitchman fought to outdo one another at advertising their goods. These mundane noises were interspersed with the soothing siren songs of prostitutes giving sophisticated concerts in brothels, the rhythmic steps of dancers, and the strains of the local opera, while playing a familiar story promising creature comforts and entertainment to its listeners. Contemporary novels such as the *Jin Ping Mei* (*Blossom in a Golden Vase*, ca. sixteenth century) used ornate descriptions and illustrations to bring China's musical tradition to life, de-

*Both figures mentioned in this quote represent legendary ideals of master craftsmen who were assigned the role of inventors of useful devices. Gongshu Ban, later also identified as Lu Ban, became the patron of carpenters and cabinetmakers. He was venerated as the inventor of the saw and the plane and was believed to have designed fantastic mechanical devices such as a bird that could fly for three days. Chui is deemed the inventor of compass and plumb line and usually placed in the court of the Chinese legendary Yellow Emperor.

scribing nostalgically how male singers, groups of musicians, and female prostitutes entertained their guests, rich merchants, and officials during the Song period.[1] But music provided much more than a melodic undertone to the ambient noise of this era. Music was a means to communicate with the heavens and harmonize the universe. It was central to state ritual and many forms of religious devotion, from ancestors' feasts to Buddhist preaching. This made it a key topic for the intellectual and politically minded scholar of the late Ming. Indeed, the middle and upper class of this period deemed themselves connoisseurs of music. Scholars inquired into sound as an issue in phonetics and philosophers investigated elements of music to reveal valuable principles and ethical patterns. The most prominent musical scholar, Prince Zhu Zaiyu (1536–1610) drew new and fruitful connections between mathematical calculation and melodic composition.[2] And the engagement was not only theoretical; following the cultural ideal of the erudite "amateur," members of the elite played the lute or the *guqin*-zither with great virtuosity.[3] Culturally and intellectually, music was central to Ming life.

Song originally planned to incorporate an essay on mathematical harmonics (*yuelü*) into *Works of Heaven*. His surviving works show he was more an inquirer into theoretical acoustics than an enthusiast of musical theory. His *On Qi* offers a philosophical survey of sound, attaching as much importance to the noise of a squeaking pig or a flying arrow as to the dulcet chords of a chimestone ensemble. For Song, all instruments—drum, flute, lute, or the suona, and gun, bow, chisel, or hammer—resonated in the world of *qi*. As the previous chapters have shown, Song's recourse to the mundane was part of his strategy to make the world explicable in the context of *qi*. In his inquiry into sound Song showed that the principles he discerned in material objects and natural phenomena were also valid for the void, the invisible sphere. This chapter explains Song's view on the sense of hearing, the voice, the physicality of sound and silence, and what harmony and resonance meant for him. Identifying *qi* as the carrier of sound, sound was the effect of a change in the status (*shi*) of void-*qi* in relation to itself. On this basis Song examined how materials, their shapes and characteristics, affected the emergence of sound and how its variances came about. He explained its volume, duration, and the distance it traveled in the context of what it revealed about *qi*. In his inquiry into sound Song progressively developed an explanatory model for the *qi* of the void as an audible phenomenon. He then systematically paralleled his findings on the audible *qi* with the principles he had found in visible phenomena. His model included a comprehensive concept of resonance and vibration and the explicit illustration of sound emanation as a wavelike movement.

Determining to what degree Song's notion of sound is related to our modern understanding of acoustics has been the incentive for scholars such as Dai Nianzu to inquire into Song's nine-part treatise on sound phenomena from the viewpoint of the history of science and technology.[4] Christopher Cullen comments, "it is possible to read Song Yingxing's work on sound as if he was giving a reductionist account of acoustics." He forewarns his reader "Song Yingxing was neither a modern physicist nor a confused thinker."[5] Following Cullen, I take Song's approach to sound on its own terms, that is, as a systematic analysis of *qi* as observed in its manifestations of sound and silence.

Premodern Chinese approaches to sound were part of cosmological considerations and natural philosophy, as were most early modern European statements about sound phenomena in the pre-Enlightenment period. Before the eighteenth century hardly any document in either culture dealt with sound or acoustics in isolation. Investigating the historical development of acoustical knowledge involves exploring the overlap of ideas about the function and application of sound phenomena. Music theory or applied acoustics were hardly ever topics in themselves but rather part and parcel of discussions on the construction of musical instruments and architecture and thus also of mathematical relations.[6] In Chinese culture ritual, philological studies, and writings on phonetics were the major fields in which music, sound, and silence were discussed.[7] The ritual context was based on the belief in a universe as a whole in which things were always affected by and reacted to each other. Sound was one expression of this reaction, exemplifying that man and heaven were connected. Music became a means to harmonize cosmic structures and keep open the lines of communication between man and heaven.[8] Ritual music functioned as a "means to turn 'the brutes' into civilized [people] and transform the commoner into a gentleman, the gentleman into a ruler, and the ruler into a sage."[9] In this regard the importance of music and the study of sound in ritual contexts cannot be overestimated for the Chinese intellectual world and the state structure. Hence, when Song made sound central to his discourse on chaos and order, he acted within a classical paradigm. His approach showed that he recognized perfectly well the significance of sound for state and society, ritual and religious beliefs and followed the idea that the universe could only be harmonized by bringing into accord its manifold things and affairs. Sound and music indicated universal principles, patterns and ways, and thus were a means to harmonize relationships. The *Discourse from the States* (one of Song's favorite texts) is one of the early classics that expressly documented the function of music for statecraft in relation to experiments with sound and pitch, melody and dissonance.[10] Song associated his ideas with this classic, thus authorizing his ideas about pitch, volume (*yinliang*), sound quality (*yinzhi*), and the physical manifestation of sound. Song stayed within traditional patterns, when he took *qi* as his epistemological anchor. From a seventeenth-century perspective, he can be identified as part of a trend that had accelerated both in quantitative and qualitative terms. Song stands out from his predecessors and contemporaries, however, because he used a down-to-earth manner to emphasize that the relationship between heaven, man, and earth rested on comprehensible schemes. It was beyond any ambiguously defined moral authority. Unlike his colleagues, Song insisted that in this context there was no difference between musical instruments and tools like hammers or hatchets. Song's trespass of epistemological boundaries can also be seen in *Works of Heaven*, where he placed his description of the production of bells in the section on hammer forging. With this Song not only acknowledged the technological similarities in the two manufacturing processes, but he also underlined that, in his view, bells and hammers were tools that divulged the workings of *qi*. Equally, in *On Qi* Song maintained that sound should be investigated within the mundane sphere, if one wanted to understand the principles that connected man and heaven. Listening to a cooking pot or a flying arrow was just as legitimate and revealing as an inquiry into the field of ritual music. While his coherent survey into *qi* harmony and dissonance verifies the existence of Chinese knowledge about what is now called theoretical acoustics, I suggest that the value of this particular part of his oeuvre is the perspective it opens on the concerns that motivated seventeenth-century Chinese inquiries into hearing and the subtleties of sound and silence.

AN ANATOMY OF SOUND

> There are drums of every variety. Their manufacture can be refined or coarse; they can be huge and stately or slender and delicate. If made to vibrate artistically, the void in their belly can emit the eight tones (*bayin*).
> *Works of Heaven*, chap. 2, no. 8, 17a

Premodern imperial Chinese orchestras often used an even number of musicians to represent harmony. For example, the famous late Ming dynastic painting of the Tang emperor Xuanzong (685–762, reign 712–56) and his favorite concubine Yang Guifei (719–756) shows eighteen elegant female musicians placed in a double crescent on ground mats in front of their elevated listeners (fig. 6.1).[11] The front row is made up of a performer with

Fig. 6.1. Xuanzong and Yang Gueifei Listening to Music, 1368–1400. Handscroll, ink and light color on silk, Worcester Art Museum.

a clapper; two fragile figures with long sleeves falling in folds to their ankles holding pear-shaped, crook-necked pipas with wooden pegs and ivory inlays; and harp and zither players flanked by large racks of chimestones. The second row features pairs of exquisite flautists, lutists, cymbalists, and tambourinists. Behind the others, in the exact center of the crescent, two percussionists stand by large drums: one is vigorously wielding two wooden mallets on a stand drum (*tanggu*), and behind her, the other is holding a flower drum and waiting for her cue. A similar group of eighteen female musicians is arranged on a fifteenth-century painting traditionally assigned to Zhou Wenju (fl. mid-tenth century). In this painting the musicians are not all in pairs. The center stage is reserved for a single large drum, covered at both ends with a skin and sheltered with a decorative umbrella.[12]

Emitting a smooth mellow tone that could both soothe the audience and demand its undivided attention the drum presided over secular music and played a dominant role in ceremonial performance.[13] Whether imposingly large or arranged in matched sets of five or six, drums were lacquered in red to symbolize the power of the ruler. The sounds of idiophones and membranophones, clappers, gongs, hand-drums, and twirling or hanging drums made of brass, bamboo, or wood, empty or filled with rice-hulls, merged together in soft waves to create subtle undertones or impressive crescendos. By varying the tension of the drum skin, usually tanned from pig or water buffalo, the drummer could produce almost every pitch. Its variety of tone was indeed remarkable and in fact indispensible, for its sound influenced the harmony of the universe. Following this logic, Chinese musicians categorized the drum with regard to its pitch, rather than its material. The general system of instrumental timbres during the seventeenth century was fourfold, distinguishing the categories metal, silk, bamboo, and stone with their cosmological qualities defined in relation to the five phases theory.[14] In *On Qi*, however, Song referred to a much older categorization found in the *Book of Documents* that distinguished eight material qualities: metal, stones (*shi*), earth (i.e., clay [*tu*]), leather (*ge*), silk (*si*), wood (*mu*), bottle-gourd (*pao*), and bamboo (*zhu*).[15] Song argued, however, that all those who believed that the importance in the making of instruments lay in the material had got it wrong. The cause and logic of sound and silence lay first and foremost in *qi*. Therefore, he took these categories as his cosmological points of reference and detached the materials from their sound-giving quality.

As shown in previous chapters, for Song the world's phenomena and things were triggered and governed by the interaction of *yin* and *yang*. The various phenomena of sound exemplified for Song how the *qi* of the void functioned; or seen another way, its function could be observed in sound. In line with traditional *qi* thinkers Song identified sound as a characteristic and innate property of a world saturated with *qi*. He insisted that sound was audible only because of *qi*, which, he argued, entailed the principle of sound in its smallest unit. Substantiating his view on *qi*, he explained that sounds must be distinguished from their contributory factors: While "humans (*ren*) as well as things (*wu*) come into being (*sheng*) [because both] receive (*shou*) *qi*, sound (*sheng*) exists (*you*) because *qi* exists (*you*)."[16] Sound was a behavioral characteristic of the void *qi*, and *qi* in conversion and not a transformational stage of *yin-yang* forming the more or less visible things and affairs. Sound was an objective reality in the world of *qi*, independent of human senses, and unaffected by man's hearing ability or his judgment.

Corresponding to Zhang Zai's concept of sound, Song saw the original state of *qi* as all-encompassing "with nothing in between (*mo huo jianzhi*)."[17] Christopher Cullen suggests that Song drew heavily on the legacy of Zhang Zai when cataloguing the various objects and forms involved in sound production.[18] Yet, for Zhang Zai, void *qi* collided with form, and form collided with *qi*. He explained diversity by identifying various interactions between form and *qi* that created sound. For Zhang Zai objects actively took part in the production of sound.[19] Although both thinkers may have started from the same point, void *qi*, their conclusions differed in essentials. For Song only the void-*qi* was significant. His argumentations led to a detachment of sound from the object; sound became a corollary of void *qi* activity. Materials such as leather, stone, or wood disturbed, split, or destroyed the integrity of the *qi* of the void, its harmonious rest, through their material properties. The object was hence involved in sound production only as a barrier that separated void *qi* entities. Furthermore any movement of the object triggered sound because it disturbed the equilibrium, and the unity of the *qi* of the void.[20] Whenever the entity of *qi* was disturbed, a sound emerged. Separated, the various entities were in tension. As they were originally a whole, the parts tried to synchronize with each other to keep their connection and be able to rejoin. The differences between Zhang Zai's and Song's ideas can be exemplified by the drum. Zhang Zai suggested that the drum created sound because the drum was hit by an object, the drumstick, and thus the *qi* was shaken and created a sound. Song focused on a different issue in his descriptions. He drew attention to the fact that sound was generated because the *qi* inside and outside the drum felt, as Song put it, "a resentment (*hen*) about the barrier (*gemo*)."[21] Part of the harmonized *yin-yang qi* of the void was caged in the belly of the drum and could not unite with the *yin-yang qi* outside. It could, however, still communicate with it. Separated by barriers into spatial entities, *qi* yearned to be and act as one body. It retained its association by "bearing the other in mind (*xiangyi xiangsi*)." If one part was agitated, the other part "resentfully" resonated to the impact in an attempt to synchronize with the separated *qi*. This synchronization produced sound and for this reason a sound emanated whenever the drum was hit.[22] If synchronization were made impossible, for example, by thickening the barrier, no sound would emanate. One could prove that the body of sound, its physicality, came from the attempt of the void *qi* to synchronize by filling a drum with soil or stuffing a bell with clay. Both would no longer be able to produce sound. As sound was produced when separation took place, or when *qi* was moved, it consequently vanished after synchronization had been achieved: either when the reunion was accomplished or after the moved *qi* (not the object!) had come to rest again. Song stated: "Deepest silence (*jingman*) prevails when *qi* is at rest (*xi*)."[23] Since Song identified silence as the natural state of *qi*, the dynamics of the system sought a return to silence.

Emphasizing sound as the result of a separated *qi* of the void that tried to synchronize or as the effect of movement, Song's terminology of *qi* motion was quite specific. Sound was produced when the *qi* of the void collided (*zhuang*). Sometimes *qi* collided (*chong*) or trembled (*zhen*), which designated the violent pressing together of two *qi*, or *qi* broke (*po*) into two or more partitions. Song suggested that "when there is silence (*jing*), then the *qi* is silent (*jing*) and thus the whole lot is without sound. When there is motion (*dong*), it is the *qi* moving (*dong*) and thus everything bears sound."[24] Song's repeated insistence on this point verifies his awareness that his dismissal of the relevance of materials was in opposition to the ideals of his era.

From a modern perspective the originality of his approach also lies in his use of examples from everyday experience, as if to demonstrate the metaphysical significance of the mundane. Shooting an arrow, playing a lute, and tearing silk, all of these actions verified the way *qi* was brought to collision by motion and thus produced sound. Activities such as clapping hands or hammering metals were his preferred illustrations to explain how a pressing (*ya*) of *qi* created sound. In all of these cases, Song insisted, through explicit statements or affirmative rhetorical questions, that the sound, though caused by objects, was subject to and conveyed by the *qi* of the void. Describing the sound produced by hammer forging (*huizhui*) in *On Qi*, he drew attention to the moment when "a fixed object receives a blow from another. Thus the *qi* emanating from the held object collides with the *qi* following (*sui*) the item being driven and this collision produces a sound.

This is it."[25] Looking at the object, Song saw a moving part of void *qi* colliding with a resting part of void *qi*.

Song's delineations explain sound's occurrence. But his actual point was that sound was a verification of *qi*: "A movement within the indistinct and vague (*weimang*) void is a verification (*zheng*) of *qi*. Every time *qi* is destroyed with the help of a form, this brings forth a sound." Sound was ephemeral and transitory, because *qi* always attempted to "revert to its state of saturation (*jingman zhi wei*) immediately after any casual collision (*ou feng bi ya*)."[26] Sound thus could "exist and vanish within the twinkling of an eye," as *yang-qi* could.[27] For Song, its impermanence proved that sound was not a transformational process, as "*qi* transforming (*hua*) into a form (*xing*) is a [process of] progressive accumulation (*jijian*) which then reverts (*gui*) to its original state. *Qi* transforming to sound is [the process] of an instant (*shana*)."[28] Sound hence had "no body (*ti*) of its own." It was also "unable to generate itself" (*ziwei sheng*). It needed a trigger (*ji*). Because of this, sound had to be carefully distinguished from the five *qi*. Sound was the observable phenomenon and representation of *qi* "at the juncture of agitation." Any movement hence brought forth sound. "In wind (*feng*) two *qi* press on each other (*ya*), and this creates a sound. A reed instrument (*luzhu*) creates a sound because human *qi* [in the form of breath] presses (*ya*) on the outside *qi* [of the void]. A flying arrow (*feishi*) produces a sound when [*qi*] collides (*chong*). Driving a wedge and lashing (*yuebian*) a whip produces sounds similarly because both cut across (*jie*) *qi*. Playing a three-stringed lute (*tan xian*) shakes (*zhen*) *qi* and thus a sound is produced. Similarly a fragmentation (*pi*) of [*qi*] produces a sound, which happens when silk is torn (*lie zeng*) apart. Uniting (*he*) *qi* by clapping one's hands produces a sound."[29]

Important to Song was that the shape of an instrument used to divert the void *qi* influenced the sound, its timbre and pitch. He also conceded that the number of sounds thus equaled the number of possible forms, multiplied by the number of materials available in man's world.[30] Yet he maintained that the object was only an influencing factor. This deduction, he claimed, was self-evident: "everyone knows this" (*ren zhi zhi*). Despite the divergent materials and objects in all of these sound-producing activities, the production of sound meant that the harmony of the *qi* of the void was disturbed. Its parts were in tension and resonated, trying to return to the original stage of unity or at least to resurrect the overall balance by synchronizing with each other. Thus Song in *On Qi* also carefully distinguished his terminology when describing the various ways of separating the *qi* of the void, giving much emphasis to its spatial arrangement. A craftsman's knife produced a sound when cutting because the "*qi* of heaven and earth responded (*ying*) to each other" and the separated partitions of *qi* communicated their relational positions to each other (*shi zhi*).[31] A drum emanating sound had *qi* inside and outside its body; or the *qi*-breath of man was distinct from the external *qi* of the void. A spatial concept thus lies at the heart of Song's view on sound. This also becomes apparent when Song delineated the impacted *qi* of the void, its volume, and direction of movement, and then drew assumptions on the length and distance of sound emanation. It is one of the points in which Song's conception contrasts essentially with that of his contemporaries who used the length, thickness, or coarseness of an object to explain variations in volume or pitch. The two *qi* Song addressed in such contexts are not those of of *yin* and *yang*, as Li Shuzeng suggests, but the harmonious *qi* of the void split in two by objects or torn apart by movement.[32] There is only one case in which Song's examples with regard to sound relate to the interaction of *yin* and *yang qi*, when he argued that there were situations in which even the limited senses of man could perceive that this interaction was also accompanied by a sound. The example he gave was a heated pot of water. He suggested when analyzing the sound *yin* water and *yang* fire produced when fusing into the void *qi*, the observer should get close to the pot. Then he would hear the fusion of *yin* and *yang qi*.[33] That Song tackled this theme is a sign of his deep conviction that principles had to be universal. Defining the functional anatomy of sound and silence as one that rested on the homogeneity and continuity of the *qi* of the void, Song argued that its dissolution, that is, the disturbance by *yin-yang* interaction, by implication also had to produce a sound. In general, however, Song's delineations on sound concentrate on the separation of the *qi* of the void into partitions. Emphasizing the principle of resonance and insisting on the quantification of *qi*, Song suggested that sound production and thus the harmonization of the universe were based on reliable patterns. In this way, "hearing" supplemented "seeing" in Song's repertoire of objectively analyzing the order of heaven.

THE HUMAN VOICE

> There are more than ten thousand sounds of things (*wu sheng*), that can be imitated by the human voice (*ren sheng*).
> *On Qi*, chap. *qisheng* 1, 64

Devoid of scenery, traditional Chinese opera worked with actors in stylized costumes and make-up who employed symbolic mime techniques to unfold their story to either soft or strident and accentuated music. Preimperial religious and public ceremonies, seventh-century Tang dynastic choral dances, and Song and Yuan dynastic styles of prosody from the thirteenth century, were all precursors to Ming opera, which merged singing, dancing, and recital into a new art form. Trained specialists employed a highly artificial style that was between singing and chanting. Men played the female and juvenile roles and had to master high falsettos. Often they added a nasal quality to the performance. Most opera singers learnt to achieve a huge vocal range, vigorously accentuating the melody and reciting the metric rhymes in different tempi and acoustic variations. Even after a lifetime of training only the best of them had a repertoire larger than a handful of plays. Each piece required hundreds of nuances in pronouncing a cough, a laugh, or a roar of anger and each locality treasured its regional style and stories.

Although cherishing their amateur status, literati were no less ambitious in their art. Some of the educated members of society spent much time and effort on improving their skill in vocal dialogues, narration, and expression of sentiment, or on successfully rendering the restrained singing style of *Kunqu* opera during "refined gatherings (*yaji*)."[34] These men and women of letters were not aspiring to showy virtuosity. Their interest in music and their ambition to achieve vocal perfection was justified by their righteous aims. Music channeled people's diverse thoughts and actions and promoted social harmony by unifying the "people's voices."[35]

An investigation of the human voice opened Song's discussion of sound in *On Qi*. Song's interest in the place of the human voice in the realm of sound followed Zhang Zai, Tan Qiao (tenth century), and Wang Chong, who all paid special attention to the relation between the mouth and *qi*.[36] Various contemporaries of Song, such as Luo Qinshun or Wang Tingxiang, also studied sound, the human voice, and *qi* in the realm of natural studies and approaches to phonetics.[37] Comparing Song's ideas with these thinkers, the idiosyncrasies of Song's theory can be seen. In line with his principal conviction that sound was produced by a separation of *qi*, Song delineated the mouth, tongue, and lips as the major elements in human sound production. Unlike his predecessors and most of his contemporaries who all emphasized physiognomic aspects, Song concentrated on the flow of *qi*. Song's idea that sound was caused by an agitated *qi* can also be found in Wang Chong's writing. In the case of the human voice, Wang spoke of the mouth enclosing *qi* and bringing it to vibration (*dongyao*). Wang further suggested that humans produced sound through the successive enclosure and release of *qi* via the oral cavity.[38] Song also used this example, but he directed his reader's attention to motion and direction, explaining "humans and things emanate sound directly from the viscera (*zangfu*). Articulated via (*diao*) the lips and tongue (*chunshe*), sound can only be produced with the help of the void's *qi* that collides and merges (*canhe*)."[39] Wang concentrated on the separation and unification of *qi*, that is, the holding and encirclement of *qi* bodies. The difference in their argumentation was a result of their different aims. Wang Chong was interested in the life-evoking force of man's *qi*, and thus he discussed voice and the emanation of breath with the aim of divulging the secrets of mortality in his text. Song, however, focused on the voice and sound as instances of a world of *qi*. Hence he focused on the interaction, in particular the collision, of two *qi* entities. Initiated by movement, the sound emanating from the human voice could be explained by *qi* collision, in that the *qi* of the "void receives (*shou*) the sound. And the wind of the belching *qi (aiqi)* makes the sound, which comes either from the mouths (*qiao*) of man and beasts, or birds and vermin. The sounds stored (*cang*) inside *qi* are ready to be skilfully pressed and pushed (*ya*) by man."[40]

In Song's view, man thus produced sound because he moved the inside *qi* that was separated from the outside *qi*. When the inside and outside *qi* tried to resonate with each other and return to a harmonious stage of rest, sound was produced. In principle Song did not distinguish man's ability from that of instruments or animals. When he elaborated on how the human voice shaped the stored *qi* to emanate sound, he explicitly correlated the activity of the human voice to that of a flute: "the Dilai bamboo flute is blown and resounds (*ming*). If I modify its body (*shen*) only slightly or for an instant, another tone (*sheng*) is created. If I blow into it without changing its body, it [sounds like] the call (*ming*) of the earthworm (*qiuyin*) which [heralds] spring, [or] the voice of the house cricket (*cuzhi*) and a winter cicada (*hanjiang*) which announces autumn."[41] Closing and opening the holes of the flute modulated the sound, just as the motion of the tongue, lips, and palate did. Song's correlation of man's voice with the flute aimed at demonstrating the universality of *qi* principles, which held true for both mouths and flutes. The anatomy of sound rested solely on *qi* entities and their communication and synchronization. Song continued this correlation between the flute and the human voice to stress that *qi* had to communicate in order to produce sound. A tightly enclosed *qi* of the void could only produce an indistinct sound. If the separation was total, it would produce no sound at all. This was the case when man's fingers held shut the holes of a flute and when the mouth and nose were closed. "Supposing one holds one's nose and mouth shut and the outside cannot penetrate and interact [with the inside] (*xianghe*), then the inside *qi*—even if it ascends up to [the region] between the gums and the palate (*ken'e*)—can only produce an indistinct sound (*moyin*)."[42] Within this process the crucial moment was when the mouth opened and *qi* was released. Striving for reunification, the enclosed mouthful of *qi* struggled against the barrier of the mouth to reunite with the *qi* outside in the atmosphere just as the *qi* in the flute's body struggled against the holes closed by man's fingers.

In the matter of man, *qi*, and sound, Song was not really concerned about hearing. For him, the ear cavities were simply a receptive organ for the accumulation and movement of *qi*. In a way similar to Tan Qiao, Song referred to the ear as an infundibulum that received sound like a valley, something unconnected to moral evaluation with no active role.[43] In the ear the "*qi* simply interweaves with the vitality center (*dan*) and then drifts into the ear cavity (*erqiao*)."[44] In this, Song's views differ from those of Zhang Zai and also from his contemporary Wang Fuzhi, who proposed conventionally that the mental distinction and classification of good and bad sounds was innate.[45] For Song, sound, and thus hearing in general, was morally neutral. It was the intuitive ability of resonance (*ying*) in things and man's aural recognition of this that gave sounds a positive or negative connotation. Thus either the *yin-yang* qualities of two things, or a man and a thing, corresponded to each other or they did not. If the qualities matched, man perceived sound as harmonious; a mismatch resulted in sound disharmonious to the ears of man. "Heaven and earth possess the [ability to] echo sounds (*shengxiang*) and the spirits (*shen*) of the body agents (*guanhai*) (which are the five senses and the hundred bodies of man) respond (*gan*) to it." Harmony was acquired when the sounds of heaven, earth, and man had a similar pitch. Within this epistemological system, music unavoidably became a cosmological instrument, a means to readjust chaos and restore harmony through a well-balanced synchronization of the colliding *qi* of the void. "Hence as we talk about the long-cut whistle and the flutes (*qinse*), their sound is melodious (*youyang*) and harmonic (*xie*), and hearing it incites pleasure and enticement. Accordingly the stream of *xue qi* is brought into harmony and releases joy (*hexie tongchang*). Music is capable of governing hearts and minds (*zhixin*). It is no wonder this is hardly a trivial matter (*xiaoshi*)."[46]

In a cosmology of *qi*, knowing how to tune and hit the drum to obtain the correct pitch meant knowing how to harmonize the universe and the heaven-human relationship. Moral behavior was the consequence of appropriately employing this principle. The rationale was simple and those who realized it and acted in accordance were morally upright. Those who denied it proved they were immoral. When one understood the elementary rules of *qi* as they were manifested in sound, man could apply the abilities and crafts bestowed by heaven in the best possible manner. Man would realize that skills such as singing were defined by man's bodily capacity, not talent or capability. The best singers were those who had a well-proportioned belly. Man could only influence this by choosing children with suitable physiques and then training them. This training was like constructing an instrument, in which man determined the shape and the thickness of the employed materials. Training one's voice meant to enlarge one's belly to the optimal size. Then one only had to learn to control the process of compressing and separating *qi* within one's body. This was similar to a craftsman mastering the form of a bell and then learning how to strike it in the best possible manner. In either case the principle was the same: it was perceived velocity and momentum that affected the invisible sphere of *qi*. In this regard the process of sound production was similar to the way *qi* affected the creation of visible matter.

VOLUME AND VELOCITY

> The sound of a bell, the foremost among metal musical instruments, summons with one strike. A big bell conveys its sound further than ten miles, even a small one can be heard more than a mile away.
> *Works of Heaven*, chap. 2, no. 8, 18a

The ritual ceremonies instigating the New Year's festival lasted several hours, with the Confucian master standing frozen like a statue next to the fragile bronze bell, waiting for his cue. When he struck its two pitches in turns, the chime of this richly ornate ritual device dictated the rhythm of the solemn spectacle in which the emperor paid his respect to heaven. As it was cast with thin walls that had a short reverberation time, it had to be struck with care, lest it emanate an unpleasant sound. Used in Daoist, Confucian, and Buddhist devotions, Chinese bells were commanding creations, embodying inexorable force. Many of them were inscribed with prayers and commemorations of donors in convex-concavo style (fig. 6.2). A layer of bee's wax and ox fat recorded sophisticated patterns, in particular at the core of the bell. These religious or decorative designs were then reproduced in a negative shape with a paste of clay and charcoal powder. Pouring the bronze into the mold, the artwork was eternalized on the bell.[47]

Made of bronze or iron, bells constituted stable elements in the Chinese world order. They represented the precision and maintenance of standards in pitch and relative scale with far-reaching cosmological and social consequences. Tuning their music to the mode of the Yellow Bell (*huang zhong*)

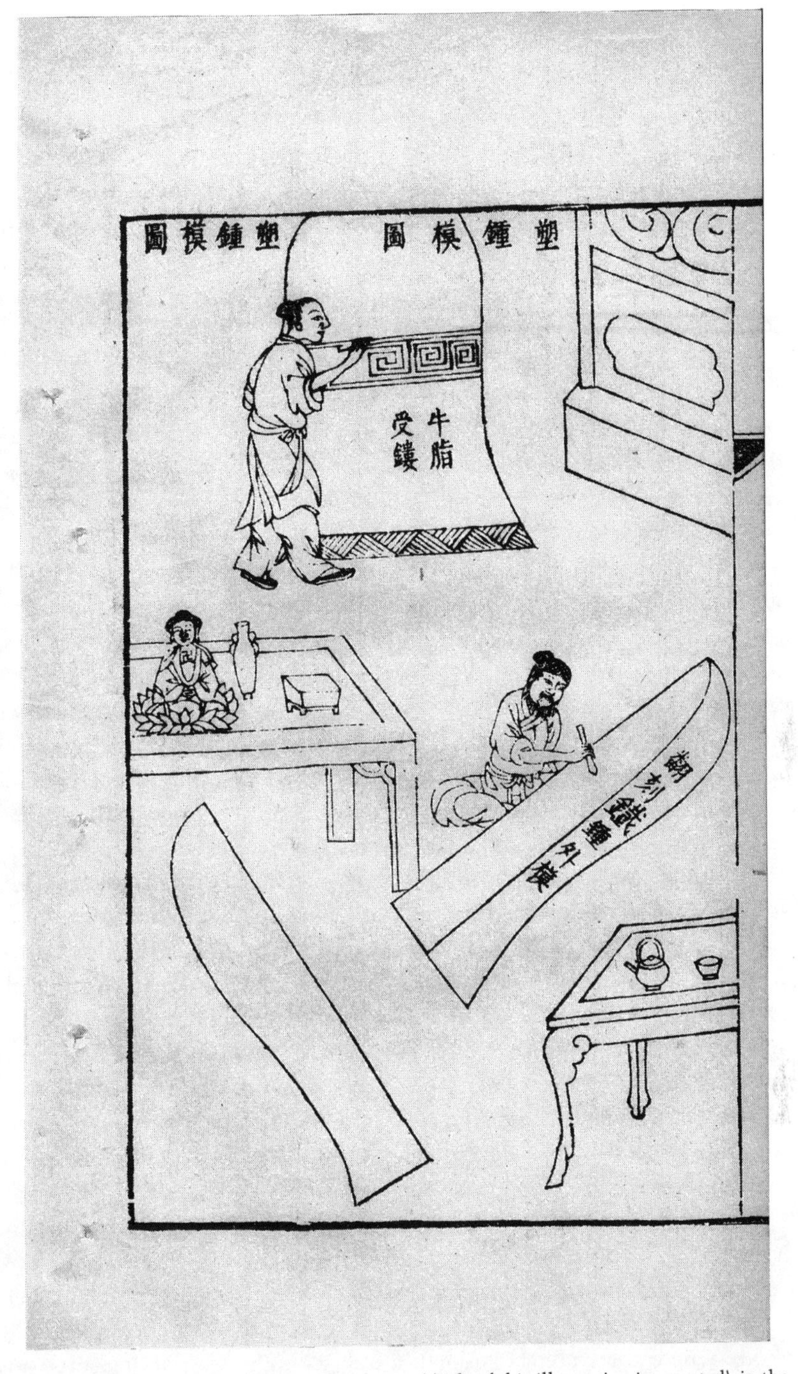

Fig. 6.2. "Modelling a bell mold" (*su zhong mo tu*) (title of this illustration is repeated); in the upper scene an artisan "beef tallow is engraved" (*niu zhi shou lou*). The man in the lower scene adds "mirror-inverted engravings to the outer mold of the iron bell" (*fan ke tiezhong waimo*). The Buddha and sacrificial wares on the table indicate the ritual purpose of bells. *Works of Heaven,* chap. 2, no. 8 *Smelting and casting,* 23a–b.

authorized a dynasty to rule all under heaven, its clang announcing imperial audiences, marking the hours during the night, and marshalling the advance and retreat of dynastic troops in the battlefield. Suspended on a rack and devoid of a clapper, the Chinese bell sounded only when struck. Song's delineations on how bells had to be shaped in order to produce sound reveal that he thought sound was an unnatural state that required causation. He defined *qi*'s volume (which was dependent on the shape of the object), momentum, and motion as significant factors structuring sound. The work of Ban Gu (AD 32–92), indicates that long before Song, theorists had noted that sound was dependent on impact, speed and power, although none of them formalized it. In *Discussions of Virtue in the White Tiger Hall* (first century), Ban Gu wrote that the bells on a carriage "will not sound when [the carriage] drives slowly, if it hastens the pace too much, the [bells] will lose their sound [too]."[48] Ban Gu was aiming at a cosmological nexus when he suggested that the movement of the carriage had to be in accord with the harmonic rhythm of the universe and the driver should, therefore, adjust the speed so the bells would ring pleasantly. Some hundred years later Tan Qiao claimed that a spirit (*shen*) played the role of an intermediary structural element and that sound was initiated because "form and *qi* avail themselves of each other (*xiang cheng*). It is not the ear that listens to sound but sound which makes its own way into the ear. It is not the valley which by itself gives out echoing sound, but sound itself fills up the entire valley."[49] Ban Gu's text was canonical, and so Song may have come across it. The same was not likely for Tan Qiao's text which was included in the *Daozang* (*Daoist Canon*), a collection of texts that was available during the Ming. Song did not embark on any of these approaches. Instead he expounded further on impact and velocity and saw them as distinctive qualities creating and shaping sound. He explained in *On Qi* that forceful chiming generated a loud sound, while caressing the bell with care would bring forth a subtle chink depending on the velocity of the movement: "If [the impact on the harmonious *qi*] happens quickly, then [a sound] is produced (*cheng*). If it is done slowly, it will fail. Hence, velocity produces [a sound], dilatoriness does not. Strength (*jin*) produces sound, weakness does not."[50]

Song demonstrated the significant role played by these factors by using the examples of water streaming out of a vessel and water falling down a cliff. In both cases the material was the same. The only variables were impact and velocity: "[Let us imagine] a high mountain with a waterfall (*pubu*), where the water falls down a cliff as high as a hundred men and pounds and roars (*jijian*) into a deep mountain pool. Hence a spectator and listener's soul is startled [by that sound] and he [nearly] loses his head in panic (*jinghun sangpo*). [Now let us think of] water leaking out of a worn-out receptacle leaning on its side, or [water] streaming through irrigation ditches (*gouqu*)—here we do not perceive any sound at all. The [amount of] water is identical (*tong*), the [movement and direction of the] flow (*zhu*) are identical [namely, downwards], but the sound configuration (*shi*) is different." The use of the idiom "to lose one's head in panic and startle the soul" (*jinghun sangpo*) reinforces his argument that sound was a disturbance of harmonious *qi*.[51]

Looking at the object in motion, Song went on to address the concept of "*qi* in a particular configuration" (*qi shi*) or, in Christopher Cullen's translation, the "advantage of position."[52] This technical term described an accumulated stage of *qi* that could be a gathering of *yin* or *yang qi* or an unbalanced distribution of *qi* in space caused by movement or pressure. The volume of sound depended on how much pressure was put on *qi* and on how densely it accumulated. The stronger the imbalance, the louder the sound would be. "*Qi* has got the advantage of position (*shi*) and thus sound is created (*sheng*). If it does not have this advantage of position (*shi*), *qi* will be hungry to the extreme (*nei shen*)."[53] "Hungry to the extreme" described a stage in which *qi* had accumulated in one place and *qi*, longing for a balanced distribution in space, tried to return to its original position. What Song described in his purely qualitative epistemology shows some similarities to contemporary European classical concepts of energy or the capacity to do work, that is, the concept of potential energy combined with the notion of kinetic energy. The core of Song's concept was, however, the idea of balance combined with that of *qi* being evenly distributed within the void. Hence, Song understood movement as an aspect of *qi* transformation, disturbing or acquiescing in the heavenly, or cosmic, order. Movement was not a concept in itself. Rather it was descriptive of a specific stimulus acting on *qi* in a distinct condition. Consequently his analysis focused on the destruction of *qi* unity, presupposing that completeness was *qi*'s ultimate intention. For this reason silence required no explanation. The degree of velocity with which *qi* returned to its point of departure determined the extent to which *qi* was dislocated or disturbed. Therefore, velocity stood in direct relation to the volume of sound.

In addition to velocity, impact and its form influenced loudness and duration. The degree of impact, and the size and form of the body that separated *qi* from the environment, were factors in the relocation of *qi* and its return to a state of rest. A flexible body such as water created a sound different from that produced by an inflexible or rigid body; an environment that allowed an open exchange of *qi* created another sound from that created by an environment where *qi* was pressurized or captured and unable to reconstitute quickly. Yet, one principle of all sound production never changed, namely, the fact that it was provisional and short-lived. As with water and fire, whose volatility depended on whether or not they were able to fuse with their complementary other, sound was short-term because *qi* was not able to hold its "advantageous position." "Exactly the same principle applies in the processing of wood, this creates the sound of a penetrating saw, the noises (*xuan*) of the military, of city life, and even the harmonious rustling of mulberry woods (*sanglin*). If you drive [a tool] into wood, numerous inches (*cun*) of shavings are shed. But it will become quiet again almost instantly. This is because *qi* cannot hold the advantage of position (*shi*); it is a fleeting phenomenon, not able to settle down. Whenever gold and silver are cut open for use, the hammer produces a sound and the wedge howls. As soon as a tiny bit is cut off and the desired result has been achieved, not even an iota of sound can be heard. The underlying sense contained in all examples is identical [with that of the waterfall]."[54]

Song noted that the principle correlating the production of sound and the return to silence was similar to that of the growth and decay of materials. The speed and power that caused the impact and created sound influenced the volume of a noise and affected its duration inversely. A high-speed powerful action produced a loud but brief sound, as the abruptly disturbed *qi* reverted just as abruptly to its natural state. Stimulating the *qi* with the appropriate action would prolong the sound, although all sounds, whether harmonious or dissonant, were always provisional, short-term occurrences and would never persist much longer than the impact. Song illustrated this vividly with the example of exploding gunpowder:

> *Qi* in its two [states of aggregation] of *yin* and of *yang*, amalgamate and form (*jiecheng*) saltpeter and sulphur. Both of them originally have a substance (*zhi*) and then they are without a substance (*wu zhi*), hence they are called divine things (*shen wu*). In fire, they unite (*he*), as they have an urgent desire to transform into void and to vanish away. When they explode they provide a forceful trigger, more unyielding (*jing*) and rapid (*ji*) than that provided by the crossbow shooting an arrow. The *qi* at ease in the void collides (*chong*) and is compressed (*bi*) [so extremely that the *qi*] can no longer find its living space. The resulting repercussion (*xiang*) is tremendous.[55]

Man perceived any movement that abruptly interrupted this harmony as bad or unpleasant sound because there was no harmonizing reaction. In extreme cases such a sound could kill, because the disturbance of the ultimate harmony of *qi* was so drastic that man's *qi* was unable to cope with it.

> The climax of a frightening sound must be like bursting and slamming cannon fire (*zhapao*) or a blazing conflagration (*feihuo*). It gives a lasting strike to the soothing (*jing*) *qi* of the void. The *qi* collides (*chongzhi*) [explosively] and is scattered (*sankai*). If it encounters a hole (*kongxue*), the *qi* consequently strolls into it (*jin*). It penetrates man's ear cavity (*ergen*) and immediately compresses the *qi* within it. Simultaneously [the intruding *qi*] also accumulates (*juji*) inside the body. Hence the gall bladder (*dan*) is tilted (*hui*) and the liver (*gan*) is injured. For this reason it is over in a moment and the person dies very soon.[56]

In this context, Song did not distinguish sound waves from pressure waves. He thought both the death of a person and the destruction of materials were results of *qi* collision. In either case the *qi* accumulation was so extreme that the inner harmony of *qi* was devastated. It was not the force of oscillation, as we would argue today, that caused death, it was the tremendous impact on the natural order of *qi* and the resulting disharmony. Interesting in this passage is also that Song explicitly pointed to waves as moving in three dimensions when he metaphorically illustrated sound. "If an item collides with *qi*, it is comparable with [an item] bouncing against water. *Qi* and water are things moving interchangeably in an analogous manner (*yi dong*). When a stone is thrown into water, the surface of the water accommodates (*ying*) the place of the stone. [The stone] makes just one single impact, but the water starts opening and undulating in successively patterned waves (*wen lang*). It moves forward in a transverse motion (*zong hengxun*) over extensive distances and does not seem to diminish. The waving *qi* (*dang qi*) acts the same way, but it is particularly subtle and is therefore imperceptible."[57]

The expression "moving in an analogous manner" (*yi dong*) has an ambivalent meaning here, implying a change and an exchange that appears as movement. As long as we assume the model of sound production to be a collision of *qi*, both sides of the ambivalence work well. Song explicitly states that *qi* set in motion was "transverse" (*zong hengxun*), thus he realized the three-dimensionality of the waves. While the correlation of sound waves to water waves was quite common, most of his colleagues were content to adduce the image of a stone thrown into a pond without further remarks on the kind of movement. Song's idea of a wave was not carried on further by later scholars who investigated *qi* in relation to sound or sound in the realm of *qi*, such as Wang Fuzhi, Fang Yizhi, his son Fang Zhongtong (1633–98), and Huang Zongxi, because Song and his colleagues explained waves—and in general why things reacted to each other—with another central concept in Chinese epistemology, the principle of stimulus-response (resonance): *ganying*. Some things resonated more, others less depending on the manifestation of diverse relationships of *yin-yang*. *Ganying* explained why sound carried, why echoes resounded, and why a bell's clang could shock another bell into oscillation.

RESONANCE AND HARMONY

> The drums on the boats cry, the sound swelling as if they were all crossing through and the boat-haulers, hearing the drum sounds echoing from the mountain rocks, unite their power.
> *Works of Heaven*, chap. 2, no. 9, 36b

Chinese scrolls of cityscapes in the style of the *Qingming shang he tu* (*Going up the River at Qingming* [Festival]) by Zhang Zeduan (fl. twelfth century) or the *Street Scenes in Times of Peace* (*Taiping jie jing tu*) by Zhu Yu (1293–1365) provide a rare window onto the everyday world of Chinese crafts. The rolls both depict wainwrights, shoemakers, and washerwomen working in the streets under the watchful eyes of their customers' servants (fig. 6.3). Many of these professions were rarely touched on in premodern Chinese art or written accounts. The lives and works of craftsmen thus remain largely hidden from the historian's eye just as they remained hidden from customers' eyes in Ming times. Artisans performed their craft behind low blackened doors, in small poky shops advertised only by a signboard. Instrument makers, for example, assembled in the courtyards and outbuildings of large compounds. According to a report on nineteenth-century Hangzhou, the seedy filth of such house entries often cunningly belied the finery behind them. In Hangzhou the community of musical instrument makers resided in a generously proportioned three-storey house around an open courtyard. Gathered under one roof, luthiers and flute makers fashioned their instruments and blacksmiths and forgers hammered brass and copper drums.[58] Anyone entering such a place would have marveled at the subtle variety of instruments that contemporary musical performances required, the shapes of stringed lutes, *erhu* fiddles, pipas, guitars, zithers, and dulcimers. Glancing at the women swathing and twisting silk for the various stringed instruments, the nobleman's servant sent to collect a purchase may have deigned to look into the section where the master gave the fin-

Fig. 6.3. *Street Scenes in Times of Peace* (Taiping jiejing tu), Yuan dynasty (1280–1368). Handscroll, ink and colors on paper, Art Institute of Chicago, 1942.112.

ishing touches to his zither, clamping filigree ivory bridges onto its oblong body. Wrinkling his nose at the acrid odor of lacquer, the priggish steward (employed by a snobbish literatus home) may have wondered how in the pandemonium of such places a craftsman could produce the masterpiece for his master's household performances.

Among the lush symphony of Ming music, the pure and subtle whisper of the zither (*qin*) stood out, unfolding its lyric for those who sought to enrich their heart and elevate their spirit. Caressing its silken strings, the literati gentleman's soft hand had long laid claim to it as an instrument of spiritual enlightenment. Capable of communicating deep feelings and metaphysical connections, each part of the instrument was identified by anthropomorphic and zoomorphic names and correlated to cosmological topics; thirteen studs (*hui*), made of mother-of-pearl, jade, or ivory, signified the heavenly and earthly stems. Two sound openings in the bottom board, represented *yin and yang*, named, respectively, phoenix pond (*fengzhao*) and dragon pond (*longchi*).[59] The rounded shape of the upper board, made of *wutong* wood (*Firmiana platanifolia*), symbolized heaven and the flat bottom board of *zi* wood (*Catalpa kaempferi*) represented earth. When played by man this instrument displayed cosmological structures, uniting in its body the three most important entities in the Chinese worldview.[60]

Chinese intellectuals of the Ming worshipped the zither for this reason, documenting its history and collecting its tablatures. Scholars played this solo instrument outdoors, situating it in nature: in mountain settings, in a garden, protected in a pavilion, or under pine trees symbolizing longevity, with serene moonlight and the air perfumed with incense. Intellectuals such as Shen Gua reported that two zithers were normally tuned to each other with the help of a paper cutout of a man's silhouette (*jian zhiran*).[61] The silhouette was placed successively on each string of the first zither, while the musician stroked the corresponding string of the second zither. When the paper silhouette began to tremble the strings were in tune with each other. Song also discussed the paper silhouettes, identifying the trembling as evidence of a world of *qi*. He explained that the instrument reacted to the agitated, trembling *qi* in its environment; it was resonating with the disturbed *qi* of the second zither. In this sense Song was convinced that the zither's music could disband the dissonances caused by dislocated *qi* and stimulate a return to order. Inducing a balanced vibration of instruments and man, sound verified the interconnectivity of all things and realms.

The phenomenon of the mutual resonance of things in relation to musical instruments can be traced to sources as early as the Han dynasty. The *Shuoyuan jiaozheng* (Collection of Speeches with Corrections and Annotations), for example, mentions a moon guitar (*luan*) resonating in tune with the imperial bell each time it was hit.[62] In connection with theories about *qi* and *yin-yang*, elite medicinal theories detected sound as a crucial factor influencing man's health.[63] Cao Shaokui, a healer at the end of the seventh and beginning of the eighth century, diagnosed a scholar's mental disorder as caused by a bell in a neighboring building. Pointing to a stone that resonated in unison (*he*) every time the bell was struck, Cao diagnosed a disorderly agitated *qi* in the surrounding of his patient as the cause for the scholar's illness. Wei Xuan (ninth century) is said to have found a similar case, tracing a monk's aching pains and the spontaneous resounding of the temple's chime stones day after day to an agitation of *qi*.[64]

Chinese explanatory models for the interrelation and mutual reaction of things and events rested on the concept of stimulus-response.[65] By the Ming dynasty the idea of reciprocal correspondence (*ying* or *ganying*) was so embedded in the Chinese moralistic paradigm that it is hard to believe that Song attempted to trace it back to its material or "natural" origin, and yet this is what he did. Perhaps he was motivated to question his cultural prerogative by his strong desire to debunk what he considered his colleagues' fallacious moral interpretation of how the world could be harmonized again. This would explain Song's great concern about harmonious resolution and resonance in the context of sound. His rhetorical method of evoking the sages substantiates this point. Song suggested that the sages, following the principles of *qi* and knowing how to order the world, had fashioned musical instruments in such a way so that separated entities of invisible *qi* were able to communicate with each other. "The sages who fashioned musical instruments, seem to have been afraid that the *qi* might not [reciprocally] interact [with each other]. Thus they worked within a hair's width. If the thickness of the bell's metal is multiplied or the drums' leather is thickened, even a vigorous strike [at the bell or the drum] will produce a very faint and negligible sound, because the spirits of the void inside and outside cannot interact with each other (*yinghe*) [in the appropriate manner]."[66]

Interaction and the idea that *qi* was inherently prone to resolve itself spatially and temporally to a stage of balance and unity provided the background for Song to argue for the advantage of round instruments. This notion subtly combined with the assumption that *qi* at rest was evenly distributed. An accumulation of *qi* in one place would in the best case cause dissonance and in the worst case, as with gunpowder, irreparable damage. Round shapes ensured that *qi* could quickly and in an orderly manner reconstitute its original condition of equilibrium. This created a pleasant and harmonious sound. Sages had manufactured mainly round (*yuan*) instruments because they had known about the principle of "the accumulation (*ju*) and inclusion (*han*) of *qi*. The *qi* of the internal emptiness can only echo the outside because everything desires to be in a state of balance (*qizhi*) and consistency (*junji*). If [instruments] are rectangular (*fang'yu*), [the *qi*] will hurry to one corner and linger in the other corner (*xi*). It will be fast (*ji*) in that corner while it is slow (*huan*) in the opposite one. In the center the *qi* is confused and wanders about unsystematically. Hence the emanated sound is inadequate."[67]

Song then likened an unbalanced accumulation of *qi* in the corners of a square object to an unbalanced relationship of water and fire agents in the atmosphere that could only dissolve to the void if they were brought together in an even manner. He argued that a pleasant sound could only be guaranteed if the *qi* was able to spread out and reconstitute its state properly. Any barrier that interrupted its flow would have an effect on the sound. In accord with his delineations in *On Qi* Song suggested in *Works of Heaven* that "all musical instruments must be round-shaped and without any solder (*han*)."[68] If it was angular, the *qi* would be, as Song put it "slanted (*pian*) in a way that the *qi* cannot reach all corners in the same way."[69]

Song also discussed the *qi* of the void in terms of quantifiable relationships. He argued that the capacity of the *qi* partitions to respond to each other was defined by their proportional relationship, as was the case with *yin-yang* forces. He maintained that *qi* was distributed across a space depending on varying forms and had different capacities for response and for that reason "its reverberation will split and be crushed"[70] if the impact was too high. If the impacting *qi* partition was small, the sound would be short. Equally, velocity had to be taken into account. Rash movement or high pressure on *qi* would create harsh and boisterous sounds. The quality of sound, as a characteristic of *qi*, should, however, not be equated with *yin-yang* forces. But as sound was subject to *qi*, it had to act and react along the same lines and in accord with the principles Song had detected.

It is for this reason that Song throughout his inquiry into sound and *qi* continuously related his theory on the behavior of the void *qi* to his ideas about the creation of the material world through *yin-yang qi*. We can see this in the way he tried to bring his ideas about material properties caused by *yin-yang* interaction in line with his theories on sound production. Embracing both the invisible and visible world in his theory, Song explained the effect of the various attributes of an object on the creation of sound by way of *qi*. He believed the various transformational processes of *qi* in the material world accounted for all the characteristics that hampered the communication of the two separated entities of invisible *qi*. Whenever *yin* water and *yang* fire *qi* fused in items that were still in their primeval shape, or some water remained, objects developed a membrane through which *qi* could no longer communicate; deprived of its ability to resonate, sound production became impossible. For this reason raw clay pots did not produce a sound, whereas fired pottery produced a harmonious tone:

> Water and soil are mingled to [produce] clay in order to form a bucket (*foupi*). The sense of fire (i.e., heat) is augmented, that is, it is dried on a sunny day. Consequently water and fire fuse and thus solidify the blank body of the piece. Knocking the outside of the bucket, the inside *qi* will not respond. No sound will emanate. In the kiln, it receives the kiss of fire (*qinhuo*) again and it changes its attitudes, discarding its resistance. This means the water agent has diffused (to the void). The spirit of fire (*huo jing*) firms the body and the soil quality changes the form. Once it is hit, [now] a most pleasing sound can be heard. It is harmonic; and in numerous combinations it can be used to make music without any dissonance (*kui*).[71]

This passage exemplifies the difference between Song's recognition of sound and our modern understanding. It shows that he was not thinking of a vibrating atmosphere or the membrane, but of a penetration of *qi* or the hindrance of its further penetration. The dried clay bucket still contained too much water *qi* and hence the *qi* on the one side could neither penetrate and unite or harmonize with the *qi* on the other side. Thus it could not produce a tone. Seen from Song's viewpoint, the ability of an object to emanate or carry the sound of *qi* was a question of how material was transformed and its *yin-yang qi* interchanged. The body of sound fully rested on *qi* qualities. The object was only involved in so far as it was able to hinder *qi* communication. Its shape aided or hindered the *qi* from returning to a balanced stage. For this reason Song suggested that "the bell-shaped percussion instrument (*zheng*), struck by a hammer or the leather stretched drum—irrespective of the fact that it can be made of copper or stretched leather—does not possess [different] attributes. The drumstick hits (*tang*) [the center] of the instrument and its booming sound coincides with the two first modes of music *gong* and *shang*. When the four edges are clattered and rattled, the modes *jiao* and *yu* resound."[72]

This description again shows that a spatial conception of interacting *qi* lies at the heart of Song's notion of *qi* in relation to sound. Sound, according to Song, always appeared at the opposite of the stimulus: "Struck from the southern direction, the *qi* north of the Qing chime stone responds. If a chime stone is struck from the east, then the *qi* west of the stone resonates." As an epistemological concept, the idea of a spatially detached but communicating *qi* was simple and all-embracing, allowing feasible assumptions about most sound phenomena: In man's voices echoing from the canyons, one *qi* echoed or answered the other *qi* and both attempted to harmonize to a unified whole. Because they were originally one, *qi*-partitions resonated with each other. Sound and its echo always corresponded: "[*qi*] echoing to *qi*. Cry out loudly, and the echo [from *qi*] is just as loud; a quiet cry, and the echo [from *qi*] is just as quiet. If the cry is urgent, the echo [from *qi*] is just as urgent."[73]

In sum, we can see that Song's treatises on sound and sound production verified his idea of a world functioning according to the principles of *qi*. Sound was the complement and counterpart to the material world in which *yin-yang* interaction shaped growth and decay. Sound allowed him to show that principles of *qi* applied to both visible and invisible phenomena. His exploration of sound added an important perspective, accomplishing the model of a world of *qi*. Song used observation to complement his theoretical knowledge. Everyday experiences, hearsay, and knowledge from medicine and alchemy supported his viewpoint, and he aligned them carefully to substantiate his major hypotheses on *qi*. When considering perceptible phenomena in his theory of sound, he refused simply to accept the conventional tenets of his era. He wanted to draw (or, he may have argued, redirect) attention to the order revealed in universal principles, the logic of *qi*. From a historian's perspective Song's universal rationale, as well as his interest in aligning natural phenomena and material inventiveness into universal and replicable generalizations, is not out of step with the intellectual trends of his time. But he reached different conclusions from Zhang Zai, who thought of sound as the intuitive ability of things to respond and even distinguished sound with regard to the different directions a collision between form and *qi* could take. Moreover, Zhang focused on the issue of the relation between this intuitive capacity and man's morals. Song approached sound as a rational matter beyond human morals. This distinguished him from contemporaries such as the Wang Yangming opponent Wu Tinghan or the natural philosopher Wang Fuzhi, who related music and sound to poetry. But even though Song harshly criticized individual features and developments in contemporary thought and had an impressive insight into natural procedures, he abstained from totally rejecting traditional concepts. This does not mean he was less than radical in his intention. To emphasize the ultimate authority of the principles of *qi* as they act on the making of things and manifestation of events, Song concluded *On Qi* by correlating the creative and destructive forces of sound to the theorem of the five phases. "That is the reason why heaven created the five phases of *qi*, with each of them owning a sound (*sheng*). The tone (*yin*) of fire and water is reflected in the *qi* of metal and earth (*tu*) and wood (*mu*). This is how to deduce [the principle] of the five tones."[74]

While Song did not reject the ideals of his culture, his all-embracing views and his attempt to make man's world subject and bound to a universal rationality not morally defined made him consider many issues in a new light. He drew interesting conclusions about physical sound phenomena and classified them within his attempt to reveal the workings of *qi* in a remarkably comprehensive, and indeed unprecedented, way. He argued, for example, that sound moved in oscillation, and that the distance it traveled and its volume depended on the advantage of position, which was then determined by the strength of the impact and the volume of interacting *qi* partitions. This, in his view, remained detached from any materialized manifestation of *qi*. In this regard Song's extraordinary essay on theoretical acoustics written for the purpose of understanding the world of *qi* provides an exceptional insight into the epistemological multiplicity of Chinese natural philosophy. In his inquiry into sound, as in his investigation of crafts, Song refused to embrace the world of delusions and depravity in which he lived, a world "made by man," emphasizing that order was provided within things and natural phenomena.

CONCLUSION

Leaving the Theater

In this way the creative transformations (*zaohua*) of heaven (*qian*) and earth (*kun*) are sometimes concealed (*yin*) and sometimes visible (*xian*) in everyday processes. This is the end of the chapter about the "Works of Heaven," that I have compiled and now release.
Works of Heaven, chap. 3, no. 14, 11b/12a

With these words, Song brought down the curtain on his masterpiece. And while the audience members gathered their belongings, the puppeteer wrapped up his figures and cleared the set. The screen was packed away, the props collected and the theatre closed, leaving the moment to history.

The intention of my book has been to unfold and separate the layers of the original production to see which factors influenced Song in his approach to knowledge and what role technology played therein. An idiosyncratic scholar has emerged who was nevertheless deeply engaged with the themes of his culture. Song's polemics reflect contemporary stylistic, terminological, and linguistic features, and his attitudes show the sociopolitical, intellectual, and practical concerns that affected seventeenth-century Chinese approaches to scientific and technological content. For Song, technology and crafts revealed a rational order that could be analyzed by observation, experimentation, and quantification. Countering the moral paradigm of his colleagues, Song suggested that material inventiveness and natural phenomena "provided by heaven" offered trust and reliability in a world of social and political chaos "made by man." While Song thus made man the arbiter of his own fate by "knowing," he did not recognize crafts: In Song's integration of technology and crafts into written discourse, what workers knew was an "epistemic object" for a higher order of knowledge that could only be deciphered by the scholar. In this regard a close analysis of Song's viewpoints has a great potential to destabilize the all-too-neat construction of linear or uniform traits in historical surveys on knowledge in the making, the interaction of practical and theoretical knowledge and the way in which crafts and technology became integrated into the world of the written account. In line with this thought, I am careful not to dismiss Song as a meticulous observer of crafts and a man knowledgeable about technological details merely on the basis of his scholarly apparel. Rather, his literati costume is a reminder that his viewpoint and attitude toward technology was the product of specific cultural and historical conditions. Appreciating the originality of his effort within this locus is more relevant to historical method than presuming to judge Song's level of technological understanding or that of the Ming period by today's criteria. This study has shown us how Song explained the technology of his time. Only when we have understood his reasoning are we in a position to reveal germane information about historical technology and Chinese ways of making knowledge.

Song's life and works echo various ambiguities inherent in the late Ming period. Encouraged from boyhood to have unrealistic ambitions for a glittering career, he concentrated his studies on a restricted set of classical texts whose validity was constantly debated. As Song's intense desire to fulfill his expected role as a sociopolitical and intellectual leader came to nothing, he became extremely frustrated. Disillusioned, Song identified his colleagues' insistence on morality as out of step, arguing that the order of the world rested on a thorough understanding of the principles of *qi*. Within this framework he developed his notion of crafts and technology as a matter of scholarly knowing. The craftsman was a performer, but not consciously aware of the principles that lay behind his actions.

Song's interest in crafts was a reflection of the state's centuries-long use of scholar-officials as the managers of craft skills. His approach to the craftsman, or rather his indifference toward their social role and abilities, was his answer to the ambiguities of a period in which the state highly venerated craft works, fostered economic developments and agriculture, and at the same time kept the scholars in the leading social and political position. In this checkered atmosphere scholars were forced by necessity to defend their leading position in "knowing." When Song took recourse to his social and political duty as a scholar he considered he was acting in absolute harmony with the ideals of his culture and its traditions. He did not repudiate seventeenth-century intellectual standards and ideals. But he did subtly redefine them. For example, his perception of farming as a gentleman's duty was in accord with his era, as was his insistence upon a hierarchical order of things which assigned oil pressing minor importance by comparison with milling. Yet, he was out of step with his contemporaries when he lumped farming together in one book with the grinding of lapis lazuli and the construction of weaponry. In such subtle but significant ways, Song redrew the intersections and borderlines within Chinese traditional views on theory and practice and ideas on fields of written knowledge.

Song's rhetoric and ordering of issues in *Works of Heaven* prove both his compliance with and manipulation of his cultural set. Referring to the sage kings, Song confirmed that his hierarchy of values rested in the cosmology of China's early beginnings. He organized topics in the sequence in which they became relevant to man's world, for they were the building blocks of civilization. Thus, the growing of grains had to be dealt with before the calcination of stones; the construction of boats preceded yeasts; dyes, putting heavenly order on display, became an addendum to the section on cloth, just as sugar, one of the five flavors, followed salt. Creation, change, and transformation were other criteria he used to classify craftwork into a sequence (*qian hou*), or "from the origins to the end" (*benmo*). In the manner of the classical order given in texts such as *The Master of Huainan,* Song argued that basic transformational processes of *qi* had to be understood first. That was why he addressed those processes in the preparation of grains and firing of ceramics before introducing more complex subjects. The subtle implications and meanings of the calcination of stones, metallurgy, and the production of vermilion and ink required a thorough understanding of the essential cosmological logic of the world of *qi.*

I see his works *On Qi* and *Talks about Heaven*, in which Song gave a more advanced explanation of the order of *qi*, as theoretical supplements to the worldview presented in *Works of Heaven. On Qi* and *Talks about Heaven* are fundamental to understand the concepts that directed Song's technical descriptions. Using the concept of *yin-yang qi* as a base he rationalized the relation between heaven, man, and earth and elaborated the subtle workings of earth, metal, and wood *qi*. Substantially informed by Chinese natural philosophy, his examples employed standard figures and took the opposition of the two *qi* of heaven and earth as axiomatic. For Song, the natural tension between the various phases of *qi*—the aspect of change (*yi*)—sufficed to explain all being. It is in these sections that Song most clearly emerges as a natural philosopher, rather than a practitioner, a passionate scholar seeking reliable features in natural and material processes in a world of political chaos and increasing social insecurity. From this viewpoint he insisted that a world of *qi* could not be unstructured or based on chaos, because *qi* itself provided the structure. The ultimate validity of *qi*, its identification as the unique substratum of change, was Song's gauge for certainty in knowledge. On this basis Song considered experimentation and the observation of the mundane world reliable methods of inquiry to shape facts and evidence in written culture. Being, life, and death, the creation and decay of things and materials, nature and man were all expressive of the order that governed the world.

In his attempt to provide a universal all-encompassing theory of *qi*, Song relied on existing theory, expanding it when it proved insufficient to explain the phenomena he so closely observed. Song often departed from traditional explanations of natural things and their transformations, but he often ended up with quite original interpretations. A good example for this is his view on the Buddhist concept of particles (*chen*) that became for Song an essential issue to explain air and man's dependence on it in *On Qi*. Throughout his works Song was concerned about consistency. He applied the same conceptual approach to reveal the workings of *qi* in the heavens and in man's world. With great clarity Song promoted his view that he saw universal principles at work in the first part of the title of the *Works of Heaven*. The second part, the *Inception of Things*, stresses his belief in the concept of change (*yi*), which Song saw as the cause and effect of everything. The original passage of the *Book of Changes* states that the inception of things and the accomplishment of affairs "advances the path of all under heaven (*mao tianxia zhidao*)."[1] We can assume that Song took this quote literally. He believed that man could gain knowledge and learn how to act properly by observing the transformations in things and affairs.

The cultural and intellectual complexities of his time make it difficult to categorize Song within Chinese scholarship. Song's approach rode a wave of contemporary literati interest in pragmatic statecraft and a growing intellectual interest in new approaches to knowledge driven by the uncertainties of this era and a changing material and cultural world. Scholars from very different backgrounds and often with very different aims pondered "knowledge and action," "the investigation of things and broadening of knowledge," and "things and affairs." Within this diversity, Song's aim was to find a reliable basis for knowing and then bring order back to the world. His approach to knowledge was comprehensive and methodologically driven by the idea one should record in writing things and affairs gained from "seeing and hearing." We can identify Song's compilation as a *biji*, a private jotting, and group him intellectually with other *li* or *qi* thinkers of his era. In particular, Song presents views similar to those of other authors with an encyclopedic approach to thinking about the "origin of things" (*wu yuan*), and authors who anchored their interests metaphysically in the *Book of Changes*. Song felt obliged to produce nothing less than the revelation of true knowledge—allowing no compromise for social bonding, ideological presuppositions, political needs, or human assessments of morality. It was the intransigency of his principles that put him in the position of an oppositionist at a time when anxiety made conformity the norm.

With regard to his theory, Song does not fit in the distinguished group of *qi* interpreters of his time because he attempted to provide an exclusive and self-sufficient system, explaining all things and events on earth on the basis of *qi*. The historians Li Shuzeng, Sun Yujie and Ren Jinjian suggest in their study that Song saw "no causative but a constituted relationship between *qi* and the void (*xu*)" and that he regarded "*qi* as the material which creates the void (*xu*)."[2] I think, however, that this gives Song's approach to *qi* an overly material emphasis. In fact, Song thought of *qi* as the core rationale of the world that was in a continuous transformational process. In that regard *qi* and the great void primarily constituted potentiality and not material existence. In sum, Song's *qi* theory was the basis for his understanding of the physical world and a critical reason for his choice of subject and explanatory content in *Works of Heaven*. For Song, technological procedures exemplified that the world and everything in it rested on *qi* change.

Song's rhetoric of knowledge reveals an individualism characterized by its rigorousness. We see from his political tract, *An Oppositionist's Deliberations*, that Song was reluctant to take part in any philosophical parley. He restricted himself almost exclusively to the description, rather than the interpretation of conceptions. His contemporaries possibly saw his pragmatic approach and consequent neglect of the philosophical backdrop as a deficiency—for them, philosophically and ethically, it was short of argument. From a historical point of view, Song's pragmatism can be seen as his way of challenging a culture that he claimed accepted only theory and ethics as keys to the revelation of knowledge. He believed this had produced misleading speculations about a morally inclined relation between heaven and man, a misconception of cyclical change, or fate, and an overall mystification of the principles that governed the world. In this context Song is revealed as a man who understood morality as the product of circumstances, namely, customs and habits. Customs together with circumstances explained how poverty could drive a farmer to riot or how a frustrated scholar could behave immorally. Considering this, Song saw morality as detached from "knowing," that is, the ability to penetrate the way of heaven (*tong tiandao*). It was exactly this ability that man required to order the social world morally. For this reason Song generally abstained from moral assessments, defining individuals as either naturally intelligent or naturally stupid. Scholars were the ones who made knowledge by observing the processes of nature, mundane activities, and craft work, all of which were equally "objects of knowledge." The craftsman was not himself knowledgeable and thus also not relevant to Song's approach to knowledge. Concentrating on craft work as a performance of general principles, Song invested no thought on the social role of craftsmen. Dividing the world into two groups, scholars and commoners, Song claimed that intelligence and talents were exclusive to the scholar. Training and experience, however, were important for all. It helped the scholar to cultivate his understanding. When he assigned importance to craft training, Song was mainly thinking of the improvement to the yield and quality of the product, not the skill as such, which has the potential to improve the social standing of the craftsman. Looking at it from a functional and skills perspective, I suggest that Song's insistence on the need for training skill had the goal of improving the effectiveness of the harnessed transformation both with regard to yield and quality.

Song furthermore claimed that while talents were always potentially present, if training was neglected and customs not controlled, man would be carried away by his emotions. And here Song closes the circle to explain the chaos of his era: his contemporaries had established customs and habits and looked for morality instead of rational patterns. Thus man had negligently allowed the world to fall into decay. His colleagues were striving after wealth and reputation and passively waiting for a sagelike man to quell the chaos of their time, instead of proactively bringing about the transition of social and moral formations by unraveling the threads of true knowing.

Throughout his writings, Song urged his colleagues to distrust any ideology or belief other than a set of well-chosen classical texts from before the third century BC, and any knowledge or facts that could be substantiated by deduction and induction within a consistent understanding of *qi*. Song deliberately abstained from referring to any other scholar, such as Zhang Zai, although his approach was clearly informed by this important thinker on the subject of *qi*. In addition, he argued that any text-based evidence had to be critically assessed through correlation with the mundane world and with common sense. Following his own principles, Song documented the myriad things in all their detail, presenting his arguments in word and image. Illustrations depicted the technical processes involved in each subject, both in scenes of work and in detailed sketches of the necessary implements and tools. When deciphered within cultural norms, they conveyed important additional arguments to the reader. For persuasive reasons, for example, the image of the loom was the most accurate in its depiction of technical details, as it represented the importance of orderly arrangements and the difficulties in achieving this order. Song also arranged the images within the text in an orderly and unified fashion. All sections gave a concise yet refined insight into the production processes; he listed raw materials and their treatment, the implements needed for each craft, details of machinery and the working process, right through to the end products, including regional diversities and special techniques. All this demonstrated that truth and knowledge lay in the orderly proceedings of the mundane world, waiting to be revealed by the knowledgeable mind of a true and insightful scholar.

Song's approach can be identified as part of an intellectual discourse that argued for the importance of thorough research by "seeing and hearing," using texts and experimentation as well as personal observation as its sources. In hindsight this observed reality may appear deeply influenced by Song's theory of *qi* and contemporary issues governing scholarly discourse, but we can assume that Song considered his methods the proper proceeding to gain knowledge. In accord with many other thinkers of his era, Song showed a healthy suspicion toward earlier documentation, in particular the findings of his colleagues. He criticized writings that provided "disorderly comments" (*luanzhu*) and offered explanations that merely "fantasize" (*wangxiang*) about processes.[3] His attitude was, however, not dogmatic. In the sections on the calcination of stones (11), vermilion and ink (16), and pearls and gems (18) of *Works of Heaven*, he draws selectively from the *Systematic Materia Medica* by Li Shizhen.[4] In his disapproval of written sources on philosophical themes, however, Song was much more obdurate than most of his colleagues. In particular he blasted the venerated iconoclastic philosopher Zhu Xi as an exemplification of the fallacies of bookish learning, denouncing him for having trusted records more than his common sense. Song argued that in Zhu's *Comprehensive Mirror for Aid in Government*,

> Zhu Xi uncritically retold an event that happened during the reign of the six dynasties. The [original source] says that two suns arose together, shifting toward the East. At the same time two moons were seen in the West. The sun [is said] to have risen at night to a height of 3 *zhang*. This is supposedly an exaggeration [which was spread] by certain people. Or the historians of the northern and southern dynasties had no common sense. If this had [really] taken place, which I highly doubt, then I

> am looking forward to the versatile and divine talent, who will investigate it as soon as it appears the next time. It is unbelievable that well-read gentlemen do not scrutinize such things and events when they happen [instead of believing in written record]. This would be truly of great benefit![5]

The scope of Song's criticism becomes apparent when one learns that Zhu Xi's interpretation of the *Comprehensive Mirror for Aid in Government* played an important role in popularizing historical studies during the Ming.[6] Such statements show Song was deeply engaged with the topics of his era. His notion of knowledge was based on universal principles and orderly proceedings, an approach that allowed no obscurities such as two suns or two moons. He called for a critical assessment of the multitude of books and materials made available to the scholar by commercial printing. Instead of blindly following the assumptions of written accounts, Song demanded reliable verification by observation of all facts at hand. For example, in the introduction to *Works of Heaven*, Song expressed his doubts about a text-based reference to the well-known story of King Zhao discovering a red flower during a boating tour. When the king asked for its name, no one knew it. Hence he sought out Confucius for advice. Confucius said, "this is the fruit of the jujube flower. You can open and eat it. This is a good omen (*hao zhaotou*)! Only the chief among several princes has the chance to catch a glimpse of (*bawang*) it."[7]

None of Song's generation had ever seen this flower. In fact no one knew if such a blossom actually existed and still certain circles speculated about its botanical specifications and read an ethical context into discussions on its existence. Song dismissed such inquiries as ingenuous and rebuked those who did not look beyond the bookish world but took its contents at face value. He condemned his colleagues for daring to speculate about the botanical specifications of a plant from a period long gone when they were unable to distinguish a peach blossom right in front of their eyes. He reprimanded "those who have only a little experience with the molds of the cooking pots (*fuxin*), yet indulge in speculations about the sacrifice vessels (*juding*) of Lü," or the painters of his era, "who prefer to depict ghosts and monsters, but scorn such common subjects as dogs or horses."[8]

Criticizing the aloofness of contemporary erudite culture, Song praised people of "broad learning" (*bowu*), historical figures such as Gongsun Jiao, a contemporary of Confucius who obtained the official title of "Erudite Literatus" (*bowuzhi*), and Zhang Hua (232–300), who analyzed natural phenomena such as thunder and subjects like wine production and salt brines in the *Comprehensive Treatise on Various Things*.[9] Such references show that Song's investigation of the physical world was anchored in a Chinese intellectual tradition of sophisticated speculations on natural phenomena and material inventiveness. These associations also again clarify that Song considered himself a scholar. To emphasize that higher knowledge was his goal, Song throughout his work evaded any association to practical experience as was common for writers of books on agriculture to express their social concern. It is for this reason that he sanctioned his interests in the glazing of tiles and uses of brimstone with reference to the sage kings, emphasizing the ancient origin of knowledge and his claim to truth.

In conclusion, Song exemplifies the characteristics of an era whose approach to knowledge does not fit today's narratives of scientific and technological endeavor. His example shows that historians must be particularly aware of the subtle preconceptions that inform their view of individual thinkers and intellectual developments in general or the major historical trends of their periods. Song may be considered a child of an era of consumerism and commodification, but he was not enthusiastic about it, nor did it raise his interest in things as mere material objects. Leaving the doorway of Ming history and entering the shadow theatre, we can see that Song's writings do not easily fit the agenda of a "world of consumerism" depicted, for example, in Timothy Brook's study.[10] Scrutinizing crafts, Song reminded officials of their responsibility to act in harmony with universal principles. In this regard he must be clearly set apart from those scholar-officials who managed and documented technology and agriculture as part of their daily duty or those who fervently discussed them as issues of pragmatic statecraft or economic growth. Neither does he belong to the world of connoisseur collectors or the nouveau riche who immersed themselves in the colorful world of luxury goods. Song's efforts also stand apart from the handbooks for users and consumers on topics such as lacquer, bronze, or porcelain wares, which guided scholars through the lavish world of consumption and claimed help the reader distinguish good from bad wares and proper reproductions from forgeries. Song disapproved of a world that was pursuing glamor and glitter instead of universal knowledge. In this context Song felt that traders were important cogs in the wheels of the state, but they should know their place. He was a man of his time, but he is not part of the current narrative that identifies seventeenth-century intellectual discussions as departing from social boundaries and waxing enthusiastic about material well-being.

We must also see that Song's priorities were not the priorities of the modern reader. This is particularly obvious in his treatment of craftsmen. Although he lived during a reign that linked crafts more closely to the state than any previous dynasty, for him the artisan remained insignificant. Previous researchers have repeatedly interpreted the title of Song's *Works of Heaven* as a pairing of human endeavor with nature, considering them structural or complementary opposites. Yet, staged in its original theatre, this juxtaposition turns out to have been irrelevant to Song. There is no allusion to a world made by man's abilities or formed through man's craft-knowledge. The title was Song's message to morally upright leaders: scholars should fulfill their duty. Performing the "works of heaven" meant identifying order and governing the state in accordance with universal principles. In Song's view, crafts and technological knowledge constituted an interface between cosmological principles and human action. This was the reason man should observe the erection of an oil press or the manifold ingredients for dyeing a cloth bright yellow or bluish-green, not because the details as such were relevant. It was from this perspective that Song included crafts and technology in the assets and meanings of premodern Chinese literature. Whether or not Song was representative of Chinese thinkers, or if he was exceptional within more global perspectives, he was without doubt a man of his time and culture. The information provided in his work on technological endeavor and scientific thinking and the role of Song's work in the making of scientific and technological knowledge can only be understood when this context is taken into account.

Finally, I find it important to add that this study should not be misunderstood as an attempt to cast doubt on the technical contents of *Works of Heaven*. While my effort to shed light on Song's silhouette did not find the fringe of his scholarly gown smudged with soil or his long sleeves rolled up, it clearly identified him as a meticulous observer of his environment and a thoughtful scholar. The fact that Song does not fit our current image of a person fascinated by crafts and practical things, the fact that he was not an engineer and thus did not approach technology or techniques or crafts, nature and culture in the way a modern mind thinks he should have is not a deficiency; it is his greatest value. Handed down to us by chance, Song's work provides an exceptional opportunity to study the thoughts and models of explanation that a former culture used efficiently and effectively for centuries to make knowledge about natural phenomena and material inventiveness. The significance of this element of my study goes beyond the question of the reliability of Song's knowledge or the informative value of Song's writing for the history of technology. It indicates that our categories and yardsticks are inadequate tools when assessing the actual dimensions of premodern Chinese views on science and technology and a tradition of knowledge making that brought forth significant results in fields of scientific and technological endeavor. Understanding their models of thought enables us to correlate their view of nature with our own. Probably even more important, this glimpse of what knowledge and practice, knowing and doing, technology and science, can mean to a specific person, at a specific place, and in a specific culture reveals the existence of a great historical range of individuals subject to change within cultures throughout time. It is through these personal ideals and cultural gauges that scientific and technological knowledge shows its history and, even more so, how history can be understood through it.

EPILOGUE

The Aftermath

> An ambitious scholar will undoubtedly toss this [book] onto his desk and give it no further thought: this book is in no way concerned with the art of advancement in officialdom.
> *Works of Heaven*, Preface, 3b

During Ming times paper was made by soaking bamboo stalks in pits on the plantation for at least a hundred days (see fig. 4.5). The fibers were then pounded and the coarse husks removed. Mixed with lime, the pulp was boiled, washed, and strained through ashes. Chemicals were added to bleach the mash before the sheets were sifted out. The papermakers did this, as *Works of Heaven* records, by "holding [the screen] with both hands and then dipping it into the water (*liang shou chi bian ru shui*)."[1] They then submerged it in the pulp tank so that the fiber suspension dispersed evenly on top of the screen. Praising its thick smooth surface, Song might have longed to have his thoughts published on the high-quality distinctive stationery (*jianzhi*) produced in Jianshan County next to his home county, Fengxin. He might also have wanted the best ink from the pine resin lampblack that, as he meticulously described, was scraped out from the last two sections of the burning chambers where the finest soot formed. Yet, the first edition of *Works of Heaven* handed down to the present was printed on standard bamboo paper with low-grade pounded ink-waste from the first chamber, cheap and shiny. Did it matter to Song? From his preface, it seems not. In fact, Song was glad that his work was published at all. The study of his life reveals, however, that his pleasure was somewhat dimmed by disappointment at his lack of career success. He remarked on his colleagues' ignorance of how the mundane cooking pots in their own households were produced, and his tone is cutting and acrimonious when he disparaged their empty discourses on ancient sacrificial vessels from a bygone era that no one had ever seen.[2] Indeed, when he considered the arrogant and self-satisfied officials who "cunningly twisted the orderly government of the state into chaos,"[3] he thought himself lucky to manage to publish with the help of his friend Tu Shaokui: "Was this turn of events not the handiwork of fate?"[4]

This chapter addresses what I call the aftermath, that is, the reception, or rather nonreception, and physical dissemination of Song's writing from the first editions up to the present day. Looking at the transmission process, I pay particular attention to the way in which the nature of knowledge is accommodated to the needs of the time and which factors affect these processes. Writings preserve the values of their moment of conception. Yet, within their transmission, written accounts are liable to epistemological fashions and individual viewpoints. Like an ancient bridge, built to link the banks of a river, texts look like stable constructs, connecting their users both technologically and socially to the knowledge and priorities of their maker's time. But, in fact, both books and bridges only survive if they are used and, this use is never neutral: just as bridges are either abandoned or frequented depending on changes or consistency in the flow of the river or trade routes, the use of a book is subject to current needs and ideals. Sometimes it is marginal; at other times it advances to a major artery. To take the metaphor further, if the bridge is no longer used for its original purpose, its stones may be used to build a new temple, or the structure itself is remodeled into a street or a row of shops. The building blocks of a text are its information. Taken out of the original construct, they can also serve new purposes. Using this perspective on knowledge transmission, I start by examining the reception of Song's work by his peers and the role of literary and intellectual sponsorship. I then go on to consider the role of Ming loyalty and the growing scholarly resistance toward Qing rulership that marks the late seventeenth century. Finally, I trace the transmission of the book as an artifact in the subsequent centuries.

BY VIRTUE OF FRIENDSHIP: LITERARY SPONSORSHIP

> I have a friend, Mr. Tu Boju [Tu Shaokui], whose perfect sincerity moves the heavens. His mind is alert, when investigating things. [He believes that] everything from the past to the present, even if there is only one laudable word or one qualified iota, however small it may in fact be, should be earnestly and assiduously considered [matched together (*qi he*) like the halves of a tally].
>
> *Works of Heaven*, preface, 3a

Paper sheets for the printing of books were generally dried on a double wall of earthen bricks between which a fire was lit to warm the stones. Song mentioned that in the north paper was often recycled to meet demand. This was known as reincarnated paper (*huanhun zhi*). But in the south, where Song lived, papermakers had access to a sufficient supply of bamboo and other plant fibers. Paper was delivered to the innumerable shops where woodcutters spent long days with their heads bowed over their work. Their fine carving inscribed administrative documents or scholarly expressions of poetry or prose into soft pine or hardwoods, depending on what the customer could afford. Song described the production of paper and ink in *Works of Heaven* because the process of making them, the transformation of materials, revealed universal principles. It was this perspective that also made him ignore the technical procedure of printing in his book. It was a technique that did not involve transformational processes. As a customer, however, he may have followed the process with interest, insisting that the expert hand rubbing and stamping the woodblock on paper reproduced the lines and drawings and work with as much delicacy as Song could afford.

Much attention has been drawn in the last decade to the growing commercialization of the book market during Song's generation, showing the close relationship between consumer attitudes and changing production modes in seventeenth-century China. The opportunity to publish books privately and disseminate one's thoughts for intellectual, political, or commercial reasons also resulted in new social attitudes toward intellectual cooperation. Scholars reconsidered the importance of the tutor-pupil relation; they changed habits of intellectual cooperation or altered modes of literary sponsorship and practices of recommendation. Literary sponsorship in seventeenth-century China could be financial or intellectual. Prominent men either provided money or lent the prestige of their name to insure publication and enhance the reputation of their protégé. They might recommend a text or book by adding a covering note or by writing a preface. In other cases, scholars simply mentioned the patron's name, as Song did in the preface to *Works of Heaven*.

As in European cultural patronage, literary sponsorship in seventeenth-century Chinese culture served many functions. Gentry or retired officials, for example, commissioned plays to promote social ideals or standards of morality. In such cases the author was of minor importance and kept in the background. The sponsor added more to the moral reputation of the sponsor.[5] In general, we can say that scholar-officials of this era used sponsorship of erudite works within an acknowledged framework to foster their professional interests or social standing. Sponsorship was also used to tighten political bonds and forge ideological links. In their complex role as political, social, and intellectual leaders, these men used the support of literary works as part of a complicated cultural process to construct, define, and maintain relationships. Compiling a preface for someone else's works or allowing an author to mention one's name was a means employed by a sophisticated society to lend credibility to ideas and writings. It was one of the many facets in the creation of political, intellectual, and social networks. Literary sponsors also helped in other matters, in particular arranging appointments or defining scholarly duties. They wrote letters of recommendation to promote their protégés to better positions. But, in contrast to their European colleagues, who considered literary sponsorship a legitimate means of recruitment, Chinese scholars of the Ming carefully avoided openly connecting their patronage in literary sponsorship to the matter of appointments. They saw the intermingling of these issues as a sign of corruption.[6]

Literary sponsorship was quite common in China. Its overall aim was to define and foster a complement or counterpoint to governmentally acknowledged frameworks of discourse. Sponsoring "private jottings" (*biji*), scholars showed their mutual concern; financing unofficial histories (*yeshi*), specialist treatises (*pulu*), or their colleagues' examination of the classics from a new angle, literati displayed their critical acuity and enhanced their scholarly reputations; promoting household encyclopedias, primers, or agricultural writings, rich and honored households displayed their humanitarian minds. Minor authors sought the sponsorship of prominent philosophers and acknowledged experts to lend credence to their work, to augment its distribution, and, if their opinions and ideas had been expressed too freely, to ward off attack. Thus literary sponsorship affected the reputation of an author and his work and, consequently, its circulation and dissemination. Sponsorship established and framed power structures and the reception of knowledge within disciplinary communities.

Song had two long-standing friends, Tu Shaokui and Chen Hongxu, who supported him and his work for most of his life. How did these friends receive Song's works? What effect did their friendship have on the contemporary reception of Song's writings? A scholarly bibliophile and Ming loyalist, Chen mentioned Song's writings in his works and included Song's efforts in his library. He did so, however, quite selectively: he did not list Song's compilation on crafts in his library catalogue. Tu Shaokui, the son of a high official, financed Song's *On Phonetics* (now lost) and *Works of Heaven* as Song noted in the preface to the latter.[7] Both friends had far more financial resources than Song and came from more reputable families. They sponsored Song thus not only materially but also through their higher social and political rank. Apart from lifelong acquaintance, mutual interests welded these three men together. All three were concerned with the military, education, and practical issues, matters of some urgency during this period of turmoil. And yet, both Tu's and Chen's reactions to the publication of *Works of Heaven* have left no trace for the historian.

Upper-middle echelon officials such as Tu usually found their protégés in their immediate intellectual environment. Sometimes these were their social equals, more often their inferiors. Often sponsor and protégé were already long-term friends, but occasionally they also became friends through the process or retained a formal relationship. Song and Tu had known each other since they studied together to pass the district examinations. They both went to the White Deer Grotto Academy, where they prepared for the metropolitan exam (see chap. 1). Throughout their acquaintance, Tu remained Song's social superior. His father, Tu Jie (*jinshi* 1571) had the influential and high-ranking position of a court censor (*yu shi*, rank 1b). As the eldest son, Shaokui inherited the family property and followed in his father's career footsteps, occupying several censorial posts at the provincial level (ranked 3–4) until his death in 1645. Like many senior officials in the late Ming, he dedicated much of his attention to strategic military issues. In addition his career responsibilities included education, mining, and minting. All sources describe Tu Shaokui as a fervent Ming loyalist, committed in a variety of ways to save the Ming.[8]

Tu seems to have supported Song's philological study with a distinct purpose in mind. From 1623 to 1632 Tu held an appointment as an education commissioner of Sichuan Province (*Sichuan tidu xuezheng*) and thus became interested in the public schooling system. The chronology of events suggests that Tu Shaokui encouraged Song to compile an educational handbook when Song was appointed to the position of teacher in Fenyi County in 1634, thus combining the personal interests and official obligations of both parties. The title *On Phonetics* is all that has survived of this book. It implies that this text concerned an important traditional branch of Chinese classical education. Presumably the book was addressed to the pupils attending the local county schools in which Song taught.

Events suggest that Song's first project was carefully prepared and that Tu's financial support for his study had been long anticipated by both friends. Starting in 1634, when Song gave up trying to pass the metropolitan exams, he seems to have invested a considerable amount of his time and Tu's money in the compilation of the study *On Phonetics*. The work was finally published in 1637 when Song still held the position of local teacher. This was five years after Tu had left the educational sector after a promotion to the position of military defense official (*bingbei dao*).

Song indicates that Tu's later financial support of *Works of Heaven* was an extension of their first joint project, mentioning that "owing to the gentleman [Tu Shaokui] the woodblocks for *On Phonetics* were prepared [for publication] last year [1636]. There is now another demand (*houming*) [by Tu Shaokui]. He has also taken up this piece of writing (*juan*) [*Works of Heaven*] and continues to raise [funds] for it."[9] As the study *On Phonetics* is lost, historians normally ignore it and presume that *Works of Heaven* was of more importance for both Tu Shaokui and Song. While this may have been true for Song, it was not necessarily so for Tu Shaokui. We can, however, assume that Tu knew about Song's interest in crafts, as Song could not have gathered the information necessary to compile a work like *Works of Heaven* in just a few months.[10]

The history of Tu's sponsorship is significant because it shows that Tu's support was initially for a traditional scholarly work on a conventional academic subject. Song mentioned humbly in his preface that his friend Tu was interested in the "investigation of things," that is, *gewu*. As mentioned in chapter 2 this reference to the *Great Learning* signposted a scholar's concern about more practically oriented forms of inquiry beyond philological studies. Song's remark, quoted at the beginning of this section, that Tu was interested in the investigation of things suggests that Tu recognized the relevance of serious and assiduous investigations to pragmatic statecraft and thought erudite studies in the tradition of *kaozheng* not up to achieving this aim. Thus, Tu was interested in practical issues, yet he associated himself intellectually and ideologically with a more acknowledged tradition than Song, namely, the *gewu qiongli* paradigm of the "learning of the way." Elman suggests that by the late Ming "*gewu* had become the key to opening the door of knowledge for literati versed in the Classics and Histories."[11] If we take into account that Song's title associated his work with the *Book of Changes* and never used the language favored by *gewu* intellectuals, his connotation may imply an intellectual dissent between the two friends about the way an interest in things and affairs should be ideologically anchored. Song tended much more toward *qi* studies, exemplifying the prime characteristic of members of this group which can be summed up as "outside and opposition." This is in strong contrast to the earnest and assiduous higher official Tu, whose career followed contemporary conservative ideals. Whatever the reason, Tu did not extend his generous support to any later works by Song. In short, his support turned out to be selective.

Tu's biography reveals that his posting to Sichuan furthered his interest in practical issues and technology, in particular in mining and minting. Reacting to the political situation, Tu also became interested in military technology. Tu's firsthand experience in these fields may have provided the final touches to Song's descriptions of mining, weaponry, and the calcination of stones in *Works of Heaven*, or Song may have shared the information he had collected with Tu.[12] Yet, despite Tu's role as Song's financial backer for *Works of Heaven* and his interest in issues such as mining, Tu never acknowledged the existence of his friend's documentary effort. There are no historical records of any contact between the two friends after Song's works were published. Possibly, Tu had willingly sponsored Song's philological survey but was not particularly enthusiastic about the second result of his patronage, the *Works of Heaven*.

The other significant sponsor in Song's life was Chen Hongxu, who supported him intellectually rather than financially. Their unbroken correspondence, documented in Chen Hongxu's anthology *Chen Shiye xiansheng ji* (*Collected Works of Master Chen Shiye*, published 1687) verifies that they were lifelong friends. Unlike Tu, Chen kept in touch with Song even after *Works of Heaven* was published.[13] Chen Hongxu was a scholastic bibliophile interested in applied knowledge with some personal experience of weaponry. He inherited a huge library from his father, Chen Daoting (d. 1628), and expanded it to 80,000 chapters (*juan*) by 1637 (many bibliophiles of this period counted their collections not in terms of books, but in chapters). He devoted much of his time to cataloguing his treasures. We know that the library offered most of the important classic and orthodox literature together with writings on irrigation, hydraulics, and botany as well as rarities such as translations of Western writings compiled by Matteo Ricci (1552–1610) and Diego de Pantoja (1571–1618).[14] Song mentioned in *Talks about Heaven* that Westerners believed the earth to be a globe and described in *Works of Heaven* the construction of European guns.[15] He may have gained the sparse information he had about Western culture and science from these sources.

Chen's library offered Song more than bookish knowledge. Chen was one of the rare Chinese bibliophiles who opened their treasure rooms to their colleagues, using the books to attract philosophical gatherings. Popular contemporary intellectuals, poets, and politicians such as Xu Shipu (1607–57), Shi Runzhang (1618–83), and Shi Kefa (1601–45), eagerly accepted Chen's invitations.[16] All of them visited Chen regularly to gain access to this exceptional private collection of books. Compared with his contemporaries, who often refused all visitors, Chen Hongxu seems to have been remarkably hospitable.[17] Chen mentioned in his writings that an important author of texts with scientific content, Fang Yizhi, had once asked for an appointment.[18] We do not know if Fang ever managed to visit, but whether or not, no connection can be drawn to Song. Song, who was in regular contact with Chen, was either not willing to use his friend's "salon" to propagate and discuss his ideas with his contemporaries, or these people remained unimpressed by his ideas and thus ignored him.

Catalogues were systematized overviews for the bibliophile and were used to advertise their possessions to their peers. Rare editions of all kinds enhanced the reputation of a scholar and were often meticulously delineated. Moreover, bibliophiles may not necessarily have included all titles in the catalogue, but only those that would have a positive effect on their social and intellectual standing.[19] So Chen's omission of *Works of Heaven* is not in itself remarkable. It is puzzling, however, when one considers that Chen recognized two other works by Song and even reviewed them. These two lost works were anthological and historical in nature. Probably they were never printed, but circulated in manuscript form. Chen described the first, *Yuan hao* (*Essential Manuscript,* ca. 1638), as a discussion of topics such as hemp production, shoemaking, economics, and administrative organization, and the second, *Chunqiu rongdi jie* (*Explanations about the Rong Slaves of the Spring and Autumn Period,* compiled ca. 1647), as an ethnographical study of northern tribes.

Perhaps Chen's particular enthusiasm for Song's ethnographical study was a consequence of his political conviction and loyalty to the Ming. Chen may have found anti-Qing material in this work as his review, dated 1644, honored Song's meticulous research on the customs and origins of the Shanrong, Baidi, and Sushen and the earlier Jurchen.[20] Can it be that Chen only catalogued works that discussed themes he was interested in, and this did not include the themes of *Works of Heaven*? But if this was the case, why does Chen's catalogue list writings on weaponry, farming, and military strategy, all themes discussed in *Works of Heaven*? Chen obviously classified Song's *Works of Heaven* in a different category. Chen's private jotting has two sections "Opinion about Irrigation" (*shuili yi*), and "On Salt Methods" (*yanfa yi*), both very reminiscent of some of Song's works. The irrigation topic appears in *Works of Heaven. An Oppositionist's Deliberations* contains a section entitled "opinion on salt politics" (*yanzheng yi*). Some parts of it are in fact almost identical in content, although Chen never refers to Song's work as a source.[21] Such plagiarism among friends was not unusual. The more interesting fact here is that Chen did not catalogue Song's *Works of Heaven,* implying that the work did not contribute significantly to Song's reputation as an author or intellectual and that Chen did not consider Song's work on crafts a writing that would enhance the reputation of his collection.

If Tu and Chen are representative of their era, scholars not only tossed *Works of Heaven* onto the table, even worse, they did not even pick it up in the first place. The examples of Tu and Chen do demonstrate, however, that this dismissal of *Works of Heaven* was not due to a lack of interest in practical matters. Chen collected literature on technical topics and Tu was very involved in the promotion of mining activities. The probability that Tu and Chen knew about *Works of Heaven* and its contents is high. Chen's refusal to comment on it and Tu's ignorance must be interpreted as deliberate acts. Taking Chen Hongxu's and Tu Shaokui's reaction as representative, we must conclude that Song's generation carefully ignored the work.

In fact it is difficult to judge how many of Song's generation knew about his works. Not one of the intimate friends and peers that Pan Jixing identified as family acquaintances of Song ever mentioned *Works of Heaven*. Some of them, such as Liu Tongsheng and the poet Xu Shituan recognized Song as the brother of Yingsheng, who was evidently more prominent than his younger brother in their lifetimes.[22] Liu Tongsheng compiled a poem when he met Song more than thirty years after their joint training for the provincial exams in 1640. He pointed out that Song had asked him for help to withdraw from his appointment in Dingzhou.[23] None of those who mentioned Song ever acknowledged him as an author, although many of their interests overlapped. None of the well-known bibliophiles we know of included any of Song's works in their catalogues, not even for the pleasure of collecting rare, bizarre, or unique compilations, although a flood of scholars engaging in similar themes would sweep through Chinese knowledge culture throughout the seventeenth and eighteenth centuries. How this reaction is linked to Song's sociopolitical standing, to his political view, and to the physical availability of his work is the topic of the next three sections.

BY VIRTUE OF POSITION: OUTSIDE AND OPPOSITION

> Each and every bright and clear thing that heaven and earth brings forth has its opposite: the dull and turbid. The damp opposes the dry and every noble thing has a vile side.
> *Works of Heaven*, chap. 3, no. 18, 53a

Living inland with limited financial resources, Song did not often have the opportunity to travel to the coast and see the divers searching for the prizes hidden in the black shells. Diving for pearls had become increasingly dangerous in the time of the Ming dynasty as the increased demand for luxury goods led to overfishing of the pearl oyster beds. By Song's lifetime divers were using air pipes and snorkeling gear to search in ever-deeper waters. Song described this trade in *Works of Heaven* in text and image, emphasizing the extreme difficulty in finding the rare treasures of nature. Song did not mention that the harvest was often frustratingly low. Even when divers were fortunate enough to discover the lustrous and perfectly formed pearls desired by the imperial household, they had to hand them over to the local authorities or the court without appropriate compensation. Most divers made more money from small pearls they sold to inland markets. They also ground imperfect pearls to powder used for medicinal and cosmetic purposes. Searching the beds of the deep sea, divers lived on the dark and turbid edge of existence.[24] In this there were parallels between them and Song: like the divers Song was below the level of the historically recognized and eked out a miserable life as a teacher. By naming his political treatise *An Oppositionist's Deliberations*, Song located his work as an outsider's view of history, a view from the "wild" (*ye*) side.[25] Song was one of the many who had been given a raw deal. How did this identification affect the reception of Song's ideas?

Within the tradition of "wild" writing, the term had originally referred to those who did not hold a court position and compiled historical documentation or wanted to stress that they offered their private opinion. The factional disturbances that rocked the Ming world in the seventeenth century had loaded it with additional meaning. The Donglin group gave "wild" a strong moral connotation as part of their fight against corrupt recommendation practices and factional alliances. In this discourse the identification of oneself as "wild," was an expression of one's duty to remain upright and vigilant and a vindication of the resignation from official service if one felt morals were in decay.[26] By the time of Song's publication campaign, discussions on inside and official vs. outside and wild, centered on the object of scholarly loyalty. Was it the state, was it the Ming emperor, or should the scholar be first and foremost loyal to moral ideals? Song had made his choice. He believed in the essentials of the social system, and he wanted to be a part of it. He deplored those who left important positions on moral grounds only to be replaced by "small men" who, as Song put it, "mulishly occupied positions" (*hui zhi bu qu*).[27]

His colleagues may have understood Song's rage but felt he was overreaching himself. Who was Song after all? He had accepted the position of a lowly teacher in a county school, a career move that was an admission of social and professional failure. Ma Tai-loi suggests in his study on the Ming schooling system that despite Ming Taizu's organizational improvements the position of teacher was "unenviable." Only the most desperate accepted the low rank and meager emolument it provided. Ma's research indicates that teachers did not even command public respect. They had no pride in their profession and often performed their administrative and ceremonial tasks perfunctorily, idling away their time. Many had to ask for presents or collect illegal rents to make a living. Consequently high officials had been trying to eliminate the position since the middle of the sixteenth century. These efforts became even more intense during the Chongzhen reign (1628–44) of Emperor Sizong (1611–44), while Song was a teacher.[28] Although Song was aware of the deficits in the education system, he opposed such policies. Song's argumentation was clearly driven by personal motives. He would have fallen victim if the proposal had been implemented. He called instead for major improvements, arguing that only local schools could locate talents and secure the necessary training.[29] From a historical perspective his efforts looked almost quixotic. The failure of the state system to provide adequate higher education was marked by the growing number and importance of private schools and academies, in one of which Song himself was trained. As these academies and schools successively gained in reputation throughout the early seventeenth century, public county schools deteriorated to little more than holding pens. The position of a teacher in a state school meant neither honor, nor privilege, nor wealth. Song realized that his colleagues would recognize his acceptance of a teaching position for what it was, a last, desperate effort to obtain an official rank.

Successful scholars who had achieved a reputable position saw Song as just one among the many dispensable scholars who had dropped out of the system. Some of his equals may have shared Song's frustration. Still, they would have seen *An Oppositionist's Deliberations* as the melodramatic reaction of an unrealistic dreamer. Many who shared Song's fate had a vocation to serve the state. Those who only achieved the lowest official appointment often felt utterly humiliated because they practically had to beg for a position—and pay for it. *An Oppositionist's Deliberations* shows Song was very angry. He realized he was an outsider and, unlike the powerful officials, he would never be asked for his opinion. With this in mind he marked *Works of Heaven* as a book "compiled in a studio of private sovereignty" (*jiashi zhi wen tang*).[30] This refers to a passage in the *Book of Changes* that obliged a virtuous man to speak for rights and virtuousness and take up civil service, saying that "not eating at home (*jiashi*) is auspicious (*ji*): it encourages righteousness (*xian*)!"[31] Song's assignment shows that he was aware of his position and that he was not given the chance to eat outside. At the same time his remark shows his condemnation of those who deliberately shut themselves off from the world as an expression of their high morals, although they could have had the opportunity to order the chaos. If he were in their position he would take action. Song twisted his outsider's position to that of a man who morally deserved to be inside.

Peter K. Bol has argued that from the southern Song dynasty onward, the local level offered a legitimate realm in which a scholar could be active.[32] For Song the local was too unimportant to meet his ambitious aims. It was not by itself worthy of his attention. Many scholars who settled down in their locality in this time of crisis kept themselves busy with tedious administrative tasks. Yet there was a vast group of scholars who refused to serve the state at the end of the Ming, even on the local level. They could say they were expressing high morals by doing so, but first and foremost they were expressing their profound personal disappointment. They dressed their individual concerns in a cloak of self-vindication. The more influential intellectuals and politicians saw the low social standing, negligible political influence, and minor intellectual reputation of men like Song as nothing more than the inconsequential dull and turbid that stood in opposition to the bright and clear, the detritus of the system which a good spill of water and the course of history would wash away. And thus Song's self-identification and his colleagues' recognition of his position was not favorable to the reception of his works.Indeed the group of the "bright and clear" who overlooked Song's works encompassed not only those who stood at the core of social and political power in China's early seventeenth century, but also included those who stood up for their moral ideals and fought for them—the Ming loyalists of the late Ming and early Qing periods. Song was not one of these.

BY VIRTUE OF LOYALTY: MORAL OBLIGATIONS

> Heaven brings forth the five grains to nourish the people. [Heaven's] beauty is contained inside. This is the sense of the yellow robes [that dress the emperor].
>
> *Works of Heaven*, chap. 1, no. 4, 53a

One of the regular offerings that the Chinese emperors had to present to heaven during the performance of the dynasty's ancestral rituals was alcohol (*jiuli*). Sprinkling wine on the ground or pouring it into vessels, the emperors manifested their association with the heavens and thus their right to rule the Chinese empire. Brewed from rice or other grains with the addition of malt and yeast, what has come to be known as Chinese rice wine was technically beer. As Song mentioned in *Works of Heaven*, alcohol was indispensable to human life, despite its occasional propensity to cause social uproar. In Chinese classical writings such as the *Rites of Zhou*, the sage-kings of high antiquity had also authorized its use for medicinal purpose. Song stressed that the ancestors personally made (*zao*) alcohol, acknowledging its essential purpose. Song argued that his colleagues ignored this when they urged the prohibition of alcohol and were just thinking of it as a means for amusement.[33] When he explained in *Works of Heaven* the variety of yeasts and fermentation methods, he pointed to the importance of appropriate tools and seasonal variations. He listed ingredients and the addition of herbs to vary flavor and offered a detailed documentation on the production of red yeast (*qu*), which he deemed an important innovation. Song was not the only one of his dynastic era who realized the benefits of wine for society and state. Indeed the founder of the Ming, Zhu Yuanzhang, had shown that alcohol could be effectively employed for societal purposes. He intentionally promoted the habit of wine-drinking ceremonies as a method to enhance community feeling among his followers and servants. About two hundred years later Wang Yangming would follow Zhu Yuanzhang's example and endorse wine drinking as a method of reward and punishment within community covenants. Offenders had to drink excessively to punish themselves, whereas those who did good deeds were encouraged to honor their accomplishments by wine-drinking rituals.[34] And while excessiveness may have been taboo among elevated scholarly circles, high officials and merchants alike were often lenient about occasional drinking bouts, acknowledging that the bonds tied with a glass of wine in hand were stronger than those without and that it created loyalty where words could not convince.

Loyalty was an important issue by the late Ming as the societal order and political power that had unified the Chinese empire for almost three hundred years gradually broke down. Loyalty was also central to Song's personal value system and must be considered an important stimulus for his approach to crafts. Loyal to his state and to his duty as a scholar, Song tried to quell the chaos of his time. Yet, whereas others were loyal to the Ming, Song was concerned about the state (*guo*). He was a maverick among the community of scholars dedicated to the Ming, including his brother Song Yingsheng, whose passionate loyalty put him on the map of contemporary political history. Lacking political conviction, Song was also an outsider to the community of Ming loyalists, thus alienating another potential audience for his works.

Ideologically, loyalty in Ming times was intimately related to the tenet of the "mandate of heaven" (*tianming*) that authorized the emperor to rule the state. Sociopolitical theories identified legitimacy on the basis of *zhengtong*, "the moral right to succession" (*zheng*) and "unified political control" (*tong*). Only a ruling house that guaranteed both was authorized to rule China. Once a house had achieved this status, the scholar-official was morally obliged to serve his dynasty and obey its orders irrespective of its familial or ethnic origin. With loyalty and the legitimacy of a dynastic rule so closely intertwined, shifts in loyalty required careful apologia. As the mandate was conditional on the holder's ability to insure the people's welfare, it entailed a responsibility to renounce loyalty should a dynastic house disregard this task. In such cases, the dynasty was illegitimate and officials were allowed—or obliged—to withdraw their loyalty. Loyalty was not blind. It demanded moral, responsible behavior on the part of the ruler. It was subject to the idea of an absolute heaven and of cosmological structures such as the *qi* theory that interpreted the actions of men according to theoretical formulas and controlled them through an omnipotent heaven. In this regard, the withdrawal of loyalty could be justified as a reaction to a heavenly force.

When the Manchu tribes in 1636 declared themselves the Qing state, they directly attacked the moral legitimacy of the Ming's right to rule the Chinese sphere, thus calling into question the officials' loyalty toward their ruler. Significantly, the concept of legitimacy in premodern China did not rest on ethnic origin. Therefore barbarian regimes could not be simply dismissed on the basis of their non-Han ethnicity. With respect to morality, they theoretically could make as good a claim to legitimate rule as a native Chinese regime—their leadership was morally superior. Intellectuals in the twelfth century received a shock when the Jurchen Jin usurped half of the Song dynastic territory, that is the territory north of the Yangtze River, and thus challenged the legitimacy of the southern Song state by claiming themselves to be the sole legitimate rulers of "all under heaven" (*tianxia*).[35] Because a unified empire could not be achieved by either the southern Song (the role models for later Ming scholars) or the Jin dynasty, each argued that it was more morally entitled to rule. Ming rulers had no problems maintaining unified political control. Yet, as the officials struggled to work within the first emperor's autocratic system, moral claims became their preferred means to get a hearing in court and affect imperial politics. By the early seventeenth century, the situation had reached the point that unified control was defined as the result of moral behavior, rather than morality and unified rule each being legitimizing factors in themselves. Eventually morality alone dominated the discussion which then resulted in an essential dilemma whenever the "kernel covered by the yellow robe," the emperor, was "devoid of beauty," that is, did not meet the officials' moral standards. If legitimacy was ultimately based on morality and ethics, how could any official justify his loyalty toward the ruling house of the Ming? If we believe traditional historiography, most emperors could be denounced on these grounds in the eyes of upright officials—particularly toward the end of each dynasty. The historian Chu Hung-lam suggests that Ming scholars consequently developed a concept of a higher loyalty toward the cultural values represented by the institution of the dynasty.[36] Thus loyalty became not only detached from territorial unity, but also from the current ruling personality. The choice was subtle, because now officials could employ their loyalty as a means to criticize the emperor, calling for an idealized measure of benevolent rule. This concept of loyalty and the inherent notion of legitimacy meant devoted service could be replaced with justified periodic withdrawal or absolute abrogation of services if the current emperor proved to be morally inept. The Ming accepted this ideal, but most scholars did not recognize its incarnation in their contemporary dynastic rulers. In this sense Ming loyalty became a sociopolitical ambition uncoupled from the existence of the real state, long before the actual fall of the Ming and lasting well into Qing times.

Eleventh-century Song dynastic scholars had developed an equally abstract perception of loyalty based on the experience of a fragmented empire. The full scope of the Ming conception was, however, novel. The first Ming loyalist literary circles emerged at the end of the sixteenth century, dedicated to the resurrection of an idealized state under Ming rule. Scholarly groups such as the Restoration Society (*Fushe*) that evolved in the 1620s wanted to uphold dynastic morals. However harsh their political criticism, they were always committed to the continuation of the dynasty's rule. Scholars did not usually allow their personal disillusionment to make them enemies of the state. Song's elder brother, Song Yingsheng, is typical of this group of Ming loyalists.

As acknowledgment for his determined efforts, Yingsheng was offered an appointment as the magistrate of two districts (*fu xianling,* rank 7b), Tongxiang County in Zhejiang Province and Gaoliang County in Guangdong Province, which he accepted. After three years of faithful service, he was promoted to the position of prefect (*zhifu,* rank 4b) in Guangzhou in 1643. The local gazetteer of Enping of the Kangxi period (1662–1722) commented on the elder brother's high moral integrity, his attempts to support the underprivileged, and his interest in literature. This positive assessment of Yingsheng is not surprising as he himself edited and revised this local gazetteer during his period of service in Guangzhou Province. In 1638, he also published the *Fang yutang quanji* (*Complete Annotations of the Fangyu Halls*), candidly expressing his aversion to the Manchus and portraying himself as a fervent Ming loyalist. In a way ethnicity did play a role for Yingsheng. In his argumentation on the issue of a just state he assumed that the uncivilized tribes of the North were incapable of being just, whatever the circumstances. Shortly after the Ming dynasty was defeated by the Qing, Song Yingsheng fell ill and then, in a state of despair, committed suicide by poisoning himself a year after the Ming defeat in 1646.[37] His story indicates the enormous weight scholars of the late Ming placed on the issue of loyalty.

Yingsheng was not the only person in Song's environment who demonstrated a deep passion for the Ming state. Liu Tongsheng, a distant relative, fervently fought for the maintenance of the Ming dynasty in both the Donglin and Fushe factions. Song's peer Tu Shaokui also joined the Fushe political faction. Tu was fighting for the Ming and against the Manchus on many levels. He became quite popular during the late 1630s because of his efforts to exploit Jiangxi's mineral resources in order to finance a military offensive against Qing troops. Liu and Tu both honored Song's eldest brother as a Ming loyalist, but never recognized Yingxing in this way.[38] Was Song thus after all a Ming loyalist?

Song was definitely not in favor of the Manchu conquest. But a close look at Song's work reveals that he believed that heaven provided the same conditions for all dynasties and that only human behavior decided the course of events. This put responsibility for the well-being of the people and the state solely into the hands of man and his activities. Song thus anchored his loyalty to the authority of intelligent behavior and benevolent leadership. He was loyal to his state, but not necessarily loyal to the Ming, for the ideals he held were not based on historical phenomena. Most scholars of Song's generation still clung to the symbolic value of an idealized moral rule represented by the de facto but dissipated and vanquished Ming dynasty. Song thus developed a quite abstract view of loyalty which many of his colleagues may not have shared. By placing himself politically on the fence, Song may have scared off an audience that would have acknowledged and promoted his work: those who understood his motives for staying outside and engaging with obscure themes, not necessarily even though they did appreciate his interest in the mundane. This included the scholars who fostered and helped his brother, such as the Fushe and Donglin party members. Song's elder brother's works were banned under the zealous Manchu Emperor Qianlong (1711–99). The political implications of being banned kept Yingsheng's work and efforts alive in the historical context—a boon that *Works of Heaven* did not receive. The fact that Song refused to affiliate with oppositional groups during his life, and at the same time attacked those in power, offers another plausible reason why all his works were left without an audience. In a sense his indifference toward the cause of the Ming may have contributed to the neglect of its technological contents by his contemporaries. The combination of these factors may have resulted in Song's prophecy being fulfilled at least in his generation: His cries were not heard, his book was not read.

One factor can be excluded: *Works of Heaven* was not officially prosecuted or banned, as previous researchers have suggested. The assumption has been that the prosecution of the works of his brother, Yingsheng, affected the reception of Song's works throughout the following century. For the Qing, it would have been easy to kill two birds with one stone and have banned the work of both brothers. The Qing dynasty court was aware of the Song family affiliation to Ming loyalists. Song expressed his disaffiliation to the new Manchu rule of the Qing in occasional remarks in *Works of Heaven*. His poetry also contains some critical remarks toward northern tribes which, in other cases, gave Qing rulers sufficient reason to prosecute authors and prohibit the dissemination of their writings.[39] But the fact that Song's works reemerge in two imperial collections suggests that the Qing considered Song's political issues irrelevant; Song's damnation of the Ming and his personal interpretation of loyalty toward the state were of little interest to them. For the Qing, Song was just a Ming scholar in a minor position with a somewhat bizarre political opinion and ideological ideals. It was the technical contents of *Works of Heaven* that secured its survival.

AN ARTIFACT IN TRANSMISSION: THE EDITIONS OF THE *WORKS OF HEAVEN*

> Fearful that the rulers of the future might increase taxes unduly or become oppressive, and to ensure that those in charge of irrigation would follow the rules, the Great Yu had all these matters inscribed and cast into the tripod. This would be more durable than books, and future generations would act in accordance with the rules.
> *Works of Heaven,* chap. 2, no. 8, 17b

Chinese craftsmen employed various techniques to bind books in the seventeenth century. Scrolls of silk and paper had long replaced the bamboo strips of ancient times, the technology of which, according to Song, was completely misunderstood by his era. The popularization of newer binding methods had meant the end of the scroll and the birth of the folded-leaf book. Since the eleventh century the butterfly binding method (*hudie zhuang*) had played a pivotal role. For this material- and space-saving format, the craftsmen folded sheets of paper in half. Then he would apply paste to the folded edges and stack the sheets together so that the folded edges formed the spine of the book. As the end product resembled the open wings of a butterfly, it was given this rather descriptive name. The binding worked efficiently with woodblock printing, as it allowed two page openings to be printed from one block. By the fifteenth century, a new binding method reversed the folded leaf and the technique of stitched binding (*xian zhuang*) evolved and spread through the various centers of commercial printing. The packaging of the work to be transmitted was adjusted to the contents. The thickness of each volume (*ce*) varied in relation to the division of chapters (*juan*), and the cover cases (*han*) were designed in a purposeful harmonic relationship to the number of volumes. These technical developments went hand in hand with changing attitudes about how to deal with books as carriers of knowledge, their dissemination, and preservation.

Chinese scholars had always been deeply concerned about the integrity of their classic textual corpus, carving the written word into stone and casting it into bronze tripods to make sure that all contents were accurately transmitted to future generations. And yet, too often written records got lost. In transmission, books, and manuscripts were artifacts subject to transcription errors or inappropriate interpretations that confused a knowledge base that had to be recovered by careful philological study and thorough philosophical analysis.[40] By the mid-Ming the growing commercialization of printing enabled scholars to rectify such losses by compiling works from various sources and copying and printing them in their private studios. Scholarly interests included the repair of books and bindings, aiming at the preservation of books as artifacts. Many bibliophiles combined restoration with foresight and selectively collected contemporary manuscripts, out of personal interest as much as scholarly obligation. Even books and tracts with nonacademic content ended up on the bookshelves of private libraries and were thus preserved. As we have seen Song's oeuvre never became the object of such efforts. His various treatises lay dormant in his home, and *Works of Heaven* seems to have submerged into the grey shadows of Chinese seventeenth- and eighteenth-century book culture.

While much emphasis has been put on the imperial prosecution of books as part of the Qianlong emperor's huge collection campaign and compilation of the *Siku quanshu* (*Complete Library in Four Branches of Literature*, 1773–82), it looks as if an even larger percentage of the written legacy of late-Ming scholars fell victim to natural causes. Books were damaged by fire or were lost in the confusion at the end of the Ming. Numerous early Qing bibliophiles reported that their or their family's private libraries were lost in rivers when fleeing from warfare. Song's friend Chen Hongxu was one of these unfortunates.[41] In most cases nothing remained other than a number of collected volumes, or the library catalogue.

Despite the far-reaching effects of imperial literary prosecution, Lynn A. Struve has pointed out that few texts were actually lost. On the contrary, proscription often drew attention to a writing and helped preserve it. Struve argues, "the test of survival for a given work lay close to the author himself; his immediate progeny and close disciples, if he had any and his situation just before and soon after his death."[42] It also lay in the attitudes of the people consuming the growing number of books, collecting objects for their status value rather than pursuing their private interests. New studies on book culture have shown that many books in China were rarities and bibliophiles often enhanced this trend. Though printed, many compilations were almost as scarce as manuscripts, initially produced in small numbers and then copied on demand. The woodblock, when stored, provided an ever-present opportunity to produce more copies. As far as we know, the two different editions of *Works of Heaven*, one published in 1637 and the other perhaps in the 1650s, were printed in only the usual run of 50 or so copies.

The preamble of the first edition of *Works of Heaven* states that Song had the book printed in the fourth month of the year 1637. Only three copies of this first edition have been discovered (see appendix 3 for overview of editions). Although *Works of Heaven* does not appear in the booklists of bibliophiles and is never mentioned by contemporaries, it seems to have been circulated. One reason to believe that the first edition was distributed to more than just a few friends is the production of a second edition in the 1650s. The second edition of *Works of Heaven* was a revised edition, with the phrase "our dynasty" (*wochao*) updated to "Ming dynasty" (*Ming chao*). Its illustrations were of poorer quality.[43] A note in the two remaining imprints of the second edition of *Works of Heaven* mentions it was published by Yang Suqing (n.d.), a private publishing house (*shulin*) in Fujian Province, a center of publishing at the end of the Ming. The Jianyang printers published for commercial reasons and were "concentrating on books that could be produced cheaply and entailed less of a financial risk."[44] Thus the book must have been well enough known to have appealed to the publisher or his principal as a potential seller. The publisher also realized that the potential readership fell into more than one category. Thus, as the remaining two copies of the second edition show, he made two covers, one addressing people who wanted to make a living and the other addressing scholars. The sentence on the cover of one imprint of the second edition (fig. 8.1) reads: "Contained inside are all kinds of money-making and ever-useful professional secrets and essential instructions about agriculture, weaving manufacture, the collection and refining of metals, and precious stones (*neizai geng, zhizao, zuolian, caijin bao yiqie shengcai beiyong michuan yaojue*)." Above the outer frame another marketing slogan was added claiming all "exceptional skills at a single glance (*yijian qineng*)."[45] These slogans suggest that the printer was a smart businessman, doing his best to sell a shelf warmer.

Like other printers in the region, the Yang Suqing publishing house may have specialized in medical works, household encyclopedia, and works for amusement that were produced for commercial gain. Assuming that merchants had the same motivation then as now, the printing house would not print a book unless they expected to make a profit. The absence of these sentences in the first edition suggests that the printing house rather than Song initiated another carving. It also suggests that it was not Song's aim to use his knowledge about technology to obtain wealth or help others to do so. It is not possible to tell whether this marketing slogan was added to one woodblock or if two different woodblocks existed. Pan Jixing suggested that the publishing house might have used one cover for exhibition and sold the simpler version, since the two second-edition books still extant are identical.

Studies on print traders from Huzhou County and Jiangxi Province who

Fig. 8.1. Two different covers of the 1637 *Works of Heaven*, edition. Right: Version in storage of the Fu Ssu-nien library of the Institute of History and Philology (IHP) of the Academia Sinica, Taipei Call number: A 640 122 with marketing sentence in the middle bar. Left version held by Bibliothèque Nationale France without marketing sentence in the middle bar.

distributed their publications throughout China have shown that household encyclopedia, primers, educational guides, and novels produced in Fujian could spread throughout the entire sphere of Ming political influence. These traders often had family connections from Nanchang in Jiangxi Province in southern China to the famous book quarter of Beijing, Liulichang in northern China, and from Hangzhou on the eastern coast to Chengdu in Sichuan Province. They were listed in craft registers and known by name.[46] Lucille Chia, who compared information on these printing houses and their connections, discovered that the private publishing house of Yang Suqing published at least one other book, *Chunqiu Zuo Zhuan gangmu dingzhu* (Annotations and Remarks to the Spring and Autumn Commentaries of the Master Zuo Tradition, systematically rearranged) supposedly compiled by Li Tingji (1542–1616).[47] The Yang Suqing house was probably a minor branch of the great Yang publishing family whose activities spread across the entire country at this time. Some parts of their enterprise can be traced back to the Song dynasty and they continued to flourish until the late Qing.[48] Like the Zou and Ma family enterprises investigated by Cynthia J. Brokaw, the Yang were family merchants (*zushang*) with a regional distribution network and clusters of printing shops, each of which was established by a branch within a descent group.[49] Backed by a strong printing house, it appears Song's *Works of Heaven* could have been distributed much farther than previously assumed, subsisting in the twilight zone of Chinese seventeenth- and eighteenth-century book consumer culture.

The publisher's initiative was the first in a line of events that secured the survival of *Works of Heaven.* This survival is characterized by a shift in the perception of Song's work from a writing contextualized within other writings and the thought of its time to an individual item, a monograph documenting crafts and technologies. Whatever package *Works of Heaven* came in, it must have excited enough interest, albeit quizzical and inadvertent, to keep it on the market since it had reached Japan and France by the eighteenth century.[50]

WRITING ABOUT PRACTICAL KNOWLEDGE IN THE CHINESE LITERATI WORLD

Song's engagement with the material world was part of his generation's intellectual effort to cope with the increasing importance of material things and decreasing social and political stability. Looked at from a historical distance, it can be seen as an attempt to maintain and create traditions around these intellectual trends. Changing views on the *Works of Heaven* were acted out on a political stage where Manchu rulers were trying to consolidate their rule over a territory administered by scholars dedicated to the heritage of the preceding dynasty. While the scholarly community hovered between aggrandizing the past and adapting to the changing times, the most active Qing rulers Kangxi, Yongzheng (1678–1735), and Qianlong, decided to actively foster cultural construction, initiating large imperial collections of literary heritage. The compilers of the *Gujin tu shu jicheng* (*Complete Collection of Illustrations and Writings from the Earliest to Current Times,* finished 1725, printed 1726), for example, not only improved the work by replacing the original illustrations with lavishly decorative ones and by adjusting the technical contents and details to meet contemporary interests and taste. Under the supervision of Chen Menglei (1650–ca. 1741), and later Jiang Tingxi (1669–1733), mid-Qing scholars stripped away the ideological texts accompanying Song's technical descriptions. They carefully excluded his preface and the introductory remarks to each of the eighteen sections of the book, when they tried to place Song's work within the larger context of Chinese literature. Their attempts show the lack of contemporary concern for Song's ideological issues. In the larger context of Chinese literature on agriculture or agronomy, hydraulics or orthopraxy, these scholars obviously also struggled with Song's placing farming and crafts in one work. Hence they rearranged the sections of Song's books, associating Song's *Works of Heaven* to two prominent and influential schemes, that of the *Artificers Record* and agricultural writings (*nongshu*). With these rearrangements, the Qing scholars pointedly ignored that Song had originally refused to use these associations. They clearly disapproved of Song's attempt to redraw the epistemological and moral boundaries that existed in Chinese scholarly culture between farming and crafts.

Following traditional schemes, the governmental officials around the chief compiler Chen Menglei used the structure of the *Artificer's Record* to rearrange all technical contents they identified as nonagricultural into their encyclopaedic effort. Their approach not only followed traditional schemes, it was also reflective of a general interest in the *Artificer's Record* that had started in the Ming. In the early fourteenth century, this text had been made a convenient reference point in political negotiations in Ming state culture. Throughout the sixteenth and seventeenth centuries the *Artificer's Record* became part and parcel of discussions on pragmatic learning and practical statecraft. Politicians used references to the text to stress the importance of tools and techniques within state and society in general and for the control over such fields in particular.[51] Up until the sixteenth century scholars rarely saw this text in connection with the growing interest in the study of material inventiveness and their natural surroundings, a development that Elman identifies as a characteristic of later eighteenth-century interests in the *Artificer's Record.* (He connects this interest to the new knowledge received from the West through Jesuit and Protestant missionaries.) Indeed scholars discussed the *Artificer's Record* increasingly within the scope of new text-critical approaches, driven by a healthy mistrust of its historicity and interpretations. Compiling an imperial collection, the state nevertheless made use of the *Artificer's Record* as an authoritative point of reference to discuss the purpose of crafts for the state.

Eighteenth-century Chinese scholars approached *Works of Heaven* intellectually and politically quite selectively, either adjusting its structure to fit traditional lines of knowledge categorization, or purposefully employing its contents to suit their needs. The association with the *Artificer's Record* emphasized issues of statecraft; another line was to associate Song's work with agricultural writings. Although Song had not made this association, the Manchurian official Ortai (1680–1745) included Song's systematic descriptions of basic agricultural themes into the *Qingding Shoushi tongkao (Compendium of Works and Days)*, a huge imperial collection of agricultural writings in seventy-eight volumes published in 1742.[52] Ortai's choice of themes was very traditional: he ignored many tasks described in *Works of Heaven*, such as wagonry, or dyeing, which could have been associated and authorized loosely to the category of agricultural writing. The category of agricultural writings by that time had vastly extended boundaries, especially with regard to many subsidiary tasks such as oil pressing, cotton processing, or transportation. The expansion of recognized agricultural themes was a result of the gentleman scholars of the late Ming use of the association to authorize their interest in beer fermentation or botanical studies, promoting it as a task beneficial to state and society.

Qing scholars hence cannibalized rather than canonized Song's work. This indicates that the Chinese elite of this era was more interested in the book's documentary purpose than its ideological and political purpose. An additional factor that paved the way for V. K. Ting (Ding Wenjiang, 1887–1936) and other nineteenth-century scholars to integrate *Works of Heaven* into a narrative of modernization and nation was that Song's other writings had faded into obscurity. Although a librarian informed Ting about the existence of Song's *An Oppositionist's Deliberations, On Qi, Talking about Heaven*, and *Yearnings* in the early 1920s, Ting abstained from officially recognizing their existence in the 1930s, as did Joseph Needham (1900–1995) who was informed by a Chinese colleague about their existence in the late 1970s. The enactment of Song's work in the twentieth century followed the changing taste of its audience, highlighting issues that helped this era's concerns and ideals.

Song's approach made use of traditions and acknowledged structures to combine issues in a new way. Not only were his descriptions much more detailed than those in official reports, he included new fields of knowledge into premodern Chinese literature. By broadening our view of the way in which Song made and propagated knowledge throughout his oeuvre, this study has attempted to show that Song's effort went far beyond the purpose and aims of authors on agricultural writings. He was not a clerk who attempted to elevate his administrative tasks by reports on the institutional structures just as he was not a scholarly enthusiast who engaged in all kinds of details and oddities. Looking at Song's oeuvre brings the sociopolitical and epistemological issues that characterize his interests in crafts to the fore. His effort emerges as the complex and systematic, albeit idiosyncratic, effort by a Chinese scholar to explain his surroundings, of which technology and crafts were one important part. Song deliberately brought together the pressing of oil with the mundane task of ironing cloth and an inquiry into the sound accompanying a flying arrow. Scrutinizing the relationship between things and affairs, he discussed the casting of bells and statues on equal terms with everyday cooking pots. And thus he gave crafts and technology, material inventiveness in its full mundane range, a presence in writing that challenged his time. While his explanations of all these issues within a framework of *qi* assigns Song's oeuvre a specific place within Chinese thinking about natural phenomena, Song's methods of content organization, his view of the knowledge contained in crafts and behind things and affairs, together with his subtle combination of forms of inquiry distinguish his work within Chinese literature. It is within this context that we can see the reception history of *Works of Heaven* as one in which a part of a knowledge concept became detached from its greater whole and was then placed into traditions and made to fit fields which the author had avoided. This refitting (and misplacement) speaks of a change in reading tradition and in the reception of the epistemic content and purpose of Song's *Works of Heaven*.

To sum up, *Works of Heaven* was handed down as a monograph, losing its sense and function as one volume in a series. As a monograph it existed in the grey area of Chinese book culture over a period of more than a hundred years from 1637 to approximately the mid-eighteenth century when it was integrated into several imperial collections. The scholars of this period were primarily concerned with the technical contents of *Works of Heaven*. The imperial collections ripped it out of its context and rearranged it to fit their encyclopedic system. This rearrangement paved the way for the later nineteenth-century reading of *Works of Heaven* as a text on agricultural themes, statecraft, and technology. The absence of *Works of Heaven* on the list of banned books in combination with the fact that it was even incorporated into an imperial collection suggests that the Manchu rulers thought the benefit his documentation offered to them outweighed his ineffectual political stance and any conceivable threat his writings might pose. It is unlikely that they overlooked Song's critical remarks about northerners entirely, as they did prosecute his brother. The incorporation of *Works of Heaven* into an imperial collection was due to its technical contents. Possibly it also resulted from the sheer fact that Song had his work illustrated with subjects that were not depicted elsewhere. Whatever kept Chinese scholars of Song's era away from *Works of Heaven* seems not to have bothered the Manchus and their Chinese allies in the eighteenth century. Three copies of the book's first edition survived in private hands in China (and two of them had left the country through missionaries and traders by the eighteenth century). This fact alone shows that there was a subculture of bookish interests in premodern China of which we know practically nothing. Its peculiar history of transmission, despite being the product of general historical traits, suggests that *Works of Heaven* embodied an irritation not ultimately dependent on its technical content and not apparent once the parts were taken out of context or looked at from a temporal distance. This irritation lay in Song's refusal to sustain the world of delusions and depravity in which he lived, a world "made by man." In his oeuvre we see *Works of Heaven* emerging in its original form: a part of his urgent call to bring knowledge into congruence with one's action by looking at the truth contained in the making of things and affairs alike. Later generations found the things and affairs eventually more worthy of preservation than Song's truth and thus preserved *Works of Heaven* despite its deviations from their ideals or the idiosyncrasy they identified in its original structure. In its reception history, Song's writings reveal not only the complexity of knowledge-making in the seventeenth-century Chinese intellectual world, it illuminates some traits of a distinctive history of knowledge transmission.

ACKNOWLEDGMENTS

Like any artifact, a book is often deemed the work of an individual, but in fact it is the product of many elements working together. Selecting the raw materials, sketching the rough outline and filling in individual lines and color, I incurred a number of personal and professional debts and relied on the knowledge, patience, and kindness of many colleagues. First there are those who provided the major techniques, tools, and attitudes to the apprentice at various stages. Dieter Kuhn was the one who drummed up my enthusiasm for Chinese studies and this project in particular. He gave me room to grow. It was Nathan Sivin who, at a crucial point, encouraged me later in my career to find my own way. Francesca Bray, her work and her sharp mind, provided so many stimuli to me.

In my search for the raw materials Professor Emeritus Pan Jixing became my mentor in China. He is one of the most accommodating and generous scholars I have ever met in my career, sharing his notes and patiently discussing interpretations. The same is true for Professor Dai Nianzu, who always warmly welcomed me with a cup of tea and an invitation to superbly delicious dinners to keep up my spirits. As I sculpted the basic form, I profited immeasurably from conversations with my colleagues, too many to all be named here: Michael Leibold, and Anne Gerritsen invested much thought on earlier versions of the manuscript. Mareile Flitsch and Dorothy Ko I thank for making me break the mold. Lucille Jia with enormous kindness added a crucial piece. Fu Daiwie, Chu Pingyi, Michael Puett, Hans Ulrich Vogel, Donald Wagner, Huang Yinong, and Ben Elman were always willing to discuss one or the other aspect of my work. Ruth Schwartz Cowan and her colleagues graciously welcomed me into the thriving atmosphere of the Department of History and Sociology of Science at the University of Pennsylvania. Once I relocated to the Max Planck Institute for the History of Science in Berlin and started to finalize the details, I was fortunate enough to receive immeasurable input from Jürgen Renn, Matthias Schemmel, William Boltz, Peter Damerow, Marcus Popplow, Martin Hofmann, and Matteo Valeriani. For the staging I received much incentive from Ursula Klein, Wolfgang Levefre, and Lorraine Daston. I thank my colleagues from the Institute for the History of Natural Sciences, Zhang Baichun and Tian Miao as well as Sun Xiaochun and his group, and Su Rongyu, for their serenity, support, and their exceptional hospitality during this period. With the tender care of a best friend and the erudition of an experienced scholar, Martina Siebert offered her critique of the manuscript at various stages. Gina Partridge Grzimek, who also joined my group in 2006, was a most dedicated reader and improved the flow and logic of the narrative immeasurably. Anna, Wan, Zhe, and Wolfgang provided the backroom support, and Falk did a great job on mapping Song Yingxing's journeys. The insightful remarks of two unknown reviewers from the University of Chicago Press gave the book its last, essential push into the final direction.

Several institutions provided the fuel (i.e. finances) necessary to keep my workshop warm for this research. The German Research Foundation (DFG) promoted the project initially, then provided additional funds to enable research stays in the Peoples' Republic of China as well as field research and an extended stay at the University of Pennsylvania. The University of Würzburg granted money for my research in Hsinchu, Taiwan. The funding project of the Bavarian Ministry of Science for outstanding talents as well as the female research scholarship initiative (HWP) also provided travel money for conferences, visits to China and Taiwan, and access to archives and libraries. The generosity of the Max Planck Society finally enabled the scope of this research to be expanded to the broader field of the history of science and technology. Without the special atmosphere of the Max Planck Institute for the History of Science, this work could not have been completed.

Laura, Leonie, and Noah are possibly those to whom I owe the biggest debt, as they now have to live with the memories of a youth veined with remarks on Song Yingxing and his world. Growing into most joyful and exuberant teenagers, they love to take advantage of a mother lost in thoughts on early China. And of course my husband Horst, who has shared all of this with me. Glad I found you.

Karen Darling has enthusiastically supported the project from its inception and painstakingly and tactfully helped it on its way. Jean Eckenfels did a great job of copyediting and Mary Gehl skillfully oversaw the process from manuscript to book. The Shanghai publishing house generously allowed original illustrations to be included and the Art Institute of Chicago (thanks to Elinor Pearlstein in particular for her generosity) and the Worcester Art Museum granted permission for the reproductions of the scrolls. Despite all this great help, all interpretations and conclusions and any blunders or imperfections are entirely my own.

APPENDIX I

Chinese Dynasties and Various Rulers

THE SAGE RULERS

Fuxi 伏羲
Shennong 神農
Huangdi 黃帝
Yao 堯
Shun 舜
Yü 虞

THE THREE ERAS OF ANTIQUITY

Xia 夏 (trad. 2205–767 BC)
Shang 商 (16th–1st century BC)
Zhou 周 (1045–21 BC)

DYNASTIES

Qin 秦 (221–07 BC)
Han 漢 (206 BC–AD 220)
  Western Han 前漢/ 西漢 (206 BC–AD 8)
  Wang Mang 王莽 (AD 9–23)
  Eastern Han 候漢/東漢 (AD 9–220)
Three Kingdoms
  Sanguo 三國 (AD 220–280)
Period of Division
  Nanbeichao 南北朝 (AD 316–589)
Sui 隨 (AD 581–618)
Tang 唐 (AD 618–907)
Five Dynasties and Ten Kingdoms
  Wudai shiguo 五代十國 (AD 907–960)

Song 宋 (AD 960–1127)
 Northern Song 北宋 (AD 960–1127)
 Southern Song 南宋 (AD 1127–1279)
Liao 遼 (AD 916–1125)
Jin 晉 (AD 1115–1234)
Yuan 元 (AD 1271–1368)

MING DYNASTY 明 REIGN PERIODS (1368–1644)

Taizu	太組	HONGWU	洪武	1368–1398
Huidi	惠帝	JIANWEN	建文	1399–1402
Chengzu	成組	YONGLE	永樂-	1403–1424
Renzong	仁宗	HONGXI	洪熙	1425
Xuanzong	宣宗	XUANDE	宣德	1426–1435
Yingzong	英宗	ZHENGTONG	正統	1436–1449
Yingzong	英宗	TIANSHUN	天順	1457–1464
Jingdi	景帝	JINGTAI	景泰	1450–1456
Xianzong	憲宗	CHENGHUA	成化	1465–1487
Xiaozong	孝宗	HONGZHI	弘治	1488–1505
Wuzong	武宗	ZHENGDE	正德	1506–1521
Shizong	世宗	JIAJING	嘉景	1522–1566
Muzong	穆宗	LONGQING	隆慶	1567–1572
Shenzong	神宗	WANLI	萬歷	1573–1620
Guangzong	光宗	TAICHANG	泰昌	1620
Xizong	熹宗	TIANQI	天啟	1621–1627
Sizong	賜宗	CHONGZHEN	崇禎	1628–1644

Qing Dynasty (1644–1911)
Republican China (1911–1949)
People's Republic of China (1949–)

APPENDIX 2

Song Yingxing Curriculum Vitae

Year	Age	Career	Reign
1587	Birth		Ming Emperor Shenzong 神宗
1593	7	traditional Confucian education	
1611	25		
1615	29	2d Degree Graduate	
1616	30	1st Attempt Metropolitan Exam	Hou Jin
1619	33	2d Attempt	Ming Emperor Guangzong 光宗
1622	36	3rd Attempt	Ming Emperor Xizong 熹宗
1625	39	4th Attempt	
1628	42	5th Attempt	Ming Emperor Zhuanglie 莊烈
1629	43	Father Song Guolin 宋國林 died	
1631	45	6th attempt (?)	
1632	46	Mother died	
1634	48	Teacher at the County School Fenyi 分宜	
1636	50	First publication, literary activity	Dynasty Proclamation of Qing
1637	51	Tiangong kaiwu 天工開物	
1640	54		
1642	56	Judge (*tuiguan* 推官)	
1645	59	Brother Yingsheng 應升 died	Qing Emperor Shizu 世祖
1662			Qing Emperor Shengzu 聖祖
1666?	80?	Died	

APPENDIX 3

Editions of the *Tiangong kaiwu* 天工開物

1637	涂本 *tuben*-Edition, in 3 *ce* 冊, printed in Fengxin or Nanchang, 3 copies stored in Beijing National Library, Tokyō National Storage, Paris Bibliothéque Nationale. Printed on Bamboo paper.
1640–1680	1. 揚本 *Yangben* 2. 揚所本 *Yangsuoben* Identical editions except for the title page by the *Yangsuo* publishing house in 3 *ce* 冊, printed in Fujian Province, 4 facsimile in Beijing (*yangben*), Taibei (*yangsuoben*), Paris (*yangben*), Tokyō (*yangsuoben*). Printed on bamboo-paper.
1771	管本 *guanben*-Edition, in 3, 6 or 9 *ce* 冊, Japanese facsimile, reprints on the basis of the *tuben* edition, various copies stored in Beijing and Japan.
1830	facsimile reprint of the *guanben* edition in Japan.
1927	陶本 *taoben* reprint of the *guanben* edition in China, by V. K. Ting, with woodblock illustrations from the *Gujin tushu jicheng* edition, in 3 *ce*.
1929	facsimile reprint of the *taoben* edition.
1930	通本 *tongben*, reprint of the *guanben*.
1933	商本 *shangben*, Shanghai: Shangwu yinshuguan in Guoxue jiben congshu and in the Wanyou Wenku. (Reprinted in 1954, same publishing house).
1936	Shanghai: Shijie shuju. *taoben* with corrections and illustrations, including V. K. Ting's Biography of Song.
1943	Saigusa Hiroto, 三枝博音 *et al.*, *Tenko kaibutsu* 天工開物 [*Tiangong kaiwu*]. Tokyo: Juichigumi shuppanbu. On the basis of the *Guanben*.

1953	Yabuuti Kiyoshi, 藪内清 *et al..*, *Tenkō kaibutsu no kenkyū*天工開物の研究 [*An Investigation of the Tiangong kaiwu*]. Tokyo: Kōseisha.
1959	Beijing tushuguan 北京圖書館. Shanghai: Zhonghua shuju. facsimile reprint of the *tuben* edition.
1965	Taibei: Shijie chubanshe. *Jiaozheng Tiangong kaiwu* 校正天工開物 [*Corrected Edition of the Tiangong kaiwu*]. Facsimile reprint in 1971 and 1981 based on *taoben* (?) edition.
1969	Yabuuti Kiyoshi, 藪内清 (transl.), *Tenkō kaibuku no kenkyu* 天工開物 [*Tiangong kaiwu*]. Tokyo: 1969 [reprinted in the year 1984]. On the basis of the *tuben* print.
1976	Guangzhou zhongshan daxue 廣州中山大學, Zhong Guangyan 鍾廣言 (transl. into Modern Chinese), *Tiangong kaiwu* 天工開物. Beijing: Zhonghua shuju. On the basis of the *tuben* print, *tuben* illustrations.
1980	Li Ch'iao-p'ing *et al.*, *T'ien-kung k'ai-wu, Exploitation of the Works of Nature.* Taibei: China Academy 1980. On the basis of the *tuben* print, illustrations of the *taoben* edition.
1987	Liu Juncan 劉君燦 (ed.), *Tiangong kaiwui daodu* - 天工開物導讀, [*A Reader's Guide to the Tiangong kaiwu, 2 Volumes*], Taibei: Jinpei chubanshe.
1988	Zhong Guangyan 鍾廣言 (transl. into Modern Chinese), *Tiangong kaiwu* 天工開物. Beijing: Zhonghua shuju [reprint 1989]. See also 1976 On the basis of the *tuben* edition.
1988	Pan Jixing 潘吉星, *Tiangong kaiwu daodu* 天工開物導讀 [*A Reader's Guide to the Tiangong kaiwu*]. Chengdu: Bashu shushe [Zhonghua wenhua yaoji daodu congshu].
1989	Pan Jixing 潘吉星, *Tiangong kaiwu jiaozhu ji yanjiu* 天工開物校注及研究 [*Critical Edition of the Tiangong kaiwu with a study*]. Chengdu: Bashu shushe. Reference to various differences between all available original editions, illustrations of the *tuben*.
1993	Pan Jixing 潘吉星, *Tiangong kaiwu yizhu* 天工開物譯注 [*An Annotated Translation of* the Tiangong kaiwu into Modern Chinese]. Shanghai: Guji chubanshe 1993.

NOTES

INTRODUCTION

1. Lu Jiying, ed., *Quanbu da Yan Song; you ming, Kaishan fu: Gailiang jingxiben* [A Complete Version of Beating Yan Song; or, Kaishan Prefecture: Corrected Beijing Opera Libretto].

2. William Dolby, "The Origins of Chinese Puppetry," *Bulletin of the School of Oriental and African Studies* 41, no. 1: 98.

3. Song Yingxing, *Huayin guizheng* [A Return to Orthodoxy] (lost), *Silian Shi* [Yearnings], *Ye Yi* [An Oppositionist's Deliberations], *Tan Tian* [Talks about Heaven], and *Lun Qi* [On Qi], in *Ye Yi; Lun Qi; Tan Tian; Silian Shi* [An Oppositionist's Deliberations; On Qi; Talks about Heaven; Yearnings]. I collated the 1976 compilation with the originals now stored in Nanchang, Jiangxi Province Library. Song Yingxing, *Tiangong kaiwu* [The Works of Heaven and the Inception of Things]. Full list of *Works of Heaven* can be found in appendix 3.

4. For an overview of this literature see Pamela H. Smith, *The Body of the Artisan: Art and Experience in the Scientific Revolution*, 18.

5. Lissa Roberts, Simon Schaffer, and Peter Dear, eds., *The Mindful Hand: Inquiry and Invention from the Late Renaissance to Early Industrialisation*, xix. For an overview of how the cultural conquered the history of science field, see Dear, "Cultural History of Science: An Overview with Reflections," *Science Technology Human Values* 20: 150–70.

6. Nathan Sivin, "Why the Scientific Revolution Did Not Take Place in China—Or Didn't It?: The Edward H. Hume Lecture, Yale University," *Chinese Science* 5: 45–66.

7. Smith, *Body of the Artisan*, 18.

8. See the writings of Francesca Bray on these themes, in particular, *Technology and Gender: Fabrics of Power in Late Imperial China*.

9. Michael J. Puett, *The Ambivalence of Creation: Debates Concerning Innovation and Artifice in Early China*, delineates these discussions for the early period around the third century.

10. Qiu Jun, *Daxue yanyi bu* [Extended Meaning of the Great Learning], chaps. 143–44; Chu Hung-lam, "Ch'iu Chün's Ta-hsüeh yen-i-pu and Its Influence in the Sixteenth and Seventeenth Centuries," *Ming Studies* 22: 1–32.

11. Smith, *Body of the Artisan*, 59–95.

12. The Ming history contains various biographies of carpenters and building masters (i.e., architects) rising in high ranks during the palace construction in Beijing of the *yongle* period, for example, Zhang Tingyu et al., *Ming shi* [Ming History], chap. 151, Biography of Liu Guan (*jinshi* 1384), vol. 14, 4184–85, chap. 160, Biography of Zhang Peng (*jinshi* 1451), 4367–69.

13. Martin J. Powers, *Pattern and Person: Ornament, Society, and Self in Classical China*, 27–64.

14. Qiu, *Extended Meaning of the Great Learning*, chap. 122, 2a–16b, 12a/b, chap. 25, 16b, chap. 78, 7a, chap. 97, 4a.

15. Qian Chaochen, *Wang Qingren yanjiu jicheng* [Anthology of Studies of Wang Qingren]. Another example is Wan Quan (1500–1585?), who compiled a compendium of case studies. Barbara Volkmar, *Die Fallgeschichten des Arztes Wan Quan (1500–1585?): Medizinisches Denken und Handeln in der Ming Zeit.*

16. Martina Siebert, *Pulu: Abhandlungen und Auflistungen zu materieller Kultur und Naturkunde im traditionellen China*, Opera Sinologica 17.

17. Kai-wing Chow, "Writing for Success: Printing, Examinations, and Intellectual Change in Late Ming China," *Late Imperial China* 17, no. 1: 126.

18. Antonio S. Cua, *The Unity of Knowledge and Action: A Study in Wang Yangming's Moral Psychology.*

19. Ge Rongjin, ed. *Zhongguo shixue sixiangshi* [A History of the Ideas of Practical Learning in China], 10.

20. Wang Fuzhi, *Chuanshan quanji* [Complete Writings of Master Quanshan], vol. 13, *Saoshou wen* [Questions of a Head-Scratcher], 9877–9906. This tract has been translated by Alison Harley Black, *Man and Nature in the Philosophical Thought of Wang Fu-chih*, 166–67.

21. The *Comprehensive Glossary* is contained in Fang Yizhi, *Fang Yizhi quanshu* [Complete Works of Fang Yizhi], 1:40–41. All translations in the following are mine except where noted. I used the 1785 facsimile of Fang Yizhi, *Wuli xiao zhi* [Notes on the Principle of Things], 12 vols. For Fang's life see also Willard J. Peterson, *Bitter Gourd: Fang I-chih and the Impetus for Intellectual Change.*

22. Fang refered to a passage in the *Yijing* [Book of Changes] of the second century BC, also known as *Zhou Yi*, edition of *Sishu wujing* [The Four Books and the Five Classics], suppl. *xici shang* [The Great Appendix, Part 1], 57, line 1b: "Only through thoroughness (*shen*) can [the sage] permeate (*tong*) the will under heaven (*tianxia zhi zhi*); only by [paying attention to] the minutest (*ji*) can he complete all matters under the heaven (*tianxi zhi wu*)."

23. *Shujing* [Book of Documents], edition of *Sishu wujing*, [The Four Books and the Five Classics], vol. 1: chap. *gao tao mo*, 16, lines 5–7.

24. *Yijing* [Book of Changes], chap. *Xici shang*, 56. The term "affairs" (*wu*) connotes an obligation, duty, or task, however, not necessarily a human agency.

25. *Shujing* [Book of Documents], chap. 1, 16a, commentary of Cai Shen. James Legge, ed. and trans., *The Chinese Classics: With a Translation, Critical and Exegetical Notes, Prolegomena, and Copious Indexes*, vol. 1, 70.

26. *Shujing*, chap. 1, 16b.

CHAPTER 1

1. Song Yingxing, *An Oppositionist's Deliberations,* 3.

2. For an overview to such issues see Roberts, Schaffer, and Dear, *Mindful Hand,* 2–10.

3. Liu Jun, Mo Fushan, and Wu Yazhi, *Zhongguo gudai de jiu yu yinjiu* [China's Ancient Wine and Wine-Drinking], 189–205, collected and annotated various historical examples. Chang Kang-I Sun, *The Late-Ming Poet Ch'en Tzu-lung: Crises of Love and Loyalism.*

4. Ray Huang, *1587, A Year of No Significance.*

5. William T. Rowe, "Success Stories: Lineage and Elite Status in Hanyang County, Hubei, c. 1368–1949," in *Chinese Local Elites and Patterns of Dominance,* ed. Joseph W. Esherick and Mary Backus Rankin, 51–81, emphasizes that practices of the northern regions varied slightly from those of the southern.

6. Song Liquan and Song Yude, *Baxiu Xinwu Ya qi Song shi zongpu* [Eighth Supplement to the Family Annals of the House of Song from Xinwu], chap. 22, 33, 43–44. Pan Jixing, *Song Yingxing pingzhuan* [A Biography of Song Yingxing], 70–72, 75, 182–86, 191–93.

7. Zhao Jie, "Ties That Bind: The Craft of Political Networking in Late Ming Chiangnan," *T'oung Pao* 86, nos. 1–3: 141.

8. Ikoma Shō, "Minsho kakyo gōkaku sha no shusshin ni kan suru ichi kōsatsu" [A Study of Birthplaces of Successful Examination Candidates during the Early Ming], in *Yamane Yukio kyōju taikyū kinen Mindaishi ronshū jō* [Publication in Honoriam of the Retirement of Yamane Yukio, Discussions about the Ming Dynasty], ed. Yamane Yukio and Mindaishi Kenkyūkai, 48.

9. John W. Dardess, *A Ming Society: T'ai-ho County, Kiangsi, Fourteenth to Seventeenth Centuries.*

10.Lü Maoxian and Shuai Fangwei, *Fengxin xianzhi* [Local Gazetteer of Fengxin], chap. *renwu zhi,* 5a–7b; Li Yinqing, Xia Congding, and Yan Shengwei, *Tongzhi Fenyi xianzhi* [Gazetteer of Fenyi county] (Jiangxi, 1871), chap. 6 (*zhiguan, wenzhi, ming zhi xian),* 6, 23b.

11. Joseph Dennis, "Between Lineage and State: Extended Family and Gazetteer Compilation in Xinchang County," *Ming Studies* 45–46: 71.

12. Song Liquan and Song Yude, *Family Annals of the House of Song,* chap. 22, pp. 24–25.

13. Keith Hazelton, "Patrilines and the Development of Localized Lineages: The Wu of Hsiu-ning City, Hui-chou, to 1528," in *Kinship Organization in Late Imperial China, 1400–1900,* ed. Patricia B. Ebrey and James L. Watson, 150–51; Kai-wing Chow, *The Rise of Confucian Ritualism in Late Imperial China: Ethics, Classics, and Lineage Discourse,* 76–79.

14. Stevan Harrell, "On the Holes in Chinese Genealogies," *Late Imperial China* 8, no. 2: 54.

15. Pan Jixing, *Biography of Song Yingxing,* 150; for a discussion on dating see 175–80.

16. Zhao Zhongwei, "Chinese Genealogies as a Source for Demographic Research: A Further Assessment of Their Reliability and Biases," *Population Studies* 55, no. 2: 183.

17. William T. Rowe, *Saving the World: Chen Hongmou and Elite Consciousness in Eighteenth-Century China,* 231–34.

18. Peter K. Bol, *Intellectual Transitions in T'ang and Sung China,* 72.

19. Song Liquan and Song Yude, *Family Annals of the House of Song*, chap. 22, 70.

20. Zhang et al., *Ming History,* 26: chap. 196, 7915; Pan, *Biography of Song Yingxing,* 76.

21. Song Liquan and Song Yude, *Family Annals of the House of Song*, chap. 17, 27, chap. 22, 71. Parts of the text can also be found in Song Yingsheng, *Fangyu tang ji* [Collection of the Fangyu Hall], microfilm, chap. 11, 1–6. Original stored at Fengxin,Yaxi.

22. Frederic E. Wakeman, Jr., *The Fall of Imperial China,* 23.

23. Pan, *Biography of Song Yingxing,* 106–7; Song Liquan and Song Yude, *Family Annals of the House of Song,* chap. 17, 27a.

24. Benjamin A. Elman, *A Cultural History of Civil Examination in Late Imperial China,* 380–420.

25. Song Yingxing, *An Oppositionist's Deliberations,* 7.

26. Frederic E. Wakeman, Jr., "China and the Seventeenth-Century Crisis," *Late Imperial China* 7, no. 1: 3, 12. At the end of the Ming era *jinshi* were even 2 to 10 years older. Ann Waltner, "Review Essay: Building on the Ladder of Success: The Ladder of Success in Imperial China and Recent Work on Social Mobility," *Ming Studies* 17: 33.

27. Lü Maoxian and Shuai Fangwei, *Local Gazetteer of Fengxin*, chap. *renwu zhi*, 5a–7b; Song Liquan and Song Yude, *Family Annals of the House of Song,* chap. 5, 113–14.

28. Mao Deqi, *Bailu shuyuan zhi* [On the White Deer Academy]; Deng Hongbo et al., *Zhongguo shuyuan zhangcheng* [Regulations of Academies in China], 144–67.

29. Xu Xiake compiled his travelogues while being on this journey in 1642. Xu Hongzu, *Xu Xiake youji* [Travelogues of Xu Xiake], ed. Chu Shaotang and Wu Yingshou, 144, 160; Julian Ward, *Xu Xiake (1587–1641): The Art of Travel Writing,* 170–73.

30. James W. Tong, *Disorder under Heaven: Collective Violence in the Ming Dynasty,* 45–55.

31. Liam Matthew Brockey, *Journey to the East: The Jesuit Mission to China, 1579–1724,* chap. 1, 50.

32. John W. Dardess, *Blood and History in China: The Donglin Faction and Its Repression, 1620–1627,* 31–71, 126–41. See also Benjamin A. Elman, "Imperial Politics and Confucian Societies in Late Imperial China: The Hanlin and Donglin Academies," *Modern China* 15, 4: 379–418 for the great impact of these discussions on scholarly culture.

33. Chang Fuyuan, *Li Zicheng Shan bei shishi yanjiu* [Research on Li Zicheng and the Historical Affairs in Northern Shanxi], 32.

34. *Lunyu* [Analects], 2:14, 16:6; Liu Baonan, Liu Gongmian, and Song Xiangfeng, eds., *Lunyu zhengyi* [Annotated Commentary to the Analects]; Chu Hsi and Lü Tsu-ch'ien [Lü Ziqian], eds., *Reflections on Things at Hand: The Neo-Confucian Anthology,* trans. Wing-Tsit Chan, 94–102; Richard John Lufrano, *Honorable Merchants: Commerce and Self-Cultivation in Late Imperial China,* 57–58, 112.

35. Liu Tongsheng, *Jinlin shiji* [Anthology of Poetry by Jin Lin], vol. 8, chap. 13, 5b.

36. Liu Zepu and Gao Bojiu, eds., *Bozhou zhi* [Local Gazetteer of Bozhou], chap. *guanwu,* 13b.

37. Zhang et al., *Ming History,* 22: chap. 258, 6672–73 describes the event summarizing the memorandum. A more elaborate version can be found in Lu Shiyi, "Ming ji Fushe

jilüe" [A Report about the Restoration Society], in *Zhongguo yeshi jicui* [Anthology of Superior Unofficial Histories of China], ed. Chen Li, 1: 540–600; Zheng Da, ed., *Yeshi wuwen* [Unofficial History Without Documents], (1962), 1: chap. 3 *Lie Huangdi yishi*, 3; see also Xu Zi, *Xiaotian jizhuan* [Biographies of an Era of Little Prosperity], 2: chap. 56 *liechuan* 49, 6a.

38. Song Yingxing, *An Oppositionist's Deliberations*, 3.

39. The event found entry into the official biographies of Zhan Erxuan (*jinshi* 1631) and Zhang Zhifa (?–1642). Zhang et al., *Ming History*, 22: chap. 258, 6672–73, as well as the biography of Jiang Yueguang in Zhang et al., *Ming History*, 23: chap. 274, 7029–31; Wen Tiren et al., "Huaizong shilu" [Veritable Records of the Reign of Emperor of the Emperor Huaizong], in *Ming shilu* [Veritable Records of the Ming Dynasty], chap. 9, 231; Tan Qian, *Guo que* [Deliberations about the State], chap. 95, 5727.

40. Donald Potter, "[Biography of] Wen T'i-jen," in *Dictionary of Ming Biography 1368–1644*, ed. L.C. Goodrich and Fang Chaoying, 1: 1474–78; Zhang et al., *Ming History*, 22: chap. 258, 6666, 26: chap. 308, 7936.

41. Qi Jiguang published in 1561 *Jixiao jinshu* [A Treatise on Military Training]. James Ferguson Millinger, "Ch'i Chi-kuang Chinese Military Official: A Study of Civil-Military Roles and Relations in the Career of a Sixteenth Century Warrior," 14–19; Joanna Waley-Cohen, *The Culture of War in China: Empire and the Military under the Qing Dynasty*, 89, distinguishes the Ming in contrast to the martial (*wu*) Qing as a rulership of "being civilized" (*wen*).

42. Zhang et al., *Ming History*, 22: chap. 258, 6666–68, 6672–74.

43. Song Yingxing, *An Oppositionist's Deliberations*, 3.

44. Ibid.

45. Chen Hongxu, *Chen Shiye xiansheng ji* [Collected Works of Master Chen Shiye], *juan* 2, 35–36.

46. Pan, *Biography of Song Yingxing*, 147–48, 232–35 See also Pan Jixing, *Tiangong kaiwu jiaozhu yu yanjiu* [Critical Edition of the Tiangong kaiwu, with a study], 136–50.

47. Song Yingxing, *Works of Heaven*, preface, 3a–b.

48. Siebert, *Pulu*, 9, 16.

49. A contemporary example for this genre is Luo Qi, *Wu yuan* [Origin of Things], Congshu jicheng jianbian 64 who praised the sage kings as those who initiated the "inception of things and construction of tools" (*kaiwu zhi qi*).

50. Song Yingxing, *Works of Heaven*, preface, 3b. For an English translation see Sun E-Tu Zen and Sun Shiou-chuan, trans., *Chinese Technology in the Seventeenth Century: T'ien-kung k'ai-wu*, xii.

51. Pan, *Biography of Song Yingxing*, 145–46, 266–67 suggests that Song later decided to incorporate the tracts into another (unpublished and lost) writing.

52. Fu Daiwie, "When Shen Kuo Encountered the 'Natural World,'" 3; Fu Daiwie, "A Contextual and Taxonomic Study of the 'Divine Marvels' and 'Strange Occurrences' in the Mengxi bitan," *Chinese Science* 11: 3–5.

53. Ming scholars referred to Song dynastic categories of Chinese *leishu* (such as the "Imperial Readings of the Taiping Era" [Taiping yulan] by Li Fang 1223–96), but also considered the Tang dynastic classification system. The categories mentioned by Fu Daiwie

slightly vary from those of Tang works such as the *Yiwen leiju* [Collection of Literature arranged in categories]. Hu Daojing, *Zhongguo gudai de leishu* [Encyclopedia of Ancient China], chaps. 4 and 5. For a major Western approach to the taxonomy of knowledge, see Michel Foucault, *The Order of Things.*

54. See the taxonomic division of "gateways, doors" (*men*) in Hu Daojing's edition of Shen Gua, *Mengxi bitan jiaozheng* [Brush Talks from the Dream Brook], 1. Hu's edition is based on the *Yangzhou zhouxue kanben*-edition by Tang Xiunan of the year 1166. Nathan Sivin, "Shen Kua," in *Dictionary of Scientific Biography,* 369–93; Sivin, "Why the Scientific Revolution Did Not Take Place in China—Or Didn't It?" 45–66.

55. Liu Yeqiu, *Lidai biji gaishu* [Survey of the Private Jottings of Past Dynasties], 185–94.

56. Zhang Hui. *Songdai biji yanjiu,* 1–2.

57. Fu, "A Contextual and Taxonomic Study," 5.

58. Song Yingxing, *Talks about Heaven,* chap. *ri shuo* 3, 105.

CHAPTER 2

1. This passage paraphrases and quotes from two of Yingxing's works: Information on boats is from *Works of Heaven,* chap. 2, no. 9, 29 a/b. Yingxing elaborates on chaos and order in his *An Oppositionist's Deliberations,* chap. *shiyun yi,* 5, chap. *luanmeng,* 44.

2. See Kidder Smith, "Sima Tan and the Invention of Daoism, 'Legalism,' 'et cetera,'" *Journal of Asian Studies* 62, no. 1: 129–56, for the difference between the contemporary view and the historical construction of isms.

3. For the changing views on nature in European thought during the same period, see Lorraine Daston, "The Nature of Nature in Early Modern Europe," *Configurations* 6: 149–72; Lorraine Daston and Katharine Park, *Wonders and the Order of Nature, 1150–1750.*

4. The thought originated from Mencius, *Mengzi jizhu* [An Annotated Collection of Mengzi], edition of *Sishu wujing ji zhu zhang ju* [The Four Books and Five Classics with commentaries by Zhu Xi], chap. 4, 10a, and chap. 7, 7a. Ming scholars pondered Zhu Xi's commentary to it, *Hui'an xiansheng Zhu Wen gong wenji* [Zhu Xi's Literary Writings], chap. 76, 23. For the general discourse on this topic see Philip J. Ivanhoe, *Ethics in the Confucian Tradition: The Thought of Mencius and Wang Yangming,* 81.

5. Cynthia Brokaw, *The Ledgers of Merit and Demerit: Social Change and Moral Order in Late Imperial China,* 26. For a general overview to ledgers of this era see 52–60.

6. Song Yingxing, *On Qi,* preface 51. Song here draws on a contemporary discussion, repeating its terminology. The term *deng tan,* for example, constantly appears in *Fu she dang'an* [Archives of the Restoration Society], Collection of Manuscripts, Reg. Nr. 25930094, Beijing Library. It originally designated Buddhist preaching. Using it for intellectual court debates held by rhetorically trained officials disgraced the discourse as being overburdened with religious aims. Lu Shiyi, "Report about the Restoration Society," in *Zhongguo yeshi jicui* [Anthology of Superior Unofficial Histories of China], ed. Chen Li, 578–83, 588.

7. Song Yingxing, *On Qi,* preface 51.

8. Timothy Brook, *Praying for Power: Buddhism and the Formation of Gentry So-*

ciety in Late-Ming China, 64, 68–74, suggests that approaches of intellectuals such as Jiao Hong were based on the assumption that eventually all three schools pursued a common end and mainly differed in application.

9. Song Yingxing, *An Oppositionist's Deliberations,* chap. *lianbing yi,* 28.

10. Tong, *Disorder under Heaven,* 100–122.

11. Zhu Xi, *Zhuzi yulei* [Classified Conversations of Master Zhu Xi], Lixue congshu, chap. 18, 14b–15a.

12. Rowe, *Saving the World,* 133–45.

13. David Bloor, *Knowledge and Social Imagery,* 5.

14. Another important representative, apart from Luo Qi's Origin of Things (mentioned in chap. 1, n. 49), for this genre is the *Shiwu jiyuan* [Record of the Origins of Things and Affairs], compiled by Gao Cheng circa 1078–85. For an overview of the compilations of this genre, see Martina Siebert, "Making Technology History."

15. Benjamin A. Elman, *On Their Own Terms: Science in China, 1550–1900,* 115–16.

16. Nathan Sivin and Geoffrey Lloyd, *The Way and the World: Science and Medicine in Early China and Greece,* 201–2.

17. Liu An, *Huainan zi* [The Master of Huainan] (120 BC), chap. *tianwen xun* 3, 1a. For an interpretation of the concepts of heaven and earth in the *Master of Huainan,* I refer to John S. Major, *Heaven and Earth in Early Han Thought: Chapters Three, Four, and Five of the Huainanzi.*

18. *Guo Yu* [Discourses from the States], attributed to Zuo Qiuming (5th century BC), chap. *Yue Yu xia,* 12a.

19. Zuo Qiuming, *Zuo Zhuan* [Commentaries of Master Zuo Traditions], edition of *Sishu wujing* [The Four Books and Five Classics], chap. *Zhao gong shang* 1, month 8.

20. Nathan Sivin, *Traditional Medicine in Contemporary China: A Partial Translation of Revised Outline of Chinese Medicine (1972) with an Introductory Study on Change in Present-day and Early Medicine,* 43–90, gives a concise and refined overview of the development of these concepts with a comparative perspective.

21. Nathan Sivin, *Traditional Medicine in Contemporary China,* 61.

22. Zhu Xi, *Classified Conversations of Master Zhu Xi,* vol. 7, chap. 65, 1a.

23. This comes near to that what Tu Wei-ming, "The Continuity of Being: Chinese Visions of Nature," in *On Nature,* ed. Leroy S. Rouner, 116 identifies as Zhang Zai's approach to *qi:* "*Ch'i* [*qi*], the psychophysiological stuff everywhere. It suffuses even the 'great void' (*t'ai-hsü*) which is the source of all beings in Chang Tsai's [Zhang Zai] philosophy. The continuous presence of *ch'i* in all modalities of being makes everything flow together as the unfolding of a single process. Nothing, not even an almighty creator [a concept that is not inherent in Chinese thought] is external to this process."

24. Charles Le Blanc, "Résonance: Une Interprétation Chinoise de la Réalité," in *Mythe et Philosophie à l'Aube de la Chine Impériale: Etudes sur le Huainan Zi,* ed. Charles Le Blanc and Rémi Mathieu, 94.

25. Steven Shapin, *A Social History of Truth: Civility and Science in Seventeenth Century England,* esp. 16–22.

26. Irene Bloom, *Knowledge Painfully Acquired: The K'un-chih chi by Lo Ch'in-shun,* 17–21.

27. Michael Lackner, Friedrich Reiman, and Michael Friedrich, eds. *Chang Tsai*

[*Zhang Zai*] *Rechtes Auflichten: Cheng-meng. Übersetzt aus dem Chinesischen mit Einleitung und Kommentar versehen*, lxxxi–civ, suggest that early reception of the *Zheng Meng* has long been concealed because of their historical delineation by thinkers such as Huang Zongxi.

28. Chen Guying, Xin Guanjie, and Ge Rongjin, eds. *Ming Qing shixue sichao shi* [The History of the Trend of Practical Learning during the Ming and Qing Dynasty], vol. 2, chap. 27, 819–44; Li Shuzeng, Sun Yujie, and Ren Jinjian, eds., *Zhongguo mingdai zhexue* [Philosophy in China during the Ming Dynasty], 1438–56, group him, for example, with Zhang Juzheng and Li Shizhen; Zhang Qizhi, *Ruxue, lixue, shixue, xinxue* [New Studies About the *Ru*-School of Confucianism, the *Li*-School of Principle, and the *Shi*-School of Practical Learning].

29. Pan Jixing, *Mingdai kexue jia Song Yingxing* [The Scientist Song Yingxing of the Ming Dynasty], 95; Christopher Cullen, "The Science/Technology Interface in Seventeenth-Century China: Song Yingxing on Qi and Wuxing," *Bulletin of the School of Oriental and African Studies* 52, no. 1: 301.

30. Li Shuzeng, Sun Yujie, and Ren Jinjian, *Philosophy in China*, 24–28, 860–62 (Wang Tingxiang); the same can be said about Luo Qinshun. Kim Youngmin, "Luo Qinshun (1465–1547) and His Intellectual Context," *T'oung Pao* 89, no. 4: 367–441.

31. Bloom, *Knowledge Painfully Acquired*, 21.

32. Dai Nianzu et al., *Zhongguo wuli xueshi da xi, shengxue shi* [A Series of Books on History of Physics in China, A History of Acoustics], 60–61.

33. Wang Chong, "Lun Heng" [Discourse Weighed in Balance], in *Lun Heng xigu* [Analysis of "Discourse Weighed in Balance" with analytic notes], ed. Zheng Wen, chap. 4, 70–107, par. *qishou* 4, 227.

34. Wang Fuzhi, Zhang Zai, *Zhang Zi Zheng Meng Zhu* [Commentary of the Discourse for Beginners of Master Zhang], chap. 3 *dongwu bian*, 6a. Wang Tingxiang, *Wang Shi jia cang ji* [Anthology of Writings Stored in the House of the Wang Clan], vol. 3, chap. 12, 6a.

35. Thomas Kuhn, *The Structure of Scientific Revolutions.*

36. Bloor, *Knowledge and Social Imagery*, 37–45; Lorraine Daston, "Objectivity versus Truth," in *Wissenschaft als kulturelle Praxis, 1750–1900*, ed. Hans Erich Bödeker and Peter Hanns Reill, 17–32.

37. For various *qi* interpretations see also F'ung Yu-lan [Feng Youlan], *A Short History of Chinese Philosophy*, ed. Derk Bodde, 2: 45; Feng Youlan emphasized the material aspect of *qi* in his study of Chinese ancient philosophy. Wing-tsit Chan, *A Source Book in Chinese Philosophy*, 757.

38. Song Yingxing, *An Oppositionist's Deliberations*, chap. *shiyun yi*, 5.

39. Timothy Brook, *The Confusion of Pleasure: Commerce and Culture in Ming China*, 8–9, 153–72, esp. 168.

40. Song Yingxing, *An Oppositionist's Deliberations*, chap. *fengsu yi*, 41–42. Yang Weizeng, *Research about the Ideas of Song Yingxing*, 15, 97. Yingxing, *Yearnings*, chap. *lianyu shi*, poem 11, 128, however, placed this in a concrete relation to Song attitudes arguing that Ming scholars treated things just the opposite way of the Song, although his contemporaries claimed to actually follow the ideal of Song scholars. Wang Zichen and

Xiong Fei, *Song Yingxing xueshu zhuzuo sizhong* [Four Treatises on Song Yingxing's Tenets], 3.

41. Tadao Sakai, "Confucianism and Popular Educational Works," in *Self and Society in Ming Thought*, ed. William Theodore de Bary, 344.

42. Song Yingxing, *An Oppositionist's Deliberations*, chap. *luanmeng yi*, 44.

43. Song Yingxing, *Talks about Heaven*, preface 99–100.

44. Sun Xiaochun and Jacob Kistemaker, *The Chinese Sky during the Han: Constellating Stars and Society*, 2–3, 102–5.

45. Song Yingxing, *Talks about Heaven*, preface 99–100.

46. Song Yingxing, *Works of Heaven*, chap. 3, no. 18, 54b.

47. Ruan Yuan and Chen Changqi et al., *Guangdong Tongzhi* [General Gazetteer of Guangdong]. Edward H. Schafer, "The Pearl Fisheries of Ho-p'u," *Journal of the American Oriental Society* 72, no. 4: 158.

48. Song Yingxing, *Works of Heaven*, chap. 3, no. 18, 53b.

49. Song Yingxing, *Talks about Heaven*, preface 99.

50. Song Yingxing, *Works of Heaven*, chap. 3, no. 18, 53b.

51. Cullen, "The Science/Technology Interface," 308.

52. Song Yingxing, *On Qi*, preface 51. For the three astronomical models see Cullen, *Astronomy and Mathematics in Ancient China*, 20–27.

53. Song Yingxing, *Talks about Heaven*, preface, 99.

54. Sun and Kistemaker, *Chinese Sky during the Han*, 25; Cullen, *Astronomy and Mathematics in Ancient China*, 35, 50–53, translates *gaitian* alternatively as "canopy heaven."

55. On the polemics on the *huntian* see Cullen, *Astronomy and Mathematics in Ancient China*, 59–61.

56. Cai Yong (132–192) mentioned this concept around AD 180. Fan Ye, *Hou Han shu* [History of the Former Han Dynasty], ed. Li Xian et al., 11: ch. 10 (*zhi*), 3217 commentary. Later texts are more detailed, implying that the theory was reinvented as a historiographic issue. Charles Hartman, "The Reluctant Historian: San Ti, Chu Hsi, and the Fall of the Northern Sung," *T'oung Pao* 89, no. 2: 100–148; Michael Loewe, "The Oracles of the Clouds and the Winds," *Bulletin of the School of Oriental and African Studies* 51, no. 3: 500–520.

57. This idea of the universe as an ocean was the closest yet to a nongeocentric model of the heavens. Xi Zezong, "Chinese Studies in the History of Astronomy, 1949–1979," *Isis* 72, no. 3: 456–70; Zhou Guidian, *Zhongguo guren luntian* [Ancient Chinese Views of the Heavens].

58. Wang Fuzhi, Zhang Zai, "Zheng Meng" [Correct Discipline for Beginners], in *Zhang Zai ji* [Collected Writings of Zhang Zai], chap. 2, 12, 1; Chen Jiujin and Yang Yi, *Zhongguo gudai de tianwen yu lifa* [Ancient Chinese Astrology and Astronomy], 18–25.

59. Song Yingxing, *Talks about Heaven*, chap. *rishuo* 4, 108.

60. Song Yingxing, *Talks about Heaven*, chap. *rishuo* 1, 100.

61. Song Yingxing, *Talks about Heaven*, chap. *rishuo* 2, 103.

62. Song Yingxing, *Talks about Heaven*, chap. *rishuo* 5, 111.

63. Song Yingxing, *Talks about Heaven*, chap. *rishuo* 1, 101.

64. Song Yingxing, *Talks about Heaven*, chap. *rishuo* 1, 101.

65. Song Yingxing, *Talks about Heaven*, chap. *rishuo* 1, 101.

66. Wang Fuzhi and Huang Zongxi, *Lizhou Chuanshu wushu* [Five Books by [Huang] Lizhou and [Wang] Chuanshan], 63–64.

67. Jessica Rawson, *Chinese Jades: From the Neolithic to the Qing*, 247–51.

68. Wang Xichan, *Xiao an xin fa* [New Methods of Xiao'an], Congshu jicheng jianbian 425, 5.

69. Nathan Sivin, "Wang Hsi-Shan," in *Science in Ancient China: Researches and Reflections*, ch. V.

70. Jiang Xiaoyuan, *Tianxue waishi* [Unofficial History of Astronomy], 54–67.

71. Song Yingxing, *Talks about Heaven*, chap. *rishuo* 4, 108.

72. Song Yingxing, *Talks about Heaven*, chap. *rishuo* 6, 112.

73. Song Yingxing, *Talks about Heaven*, chap. *rishuo* 3, 105.

74. Song Yingxing, *Talks about Heaven*, chap. *rishuo* 2, 103–4.

75. Song here omitted the middle part of Zhu Xi's commentary, quoting only the first and last sentence. Zhu Xi had argued that goodness would discard tyranny and that as long as the *yang qi* was prospering no eclipse would occur. Zhu Xi, *Yupi zizhi tongjian gangmu* [Comprehensive Mirror for Aid in Government], complement in *Siku quanshu zhenben*-edition, chap. 7, 12a.

76. Song Yingxing, *Talks about Heaven*, chap. *rishuo* 3, 106. See also Wang and Xiong, *Song Yingxing Tenets*, 108–9 n.

77. Song Yingxing, *Talks about Heaven*, chap. *rishuo* 3, 106.

78. Song Yingxing, *Talks about Heaven*, chap. *rishuo* 3, 106.

79. Song Yingxing, *Talks about Heaven*, chap. *rishuo* 5, 111.

80. Song Yingxing, *Works of Heaven*, part 1, no. 3, 49a.

81. Song Yingxing, *Works of Heaven*, part 1, no. 3, 49b; Ogawa Shogo, *Kinsei shokusengaku kōyō* [Systematic Studies About Dyeing in the Early Modern Period]. He specifies terminological issues with a reference to Teiji Okanobori, *Senshoku seigi* [Correct Meanings in Dyeing].

82, Song Yingxing, *Works of Heaven*, part 1, no. 3, 49b; Reference to *Yijing* [*Book of Changes*], chap. *xici shang*, 11a.

83. Song Yingxing, *Works of Heaven*, part 3, no. 16, 40a.

84. Puett, *Ambivalence of Creation*, 65–91, 72.

85. Chiara Crisciani, "History, Novelty and Progress in Scholastic Medicine," *Osiris*, 2nd ser., 6 (1990): 118–39.

86. Shapin, *Social History of Truth*, esp. 234.

87. Jessica Rawson, "The Many Meanings of the Past in China," in *Perception of Antiquity in Chinese Civilization*, ed. Dieter Kuhn and Helga Stahl, 397–421.

88. Frederick W. Mote, *Imperial China, 900–1800*, 99.

89. Ibid.

90. See also Hu Shi's (1891–1962) commentary in Cui Shu, "Kexue de gushijia" [Cui Shu as a Scientific Historian of Antiquity], in *Cui Dongbi yi shu* [An Epitaph of Cui Shi], ed. Gu Jiegang, 953. See also Michael Quirin, "Scholarship, Value, Method, and Hermeneutics in Kaozheng: Some Reflections on Cui Shu (1740–1816) and the Confucian Clas-

sics," *History and Theory* 35, no. 4 (1996): 34–53. See also his dissertation (1994) available through the University of Michigan doctoral dissertation.

91. Joshua A. Fogel, *The Cultural Dimensions of Sino-Japanese Relations: Essays on the Nineteenth and Twentieth Centuries*, 9.

92. Jia Sixie, *Qimin Yaoshu* [Essential Techniques for the Peasantry], annotated and ed. Miao Qiyu, 169, 174, 176.

93. Li Shizhen, *Bencao gangmu tongshi* [General Explanation of the Systematic Materia Medica] (1596), 1: preface, 2–5, 334.

94. The book itself is lost, but almost two thirds of it are at present still available in Yang Hui's *Xiang Jie Jiu Zhang Suan Fa* [Detailed Analysis of the Mathematical Rules in the "Nine Chapters"]; Guo Shuchun, "Jia Xian <Huangdi jiu zhang suan jing xi cao> chu tan" [An Initial Inquiry into Jia Xian's "Detailed Solutions for the Problems" in the Nine Mathematics Chapters of Huang Di], *Ziran kexue shi yanjiu* 7, no. 4 (1988): 328–34.

95. Song Yingxing, *Works of Heaven*, chap. 1, no. 1, 1a.

96. Daniel K. Gardner, "Chu Hsi [Zhu Xi] and the Transformation of the Confucian Tradition," in *Learning to Be a Sage: Selections from the Conversations of Master Chu, Arranged Topically*, by Chu Hs'i [Zhu Xi], translated with commentary by Daniel Gardner, 57–81; Zhu Xi, *Classified Conversations of Master Zhi Xi*, chap. 78. *Shujing* [Book of Documents], chap. 1 *Da Yu mo*, 3b/4a.

97. Song Yingxing, *Works of Heaven*, chap. 1, no. 1, 1a. Hou Ji was a prime representative for laborious agricultural works. See Zhong Guangyan's annotation in Song Yingxing, *Tiangong kaiwu*, ed. Zhong Guangyan, chap. 1, 9.

98. Song Yingxing, *Works of Heaven*, chap. 2, no. 9, 29a.

99. Ibid., 34b.

100. Song Yingxing, *An Oppositionist's Deliberations*, chap. *minzai yi*, 11.

101. Song Yingxing, *An Oppositionist's Deliberations*, chap. *shiqi yi*, 13.

102. Song Yingxing, *Works of Heaven*, chap. 3, no. 18, 53a.

103. Song Yingxing, *Works of Heaven*, chap. 3, no. 15, 26a.

CHAPTER 3

1. Song Yingxing, *Works of Heaven*, chap. 3, no. 5, 66a.

2. *Lunyu*, 1:12. Liu Baonan, Liu Gongmian, and Song Xiangfeng, eds., *Lunyu zhengyi* [Annotated Commentary to the Analects], 2: 19, 63.

3. Shapin, *A Social History of Truth*, 66–68, 122–24.

4. Jean Francois Gauvin, "Artisans, Machines, and Descartes's Organon," *History of Science* 44 (2006): 190.

5. I use "epistemic objects" in the sense of Hans-Jörg Rheinberger, *Toward a History of Epistemic Things: Synthesizing Proteins in the Test Tube*, 32.

6. Richard John Lufrano, *Honorable Merchants: Commerce and Self-Cultivation in Late Imperial China*, 46. For more examples see Hu Jichuang, *Zhongguo jingji sixiangshi jianbian* [A Concise History of Chinese Economic Thought], 458.

7. Guo Jinhai, "Mingdai Nanjing chengqiang zhuan mingwen luelun" [A Concise In-

troduction of the Inscriptions on the Bricks of City Wall of Nanjing in the Ming Dynasty], *Dongnan wenhua* (2001): 75–78; Xia Minghua, "Jingzhou gucheng leming zhuan yu wule gongming" [The Craftsmen Mark System and the Engraved Bricks of Xingzhou City Walls], *Jianghan kaogu* 87, no. 2 (2003): 66–72.

8. Thomas T. Allsen, *Commodity and Exchange in the Mongol Empire: A Cultural History of Islamic Textiles,* Cambridge Studies in Islamic Civilization, 30–34, esp. 32; Jack Weatherford, *Genghis Khan and the Making of the Modern World.* For the Mongolian influence on Chinese ideas about expertise see Nathan Sivin, *Granting the Seasons: The Chinese Astronomical Reform of 1280,* 19–33.

9. Shen Shixing and Li Dongyang, *Da Ming huidian* [Collected Statutes of the Great Ming], chap. 181, 201. Zhang et al., *Ming History,* vol. 6, chap. 72, pp. 1729–31, 1760. Li Shaoqiang and Xu Jianqing, eds., *Zhongguo shougong ye jingji tongshi, Ming Qing juan* [A Comprehensive History of Chinese Handicraft Economy, Ming and Qing].

10. Dagmar Schäfer, *Des Kaisers seidene Kleider: Staatliche Seidenmanufakturen in der Ming-Zeit 1368–1644,* 210–12.

11. Hildegard Scheid, "Die Entwicklung der Staatlichen Seidenweberei in der Ming-Dynastie (1368–1644)." The same sort of interference from eunuchs was common in the military sector and other administrative activities relating to the inner court. Shih-shan Henry Tsai, *The Eunuchs in the Ming Dynasty,* 69–81, 106–10; Fan Jinmin, ed., *Ming Qing Jiang nan shang ye de fa zhan* [The Development of the Economy in Jiangnan during the Ming and Qing Dynasties].

12. Chen Juanjuan, *Sichou shihua* [Silk Manufacture and Trade], 9–15, 39.

13. Michael Marmé, *Suzhou: Where the Goods of All the Provinces Converge,* 108–26.

14. An overview of works on porcelain manufacture is given by Rose Kerr and Nigel Wood, *Ceramic Technology,* pt. 12 of *Chemistry and Chemical Technology,* vol. 5 in Science and Civilisation in China, 24–28. The most detailed reports on this issue originate from Tang Ying (1682–1756) who assessed Ming developments from a Qing production perspective. See Tang Ying's *Tao ye tupian ci* [Illustrated Explanation of Ceramic Production] from 1743 in Tang Ying, *Tang Ying ji* [Collected Writings of Tang Ying], ed. Zhang Faying and Diao Yunzhan, 950–67 or Zhu Yan's *Tao Shuo* [Description of Ceramics] from 1774, ed. Bianzuan wuyuan hui. Xuxiu siku quanshu.

15. Lothar Ledderose, *Ten Thousand Things: Module and Mass Production in Chinese Art,* preface, 5.

16. Zhang Juzheng, Lü Diaoyang et al., *Shizong shilu* [Veritable Records from the Reign of the Emperor Shizong] (1577), 8: chap. 117, 8153 (5a); Schäfer, *Des Kaisers seidene Kleider,* 60–61.

17. Liang Fang-chung, *The Single-Whip Method (I-t'iao-pien-fa) of Taxation in China,* trans. Wang Yü-ch'uan, 55.

18. Liang, *Single-Whip Method* shows that the tax was rather a regulation than a rule and thus was locally adjusted, as described by Fei Siyen, "We must be taxed: A case of Populist Urban Fiscal Reform in Ming Nanjing (1368–1644)," *Late Imperial China* 28, no. 2 (2007): 4.

19. Frederic E. Wakeman, "Localism and Loyalism during the Qing Conquest of Jiangnan: The Tragedy of Jiangyin," in *Telling Chinese History: A Selection of Essays,* ed. Lea H. Wakeman, 178.

20. Chiu Pengsheng, "The Discourse on Insolvency and Negligence in Eighteenth-Century China," in *Writing and Law in Late Imperial China: Crime Conflict, and Judgment,* ed. Robert E. Hegel and Katherine Carlitz, Asian Law Series 18, 125–42.

21. Andreas Janousch, "Salt Production Methods and Salt Cults at Xiechi Salt Lake in Southern Shanxi" (conference paper, Baltimore, ISHEASTM, 2008).

22. Ina Asim, "The Merchant Wang Zhen, 1424–1495," in *The Human Tradition in Premodern China,* ed. Kenneth J. Hammond, 157–64.

23. Zhang Juzheng, *Shizong shilu,* vol. 9, chap. 172 *jiajing* 14, 8406 (4a–5b).

24. David M. Robinson, *Bandits, Eunuchs, and the Son of Heaven: Rebellion and the Economy of Violence in Mid-Ming China,* 25–57.

25. Song Yingxing, *Works of Heaven,* chap. 1, no. 2, 30a–b.

26. Song Yingxing, *Works of Heaven,* chap. 3, no. 15, 27b, 28b.

27. Sun Laichen, "Military Technology Transfers from Ming China and the Emergence of Northern Mainland Southeast Asia (ca. 1390–1527)," *Journal of Southeast Asian Studies* 34, no. 3 (2003): 495–517. For the development of gunpowder see Joseph Needham, *Military Technology: The Gunpowder Epic,* pt. 7 of *Chemistry and Chemical Technology,* vol. 5 in Science and Civilisation in China, 111–126. Thomas T. Alsen, "The Circulation of Military Technology in the Mongolian Empire," in *Warfare in Inner Asian History (500–1800),* ed. Nicola Di Cosmo, 279–93.

28. Zhang Yunming, "Ancient Chinese Sulphur Manufacturing Processes," *Isis* 77, no. 3 (1986): 490–91.

29. Song Yingxing, *Works of Heaven,* chap. 2, no. 11, 60b.

30. Song Yingxing, *Works of Heaven,* chap. 3, no. 15, 35a–b.

31. Song Yingxing, *Works of Heaven,* chap. 2, no. 11, 60b.

32. Antonio S. Cua, *The Unity of Knowledge and Action: A Study in Wang Yang-ming's Moral Psychology,* 4–9, 102–4 on self.

33. Wang Yang-ming [Yangming], *Instructions for Practical Living and Other Neo-Confucian Writings,* trans. Wing-tsit Chan, chap. 1:3, 7. In my interpretation of this passage and the application of intuitive and innate, I follow the epistemological study of Wang Yangming about the divergence of intuition in cultural contexts. An Yanming, "Liang Shuming and Henri Bergson on Intuition: Cultural Context and the Evolution of Terms," *Philosophy East and West* 47, no. 3 (1997): 337–62; Jig-chuen Lee, "Wang Yang-ming, Chu Hsi, and the Investigation of Things," *Philosophy East and West* 37, no. 1 (1987): 24–35.

34. Wang, *Instructions for Practical Living,* note 139.

35. Song Yingxing, *Works of Heaven,* preface 2a.

36. Song Yingxing, *On Qi,* preface, 51.

37. Zhang Juzheng, *Zhang Juzheng zhu sishu jizhu* [Zhang Juzheng Commentary on the Four Books, with annotations] *Siku quanshu*-edition referred to the text *Mengzi,* chap. 36: 819. For a discussion see Wei Qingyuan, *Zhang Juzheng he Ming dai zhonghou qi zhengju* [Zhang Juzheng and the Policy Circumstances of the Mid- and Late Ming Period], chap. 8, 292–298.

38. Song Yingxing, *An Oppositionist's Deliberations,* chap. *lianbing yi,* 27–28.

39. Song Yingxing, *An Oppositionist's Deliberations,* chap. *lianbing yi,* 25.

40. Song Yingxing, *On Qi,* chap. *qisheng* 1, 64–65. Treatises on warfare also use the concept of *shengqi* arguing that it psychologically influenced the soldiers. Zuo, *Commen-*

taries of Master Zuo Traditions, chap. 6 *xigong* 22, 185. This text specifies it also as the exhaled *Qi* for the training of voices.

41. Nathan Sivin, *Traditional Medicine in Contemporary China*, 126, 155.

42. Song Yingxing, *An Oppositionist's Deliberations*, chap. *lianbing yi*, 25.

43. Wang, *Instructions for Practical Living*, note 3.

44. Song Yingxing, *An Oppositionist's Deliberations*, chap. *lianbing yi*, 25.

45. Yue Fei is a storybook hero of the Song Dynasty. Deng Guangming, *Yue Fei zhuan* [Biography of Yue Fei]. John E. Wills, Jr., *Mountain of Fame: Portraits in Chinese History*, 168–80.

46. Song Liquan and Song Yude, *Family Annals of the House of Song*, chap. 22, 71. The text is copied in Pan Jixing, *Biography of Song Yingxing*, 151. The local monograph of Bozhou only mentions Song's brief appointment. Liu, *Local Gazetteer of Bozhou*, chap. *renwu*, 5b.

47. Song Yingxing, *Works of Heaven*, chap. 2, no. 12, 66a.

48. Song Yingxing, *Works of Heaven*, chap. 2, no. 12, 66a.

49. Song Yingxing, *An Oppositionist's Deliberations*, chap. *lianbing yi*, 27. In the original text Song addresses the Tang dynastic intellectual Han Yu, as Chang Li. Han Yu was an emphatic opponent of Buddhism and very much interested in the proper usage and understanding of words. His style was clear and straightforward. Charles Hartman, *Han Yü and the T'ang Search for Unity*, 13; Su Dongpo, *Dongpo zhilin* [The Forest of Records by Su Shi] punctuated by Wang Songling.

50. Song Yingxing, *An Oppositionist's Deliberations*, chap. *lianbing yi*, 28.

51. Ibid.

52. Ibid. The expression *kaiming* is used in Sima, *Records of the Grand Historian*, chap. 15 *wudi benji*, 20 in the sense of understanding, elucidate, illuminate, clarify. It is the countersubject to *ye* in the sense of wild, uncultivated or without civilization. This should not be confused with the European movement of enlightenment.

53. Mencius, *An Annotated Collection of Mengzi*, 13: 13; Tu Wei-ming [Weiming], *Centrality and Commonality: An Essay on Chung-yung*, 51, 160–62; Yong Huang, "A Neo-Confucian Conception of Wisdom: Wang Yangming on the Innate Moral Knowledge (*liangzhi*)," *Journal of Chinese Philosophy* 33, no. 3 (2006): 393–408.

54. Song Yingxing, *An Oppositionist's Deliberations*, chap. *lianbing yi*, 27.

55. Ibid.

56. Song Yingxing, *Works of Heaven*, chap. 1, no. 1, 1a/b.

57. Ibid., chap. 2, no. 7, 12b/13a.

58. Song Yingxing, *Works of Heaven*, chap. 1, no. 2, 28b.

59. Ibid., chap. 2, no. 9, 29a.

60. Song Yingxing, *An Oppositionist's Deliberations*, chap. *yanzheng yi*, 36–38.

61. Ibid., 36.

62. Mote, *Imperial China 900–1800*, 769; Chiu, *Discourse on Insolvency*, 127.

63. Frederic E. Wakeman, "Boundaries of the Public Sphere in Ming and Qing China," *Daedalus* 127: 1–14.

64. Frederick W. Mote and Denis Twitchett, eds., *The Cambridge History of China*, vols. 7 and 8, *The Ming Dynasty, 1368–1644*, 29–31.

65. Song Yingxing, *An Oppositionist's Deliberations*, chap. *fengsu yi*, 42.

66. Chun-Shu Chang and Shelley Hsueh-lun Chang, *Crisis and Transformation in Seventeenth-century China: Society, Culture, and Modernity in Li Yü's World*, 152, 174.

67. Zuo, *Commentaries of Master Zuo Traditions*, chap. 12 Zhaogong 7, 428. For a translation see Legge, *Chinese Classics*, vol. 5, *Ch'un ts'ew*, 616, vol. 1.

68. Song Yingxing, *Works of Heaven*, chap. 3, no. 14, 1a.

69. Ibid.

70. Song Yingxing, *An Oppositionist's Deliberations*, chap. *xuezheng yi*, 32–33.

71. Ibid., chap. *fengsu yi*, 40–41.

72. Ibid., chap. *yanzheng yi*, 35. The term *hangyan* includes the whole salt monopoly, which means also transport and trading.

73. Ibid., chap. *shiqi yi*, 13.

74. Ibid., chap. *fengsu yi*, 40.

75. Ibid., 43.

CHAPTER 4

1. Paraphrased from Yingxing's *Works of Heaven*, chap. 1, no. 2, 20–24 and *Yearnings*, chap. *simei shi*, 122 (poem 6) and chap. *lianyu shi*, 127 (poem 6).

2. Steven Shapin and Simon Schaffer, *Leviathan and the Air Pump: Hobbes, Boyle, and the Experimental Life*, 22–26, 55; see also Barbara J. Shapiro, *A Culture of Fact: England, 1550–1720*, 112–19.

3. Peter Dear, "Totius in Verba: Rhetoric and Authority in the Early Modern Royal Society," *Isis* 76, no. 2: 145–61.

4. William Gilbert, *Tractus, sive, physiologica nova de magnete, magneticisque corporibus et de magneete tellure: sex libris comprehensus*, 170–72.

5. Bloom, *Knowledge Painfully Acquired*, 12.

6. Joanna Handlin, *Action in Late Ming Thought: The Reorientation of Lu K'un and Other Scholar-Officials*, 4; see also her view on changes in knowledge culture, 6–20, and the correction of faults, 193–212. For the original source see Lü Weiqi, *Mingde xiansheng wenji* [The Collected Works of Lü Weiqi], 23: 6a.

7. Wang Shimao, *Minbu shu* [Memorials about Fujian Province]. William Charles Milne, *A Retrospect of the First Ten Years of the Protestant Mission to China*, 353. Yingxing, however, liked to use pottery metaphorically in his poetry. Song Yingxing, *Yearnings*, chap. *Simei shi*, poem 1, 119.

8. Shen Gua, "Hun yi yi" [On the Armillary Sphere], in *Lidai tianwen lü lideng zhi huibian* [A Collection of Historical Documents on Astronomy], ed. Zhonghua shuju bianji bu, vol. 3, 802. See also the original tract in Tuo Tuo, *Song shi* [Song History], chap. 48, Tianwenzhi-Yixiang, 954–62.

9. Fabrizio Pregadio, *Great Clarity: Daoism and Alchemy in Early Medieval China*, Asian Religions and Cultures, 13–16.

10. This is particularly obvious in the *pulu* genre. Siebert, *Pulu*, iii–v.

11. Zheng Qiao, *Tongzhi yiwen lue* [Essentials on the Arts of the Tongzhi], 6 vols.

1161. For a discussion of the Song dynasty *biji* culture see chap. 2 of this book, Stephen Owen, *Readings in Chinese Literary Thought,* 272–77, and Fu Daiwie and Lei Xianglin, "Mengqi shen de yuyan yu xiangsixing- dui Mengqi bitan zhong renmingyun zhi yuzhi ji shenqi, bishi ermen de yanjiu" [Language and Similarity in the Dream Brook: A Study of Prognostication, Divine Oddities, and Strange Events in 'Mengqi bitan'], *Ts'ing-hua Hsueh-bao* 3: 35.

12. Song Yingxing, *Works of Heaven*, preface, 2b.

13. Zheng Qiao, "Ji Fang libu shu," in *Tongzhi ershi lüe* [Selections of Twenty Essential Treatises of the *tongzhi*], ed. Wang Shumin, vol. 1, 15.

14. Song Yingxing, *An Oppositionist's Deliberations,* chap. *jinshen yi*, 8.

15. Song Yingxing, *Works of Heaven*, preface, 2b.

16. Chen Shou, *Sanguo zhi* [Record of the Three States], chap. *weishu* 9, *xiahou xuanzhuan*, 208. Cao Zhi was idolized mainly for his poetry. His most prominent poem is entitled "Bai ma bian" [On the White Horse]. It idolizes a young warrior fearlessly fighting for his country. Bai Ma was the princely title of Cao Zhi's brother Cao Biao. Cao Zhi, *Cao Zhi ji zhu zi suoyin* [Concordance to the Works of Cao Zhi], ed. D. C. Lau, Chen Fangzheng, and He Zhihua 57 (no. 234). Cao Zhi's poetry was celebrated in the Ming. Robert Joe Cutter, "Cao Zhi's (192–232) Symposium Poems," *Chinese Literature: Essays, Articles, Reviews* 6, no. 1/2: 12.

17. Geng Yinlou achieved his *jinshi* in 1625 and was nominated a commissioner (*zhixian*) of Linxi and Shouguang in Shandong Province. Zhang et al., *Ming History,* 22: chap. 267, 6878–79. Xu Guangqi, *Nongzheng quanshu* [Comprehensive Treatise on Agricultural Administration] (1639), preface and chap. 10, 90 expressed similar ideas.

18. Francesca Bray, "Agricultural Illustrations: Blueprint or Icon," in *Graphics and Text in the Production of Technical Knowledge in China: The Warp and the Weft,* 521–22.

19. The reverse of this character combination *xiangxing* refers to one of the six groups into which Chinese characters are divided, namely, the group that is interpreted as representative of objects or phenomena.

20. Sun and Sun, *Chinese Technology in the Seventeenth Century,* preface viii.

21. Peter Golas, "Like Obtaining a Great Treasure: The Illustrations in Song Yingxing's The Exploitation of the Works of Nature," and Donald Wagner, "Song Yingxing's Illustrations of Iron Production," in *Graphics and Text in the Production of Technical Knowledge in China: The Warp and the Weft,* 569–634.

22. Song Yingxing, *Works of Heaven*, preface, 2b.

23. Francesca Bray, "Agricultural Illustrations," 522–23.

24. Ibid., 521.

25. Bray, "Introduction," *Graphics and Text,* 4.

26. Zheng Qiao, "Tu pu lue" [Treatise on Images and Tables], in *Tongzhi yiwen lue* [Essentials on the Arts and Literature of the General Treatises], completed in 1162, vol. 5, chap. 71, 6.

27. Pan Jixing, *The Scientist Song Yingxing,* 48. Golas, *Like Obtaining a Great Treasure,* 572.

28. Robert E. Hegel, *Reading Illustrated Fiction in Late Imperial China,* 220–24.

29. Chai Jiguang, "Guanyu Song Yingxing Tiangong kaiwu zhong 'chiyan' bufen yixie

wenti den bianshi" [A Discrimination of Some Questions in the Part of Salt Lake in the Tiangong kaiwu], *Yanye shi yanjiu* 1: 32 also implies a purely technical content.

30. Song Yingxing, *Works of Heaven*, chap. 3, no. 15, 30b.

31. Song Yingxing, *Works of Heaven*, chap. 3, no. 15, 27a.

32. Donald Wagner suggests that this man interspersed clay with saltpeter components into the basin where iron was melted (*chao ni hui*). Wagner, "Song Yingxing's Illustrations of Iron Production," 618–21.

33. Wang Zhen, *Nongshu* [Book of Agriculture], ed. Wang Yuhu (1313); see, for example, the images on pages 212, and 234.

34. Song Yingxing, *Works of Heaven*, chap. 3, no. 15, 34b.

35. E. H. Gombrich, "Standards of Truth: The Arrested Image and the Moving Eye," *Critical Inquiry* 7, no. 2: 240.

36. Song Yingxing, *Works of Heaven*, chap. 3, no. 14, 15b. *Qianfu* is here used as a compound and applied as an abbreviation of *qianbo qingfu*, "floating and light," which Song also mentions in *On Qi*, chap. *xing qi* 1, 52–53. The cold caverns (*huxue*) here refer to the idea of a reversion to the great void that, in this case, is situated in the earth. My thanks to Nathan Sivin for bringing this to my attention.

37. Song Yingxing, *Works of Heaven*, chap. 3, no. 14, 15b–16a. Peter Golas, *Mining*, pt. 13 of *Chemistry and Chemical Technology*, vol. 5 in Science and Civilisation in China, ed. Joseph Needham, 166, takes *pingyang* to be a descriptive term rather than a place-name, even though the prefecture of Pingyang, Shanxi, is mentioned further on in the text. The commentaries note that it may either refer to Pingdan or to Pingyang. It does not mean sunny hill, but plain grounds and hilly regions. Zhong, *Works of Heaven* [annotated], 209–10.

38. Song Yingxing, *Works of Heaven*, chap. 3, no. 14, 17b. For reference see also picture 2–56 with comments in Pan Jixing's compilation of Song's works on crafts. Pan, *Investigation of the Tiangong kaiwu*, 367.

39. Song Yingxing, *Works of Heaven*, chap. 1, no. 14, 16 a/b.

40. Song Yingxing, *Works of Heaven*, chap. 3, no. 16, 40a.

41. Song Yingxing, *On Qi*, chap. *qisheng* 2, 66, chap. *shui chen* 1, 86, chap. *shui huo* 2, 82.

42. For an overview of the early developments of anomaly accounts see Robert Ford Campany, *Strange Writing: Anomaly Accounts in Early Medieval China*, SUNY Series in Chinese Philosophy and Culture.

43. Song Yingxing, *On Qi*, chap. *qisheng* 2, 66. Song refers to a passage of the *Master of Huainan*, that emphasizes the characteristic of unity in *hundun*, rather than chaos. Liu, *Master* of *Huainan*, 2:14, suggests "heaven and earth were perfectly joined, all was chaotically unformed (*hundun wei pu*); and things were complete (*cheng*), yet not created (*cheng*)"; translated by Norman J. Girardot, *Myth and Meaning in Early Taoism: The Theme of Chaos (hun-tun)*, 134.

44. Song Yingxing, *On Qi*, chap. *xing qi hua* 2, 55.

45. Song Yingxing, *Works of Heaven*, chap. 2, no. 12, 66a.

46. Song Yingxing, *On Qi*, chap. *qisheng* 2, 66. Zhang Zai, "Correct Discipline for Beginners," chap. 1 *taihe*, 7,1, lines 5–8; chap. *Hengju yishuo*, and his commentary to the *Book of Changes*, suppl., *xici shang*, 206.

47. Song Yingxing, *On Qi*, chap. *qisheng* 2, 67.

48. Song Yingxing, *On Qi*, chap. *shui feisheng huo shuo* 1, 80. For reference see Zhang Zai, "Correct Discipline for Beginners," chap. 1 *taihe*, 8, lines 2–3; chap. 2 *canliang*, 11.

49. Li Shuzeng, Sun Yujie, and Ren Jinjian, *Philosophy in China*, 1439–41; Pan Jixing pointed to this relationship several times in a personal meeting in 2002. In his *Biography of Song Yingxing*, 448–53, Pan refers to the section on acoustics.

50. Julia Ching, *The Religious Thought of Zhang Zai*, 59–60. She refers to Zhang Zai, "Correct Discipline for Beginners," chap. *chengming*, 34.

51. Girardot, *Myth and Meaning in Early Taoism*. For a Ming approach identifying *hundun* as "diffuse," see Wang Tingxiang, *Anthology of Writings Stored in the House of the Wang Clan*, vol. 3, chap. *taiji bian* 33, 597.

52. Wang Tingxiang, "Yashu, shangbian," in Wang Tingxiang, *Anthology of Writings Stored in the House of the Wang Clan*, vol. 3, 1334. See also chap. "zhenyan, wu xing bian" of this work. Michael Leibold, *Die handhabbare Welt: Der pragmatische Konfuzianismus Wang Tingxiang's (1474–1544)*, 43, 293.

53. Song Yingxing, *On Qi*, chap. *shui huo* 4, 85.

54. Ibid., chap. *xing qi* 2, 56. The expression of *aoshi* is an abbreviation of a location term *yinyang aoshi*, "dark room of *yin* and *yang*." This term designates the backroom of the palace, where the concubines reside. Fan, *History of the Former Han Dynasty*, 4: chap. 34, 1181. The *Book of Changes* interprets *ao* as the darkest corner of a room. It describes a typical *yin* attitude. It corresponds to hiding or concealing. *Book of Changes*, *xici shang* 7, 42.

55. Song Yingxing, *On Qi*, chap. *shui huo* 4, 85.

56. Ibid., 4, 81. Shikuang is documented to have been a musical teacher in the state of Jin during the Spring and Autumn period.

57. Ibid., chap. *qisheng* 5, 71–72. The literal translation of *yunjie* is "incorporated in something like a seed." It is again a technical term in the *Book of Changes*, suppl. *xizi shang* 3, 43.

58. Li Shuzeng, Sun Yujie, and Ren Jinjian, *Philosophy in China*, 1439.

59. Song Yingxing, *On Qi*, chap. *shui huo* 2, 82.

60. Ibid., 1, 80.

61. Song Yingxing, *Works of Heaven*, chap. 2, no. 11, 53b.

62. *Shujing* [Book of Documents], chap. *hong fan*, 77.

63. Ho Peng Yoke, *Li, Qi, and Shu: An Introduction to Science and Civilization in China*, 19–20, suggests that the earliest documented argumentation of this concept by Zou Yan (350–270 BC) was based on observations of physical attitudes, as "wood conquers earth, because wood is harder than earth; metal conquers wood, because one can cut wood with an axe made from metal; fire conquers metal because fire can melt metal; *yin*-water conquers fire, because *yin*-water can extinguish the fire." Zou Yan does not take this issue further. Zou Yan was the first to provide the opposed cycle of mutual conquest (*xiangke*).

64. Brokaw, *The Ledgers of Merit and Demerit*, 29–30.

65. Wei Qingyuan, *Zhang Juzheng and the Policy Circumstances*, 20: 11, 810–15.

66. Song Yingxing, *On Qi*, chap. *xing qi hua* 5, 62–63.

67. Song Yingxing, *Works of Heaven*, chap. 2, no. 11, 55b.

68. Ibid., no. 10, 44b–45a.

69. Ibid., no. 8, 17a.

70. *Works of Heaven*, chap. 3, no. 18, *Metals*, 8a/b, 60 b.

71. Ibid., chap. 3, no. 14, 16a.

72. Song Yingxing, *On Qi*, chap. *xing qi hua* 4, 59–60. The expression *zhenhuo* refers to the so-called original fires, which are said to have created life and thereafter generated a strong primordial *qi*. Both senses are certainly interrelated and can be seen as an abbreviation of the expression *tiandi zhenhuo*, "the true fires of heaven and earth." In the discipline of *Nei dan shu* (internal alchemy—literally often translated as "arts of the inner elixir") *zhenhuo* locates the shaken (*dong*) *qi* next to the kidney. According to the explanation of the medical classics, it is a fire or energy that cannot be observed. This implies that it is existent but its characteristics like heat and light are invisible.

73. Du Wan, *Yunlin shipu* [Stone Catalogue of Cloudy Forest], ed. Zhu Jiuding and Gao Zhao (1720), preface ii; see also Edward H. Schafer, *Tu Wan's Stone Catalogue of Cloudy Forest*, translation of the preface, 12, 34.

74. Song Yingxing, *On Qi*, chap. *xing qi hua* 5, 62–63.

75. Zhang Zai, "Correct Discipline for Beginners," chap. *canliang bian* 2, 12, line 9, 10.

76. Song Yingxing, *Works of Heaven*, chap. 2, no. 10, 44b, 45a.

CHAPTER 5

1. Song Yingxing, *Works of Heaven*, chap. 3, no. 14, 11b–12a; Song Yingxing, *On Qi*, chap. *shui huo* 1, 80. For counterfeiting in the Ming I refer to Richard von Glahn, *Fountain of Fortune: Money and Monetary Policy in China, 1000–1700*, 84–94.

2. The *Aobo tu* [Sketches on Aobo] by Chen Chun, published ca. 1333, delineates the various methods used for salt production up until Ming times. An English translation has been published by Yoshida Tora and Hans Ulrich Vogel, eds., *Salt Production Techniques in Ancient China: The Aobo tu*; for a survey on salt production in China see pp. 4–25. Yingxing mentions Yunnan in his poetry, though. Song Yingxing, *Yearnings*, chap. *lianyu shi*, poem 11, 128.

3. Song Yingxing, *Works of Heaven*, chap. 1, no. 5, 69b.

4. Song Yingxing, *On Qi*, chap. *xing qi hua* 1, 52.

5. Ibid., 3, 57.

6. All previous quotes are from Song Yingxing, *On Qi*, chap. *shui huo* 1, 80.

7. Song Yingxing, *On Qi*, chap. *xing qi hua* 2, 55. I use *skana* here in the sense of the smallest time unit as proposed by Yang Weizeng, *Research about the Ideas of Song Yingxing*, 168.

8. Song Yingxing, *On Qi*, chap. *shui huo* 1, 80–81.

9. Ibid., 4, 85.

10. Ibid., 1, 80–81.

11. All quotations in this paragraph are from Song Yingxing, *On Qi*, chap. *shui huo* 1, 80, 81.

12. Song Yingxing, *On Qi*, chap. *hanre*, 94.

13. All quotations in this paragraph are from Song Yingxing, *On Qi*, chap. *shui huo* 1, 80–81.

14. Song Yingxing, *On Qi*, chap. *xing qi* 2, 56.

15. Ibid., 1, 52.

16. Ibid., 2, 55.

17. *Book of Changes*, suppl. *xici shang* 7, 68 commentary Zheng Xuan. The expression *guan qi shou* originates from the Lao Zi, *Laozi xinbian jiaoshi* [New Text Laozi, with Explanatory Commentary], ed. Wang Xing, chap. 43 (former 44), 202. It stresses the dualism between earth and the heavens as *qian* and *kun*, in *yin* and *yang*.

18. Both quotations are from Song Yingxing, *On Qi*, chap. *xing qi hua* 2, 55.

19. Song Yingxing, *On Qi*, chap. *shui huo* 4, 85. The expression rapid (*ji*), used to describe the attitude of waters, correlates to the attitude of sound vibration that Song also describes as rapid. For reference see next chapter.

20. Ibid., chap. *xing qi hua* 2, 55.

21. Wang Anshi, *Wang Wengong wenji* [Collected Writings of Wang Anshi], 31: 442.

22. Song Yingxing, *On Qi*, chap. *shui huo* 2, 82.

23. Zhang Zai, "Correct Discipline for Beginners," chap. *canliang bian* 2, 12, line 9; see also chap. 3, Dongwu bian, 19, line 1. 6; Ira Kasoff, *The Thought of Chang Tsai (1020–1077)*, 44, translates this sentence with a figurative object of *qi*: "When *yin* causes it [that is *qi*] to condense, *yang* must cause it to disperse."

24. Song Yingxing, *On Qi*, chap. *xing qi hua* 1, 53; Wang Zichen and Xiong Fei, *Song Yingxing's Tenets*, 62, corrects the misprinted characters of the 1976 reprint in this passage.

25. Song Yingxing, *On Qi*, chap. *shui huo* 4, 85. For reference see also Song Yingxing, *On Qi*, chap. *qisheng* 9, 79 as to the sound of thunder.

26. All quotations in the paragraph are from Song Yingxing, *On Qi*, chap. *shui huo* 4, 85.

27. Ibid.

28. All quotations in this paragraph are from Song Yingxing, *On Qi*, chap. *shui feng guizang*, 91–92.

29. Song Yingxing, *Works of Heaven*, chap. 1, no. 5, 67a.

30. Ibid., no. 1, 4b.

31. Zhang Luxiang, *Bu [Shenshi] Nongshu* [Supplement to Shen's Agricultural Book], preface ii.

32. Song Yingxing, *On Qi*, chap. *xing qi hua* 4, 59.

33. Ibid., 5, 62. The expression *shanzhi*, "to mingle clay for developing implements (qi)" refers to Lao Zi, *New Text Laozi*, par. 11; Yang Weizeng, *A Discussion of Song Yingxing's Thoughts*, 164, explains the word *shan* as a technical process of blending water and soil.

34. Song Yingxing, *On Qi*, chap. *xing qi hua* 1, 52.

35. Ibid., 53–54. See Yang Weizeng, *A Discussion of Song Yingxing's Thoughts*, 165, for explanations on misprinted characters in this section.

36. Song Yingxing, *On Qi*, chap. *xing qi hua* 5, 62.

37. Ibid., 1, 52–53.

38. Sima Qian, *Records of the Grand Historian*, 8: chap. *liezhuan* 105, 2785–2820. Lu

Gwei-Djen and Joseph Needham, *Celestial Lancets: A History and Rationale of Acupuncture and Moxa,* 3–4, 106–13 for an introduction to Chunyu Yi [Shunyü I] (here introduced with different date, 216–145 BC) and his clinical cases records.

39. Song Yingxing, *On Qi*, chap. *xing qi hua* 3, 57.

40. Ibid., 58.

41. Ibid., 57.

42. Ibid., chap. *shui huo* 3, 83–84.

43. Ibid.

44. Ibid., chap. *xing qi hua* 1, 53–54.

45. Song Yingxing, *Works of Heaven*, chap. 3, no. 14, 16b–17a.

46. Ibid., chap. 2, no. 7, 3b–4a.

47. Song Yingxing, *On Qi*, chap. *xing qi hua* 4, 58.

48. Ibid., 59.

49. Ibid., chap. *shui huo* 3, 83.

50. Thanks to Nathan Sivin who brought this to my attention.

51. Song Yingxing, *On Qi*, chap. *xing qi hua* 1, 53, chap. *qisheng* 5, 71–72; Song Yingxing, *Works of Heaven*, chap. 3, no. 18, 65a.

52. Song Yingxing, *On Qi*, chap. *xing qi hua* 4, 59.

53. Nathan Sivin, *Chinese Alchemy: Preliminary Studies,* 6–7. See also his appendix G, in which he identifies *xiongci shi* as Schwefelarsenik (realgar). *Kui* could also mean "large clam," adding that clamshells would leave a residue when fired at a high temperature. Song was probably aware of this because clamshell is used to produce lime. He roughly delineates lime production in Song Yingxing, *Works of Heaven*, chap. 3, no. 16, 12a, however, without mentioning clamshells.

54. Song Yingxing, *On Qi*, chap. *xing qi hua* 4, 59–60.

55. Song Yingxing, *Works of Heaven*, chap. 2, no. 12, 63b. Sesame oil was preferred, yet this oil was extremely expensive and rare, and thus mostly used for cooking. Only court and high officials could afford it, and then they mostly used it for cooking. H. T. Huang, *Fermentation and Food Science,* pt. 5 of *Biology and Biological Technology,* vol. 6 in Science and Civilisation in China, edited by Joseph Needham, 102–7, 440.

56. José Ignacio Cabezón, "Buddhism and Science: On the Nature of the Dialogue," in Allan B. Wallace, ed., *Buddhism and Science: Breaking New Ground,* 60, note 38 (page 65); Yang Chung-Chieh, "Fojing linxu chen—zuizhong jiben lizi, zhenkong ji liangzi zhi yuan?" [Buddhistic Linshitron: The Ultimate Fundamental Particle, Vaccum, and the Origin of Quantum?], *Fojiao yu kexue* 7, no. 1: 34–45.

57. Song Yingxing, *On Qi*, chap. *shui chen* 1, 86; here Song focused on the quality of particles. For a different interpretation, see Li Shuzeng, Sun Yujie, and Ren Jinjian, *Philosophy in China,* 1438–39, who suggests that Song described in this passage the quality (*zhi*) of *qi*. Yang Weizeng, *Research about the Ideas of Song Yingxing,* 202.

58. Song Yingxing, *On Qi*, chap. *shui chen* 2, 88.

59. Ibid., 87.

60. Ibid., 86.

61. All quotations in this paragraph are from Song Yingxing, *On Qi*, chap. *shui chen* 2, 87.

62. Song Yingxing, *On Qi*, chap. *qisheng* 5, 71.

63. Ibid., 72. Song also mentions it in a slightly different way in chap. *xing qi hua* 2, 54.

64. Song Yingxing, *On Qi*, chap. *shui huo* 1, 80.

65. Ibid., chap. *xing qi hua* 1, 52. For a correction of characters, see Yang Weizeng, *Research about the Ideas of Song Yingxing*, 164.

66. Song Yingxing, *On Qi*, chap. *xing qi hua* 5, 62.

67. Ibid., 61.

CHAPTER 6

1. Lanling Xiaoxiao Sheng, *Zhang Zhupo piping Jinping mei* [Zhang Zhupo annotating Blossom in a Golden Vase], ed. Wang Youmei. The name of the author is a pseudonym, his identity is unknown. Zhang Weihua, "Music in Ming Daily Life, as Portrayed in the Narrative Jin Ping Mei," *Asian Music* 23, no. 2: 107.

2. Joseph S. C. Lam, *State Sacrifices and Music in Ming China: Orthodoxy, Creativity, and Expressiveness*, and "Huizong's Ritual and Musical Insignia," *Journal of Ritual Studies* 19, no. 1: 4; For connoisseurship and amateur ideals, see Craig Clunas, *Fruitful Sites: Garden Culture in Ming Dynasty China*, 38, 152–54; Zhu Zaiyu compiled several works, three of which discussed ideas of equal temperament. In 1596 he published the *Lüxue xinshuo* [A New Account of the Study of Lü], ca. 1606 the *Lülü jingyi* [The Essential Meaning of Lülü], and ca. 1603 the *Sanxue Xinshuo* [New Ideas on the Three Schools]. Chen C-Y, "A Re-visit of the Work of Zhu Zaiyu in Acoustics," in *Current Perspectives in the History of Science in East Asia*, ed. Kim Yung Sik and Francesca Bray, 332–36. For the general approach of this era to "things" I refer to Craig Clunas, *Superfluous Things: Culture and Social Status in Early Modern China*, 166–67.

3. Robert H. van Gulik, "The Lore of the Chinese Lute: An Essay in Ch'in Ideology [Continued]," *Monumenta Nipponica* 2, no. 1: 82.

4. Dai Nianzu et al., "Shengxue shi"[A History of Acoustics], in *Zhongguo wuli xueshi da xi* [A Series of Books on History of Physics in China,] ed. Dai Nianzu,17–47.

5. Cullen, "The Science/Technology Interface," 306.

6. Frederick Vinton Hunt, *Origins in Acoustics: The Science of Sound from Antiquity to the Age of Newton*, 3.

7. Walter Kaufmann, *Musical References in the Chinese Classics*, summarizes quotations concerning music in ancient classics of China. John Emerson, "Yang Chu's Discovery of the Body," *Philosophy East and West* 46, no. 4: 533–66, discusses the correlation of music, ritual, and harmonization in early thought (to third century BC).

8. Kenneth DeWoskin, *A Song for One or Two*, chap. 4 includes early Chinese documents discussing the physics of music. Charles Le Blanc, "From Cosmology to Ontology through Resonance: A Chinese Interpretation of Reality," in *Beyond Textuality: Asceticism and Violence in Anthropological Interpretation*, ed. Gilles Bibeau and Ellen Corin, 57–78, illustrates this using the example of the idea of resonance in the *Huainan zi*. Roel Sterckx, "Transforming the Beast: Animals and Music in Early China," *T'oung Pao* 136: 4–5, brings to attention the role of music for transformative processes and describes Chinese ideals of music as a civilizing instrument.

9. *Yueji* [Record of Music], compiled in fifth century BC, incorporated in the *Liji* [Book of Rites] in the first century BC. See also *Shujing* [Book of Documents], *Sishu wujing* edition, chap. 1 Wushu 5, 18; Sima Qian, *Records of the Grand Historian*, chap. *yue shu* 24, zhang 2, 1214.

10. *Guo Yu* [Discourses from the States], 5 Zhouyu xia, 4b–5b.

11. *Ming Huang and Yang Gueifei Listening to Music, 1368–1400* (Worcester Art Museum, 1936.4, handscroll; ink and light color on silk with three interpolated seals of Qian Zuan [about 1235–after 1300]). The scroll is discussed by George A. Rowley, "A Chinese Scroll of the Ming Dynasty: Ming Huang and Yang Kuei-fei Listening to Music," *Attribus Asiae*, 31, no. 1: 13–19.

12. Title of the Scroll: *Female Musicians Playing before the Emperor*, Ming dynasty (1368–1644), early fifteenth century, handscroll, ink and colors on silk, the Art Institute of Chicago, ID 1950.1370. It is traditionally attributed to Zhou Wenju of the tenth century, although the existing version originates from circa 1401–33. *Zhongguo lidai huihua: Gugong bowuyuan canghua ji* [Chinese Paintings from Successive Dynasties of the Beijing Palace Museum Collection], vol. 1, 91, shows details of the women musicians on this scroll.

13. For the organization of a Chinese orchestra see Han Kuo-Huang and Judith Gray, "The Modern Chinese Orchestra," in *Asian Music* 11, no. 1: esp. 23–25.

14. Wu Guodong, *Zhongguo yinyue* [China's Music], 330–52; John S. Major and Jenny F. So, "Music in Late Bronze Age," in *Music in the Age of Confucius*, ed. Jenny F. So, 13–33.

15. *Shujing* [Book of Documents], chap. 1 *wushu* 5, 18; Sima Qian, *Records of the Grand Historian*, chap. *yueshu* 24, zhang 2, 1214. Joseph Needham, *Physics*, 140–41.

16. Song Yingxing, *On Qi*, chap. *qisheng* 1, 64.

17. Ibid., 4, 69; Zhang Zai, "Correct Discipline for Beginners," chap. 3 *dongwu*, 20.; Wang Fuzhi gives a rather lengthy commentary on this passage in his *Commentary of the Discourse for Beginners of Master Zhang*, chap. *dongwu*, 203.

18. Cullen, "The Science/Technology Interface," 302, see also 313.

19. Zhang Zai, "Correct Discipline for Beginners," chap. *dongwu* 5, 20, line 6.

20. Song Yingxing, *On Qi*, chap. *qisheng* 2, 66–67.

21. Song Yingxing, *On Qi*, chap. *qisheng* 4, 69. The expression *gemo* Song uses in this context was a *terminus technicus* for a separating skin or a membrane.

22. Song Yingxing, *On Qi*, chap. *qisheng* 4, 69.

23. Ibid., 2, 66.

24. Ibid., 7, 75–76.

25. Ibid., 66–67.

26. Ibid., 66.

27. Ibid., chap. *xing qi hua* 2, 55. The Chinese expression *Shana* of the Sanskrit *skana* describes the smallest unit of time used by Buddhists. It was adopted into the Chinese language during the Tang dynasty.

28. Song Yingxing, *On Qi*, chap. *qisheng* 1, 64.

29. Ibid., 2, 66.

30. Ibid., 8, 77–78.

31. Ibid., 3, 68.

32. Li Shuzeng, Sun Yujie, and Ren Jinjian, *Philosophy in China*, 1439. Yang Weizeng,

Research about the Ideas of Song Yingxing, 179ff., also explains the idea of two primarily with regard to a mutual collision of *qi* without further specifying on this point. Huang Mingtong, "Cong Lun Qi kan Song Yingxing de ziran guan" [Song Yingxing's Notion of Nature according to his Treatise *On Qi*], *Huanan shifan xuebao shehui kexueban* 4, no. 6: 27 gives the same explanation.

33. Song Yingxing, *On Qi*, chap. *xing qi hua* 2, 55.

34. Joseph R. Levenson, "The Amateur Ideal in Ming and Early Ch'ing Society: Evidence from Painting," in *Chinese Thought and Institutions*, ed. John K. Fairbank, 321–23.

35. "Yueji," in the *Liji* [Book of Rites], chap. 7, 6b–7a (204).

36. Tan Qiao, *Hua shu* [A Book about Transformation], ed. Ding Zhenyan and Li Sizhen, chap. *shu hua*, entry *shengqi*, 27. See also Needham, *Physics*, 208; Wang Chong, "Discourse Weighed in Balance," chap. 20, *lunsi* 62, 808; for Wang Chong's idea of the human body see also pages 76–80, 108–13.

37. Luo Qinshun, *Kunzhi ji* [Record on Knowledge Painfully Acquired], pt. II, 12–13; Bloom, *Knowledge Painfully Acquired*, 34.

38. Wang Chong, *Discourse Weighed in Balance*, chap. *bianxu* 4, 341–42.

39. Song Yingxing, *On Qi*, chap. *qisheng 1*, 64.

40. Song Yingxing, *On Qi*, chap. *qisheng* 9, 79. The voice of the house cricket and the winter cicada is a commonly used metaphor for stillness and a return to silence. It indicates the period when the weather is getting cold.

41. Ibid.

42. Ibid., 1, 64.

43. Tan Qiao, *Book about Transformation*, chap. *dao hua*, entry *da han*, 15.

44. Song Yingxing, *On Qi*, chap. *qisheng* 8, 77.

45. Zhang Zai, "Correct Discipline for Beginners," chap. *dongwu* 3, 20. For Wang Fuzhi's commentary see *Zhang Zi Zheng Meng zhu* [Commentary of the Discourse for Beginners of Master Zhang], chap. *dongwu* 3, 52.

46. Quotations in the paragraph are from Song Yingxing, *On Qi*, chap. *qisheng* 8, 77–78.

47. Song Yingxing, *Works of Heaven*, chap. 2, no. 8, 18b–19a. For a technical analysis see William Rostoker, Bennet Bronson, and James Dvorak, "The Cast-Iron Bells of China," *Technology and Culture* 25/4: 752; Lothar von Falkenhausen, *Suspended Music: Chime-Bells in the Culture of Bronze Age China*, 98–125, 190–93, for the changes in technology.

48. Fabrizio Pregadio, *The Encyclopedia of Daoism*, vol. 1, 30–31.

49. Tan Qiao, *Book of Transformation*, chap. *dao hua*, entry *da han*, 15.

50. Ban Gu, *Baihu tong de lun* [Discussions of Virtue in the White Tiger Hall], *Baizi quanji* [Complete Anthology of the Hundred Masters], chap. 2 shanfu, 2b.

51. Song Yingxing, *On Qi*, chap. *qisheng* 2, 66.

52. Ibid., 3, 68. The expression *jinghun sangpo* literally means that the body soul and the mental soul are disturbed, which means man's health is crucially affected. Referring to man's life and death Song uses the same terms as Zhang Zai, "Correct Discipline for Beginners," chap. *dongwu* 5, 19, line 5.

53. Cullen, "The Science/Technology Interface," 308, delineates the initial idea of this compound in the context of the *Huainan zi* and the *Binglue xuan*.

54. Song Yingxing, *On Qi*, chap. *qisheng* 3, 68.

55. Ibid. The term *sanling* used in this passage is ambiguous and can mean both the mulberry trees or in the context of music a special type of court music that was played during the Yin reign and is described in the *Commentaries of Master Zuo Traditions*, chap. 1 *yingong* 3, 45, chap. 11 *xianggong* 10, 350, commentary of Kong Yingda. See Kenneth J. DeWoskin, "Picturing Performance: The Suite of Evidence for Music Culture in Warring States China," in *La pluridisciplinarité en archéologie musicale: IVe-rencontres internationales du Groupe d'études sur l'archéologie musicale de l'ICTM (8–12 Octobre 1990)*, ed. Catherine Homo-Lechner and Annie Bélis, esp. 354.

56. Song Yingxing, *On Qi*, chap. *qisheng* 7, 75–76.

57. Ibid., 8, 77–78.

58. Ibid., 7, 75.

59. Lu Jiansan, "Chuchu dou zhu zu yiyi bu neng qu—Yuan, Ming, Qing Hangzhou de lüyou" [Tourism in Hangzhou Thirteenth to Nineteenth Century], in *Yuan, Ming, Qing Mingcheng Hangzhou* [A Famous City in the Yuan, Ming, and Qing Dynasties: Hangzhou], ed. Zhou Feng and Xiang Xiuwen, 279–90; Zheng Yun, *Hangzhou Fuzhi* [Hangzhou Gazetteer], chap. *shi huo*, 15a.

60. Bell Yung, ed., *Celestial Airs of Antiquity: Music of the Seven-String Zither of China*, 2–9.

61. Huang Xiangpeng, *Zhongguoren de yinyue he yinyue xue* [Music and Musicology of the Chinese], vol. 1, 51.

62. Shen Gua (1031–95), *Brush Talks from the Dream Brook*, chap. *bu bitan* V, 382.

63. *Shuoyuan jiaozheng* [Collection of Speeches with Corrections and Annotations], ed. Xiang Zonglu, chap. 16, 108.

64. Wang Dang (fl. 1101–10), *Tang Yulin: Fujiao kanji* [A Forest of Sayings of the Tang Dynasty, Collated and Proofread], chap. 5 *buyi*, 162.

65. Liu Yuxi (772–842), *Liu Binke jiahua lu* [Record about Liu Binke's Auspicious Words], ed. Wei Xuan, Zhao Lin, and Su E, chap. *Gushi wenfang xiaoshuo* 3a.

66. Song Yingxing, *On Qi*, chap. *qisheng* 4, 69.

67. Ibid., 6, 73.

68. Song Yingxing, *Works of Heaven*, chap. 2, no. 10, 51a.

69. Song Yingxing, *On Qi*, chap. *qisheng* 6, 73.

70. Ibid.

71. Ibid., 5, 71. The expression *qingshen liaoyun* refers to classifying tones, in this case the class of earth agent (*tu*) tones.

72. Ibid., 6, 73. A *tang* belongs to the category of *jinge* instruments. Like *gong* and *shang*, Song here does not consider *jiao* and *yu* as definite tones but is just describing a sound quality. He refers to the entries for the musical officials in the *Zhouli* [Rites of Zhou], entry *Diantong* chap. *zhengyi* 46, 1b–4a. The *Kaogong ji* [Artificer's Record], chap. 4 in *Zhouli, Sishu wujing* edition, 4a discusses material-sound quality in the section on bell making and states, "bells with thick (*hou*) walls produce heavy (*shi*) sounds, whereas those with thin walls produce scattered (*bo*) sounds."

73. Song Yingxing, *On Qi*, chap. *qisheng* 4, 69.

74. Ibid., 5, 72.

CONCLUSION

1. *Book of Changes*, vol. 1, suppl. *xici shang*, 56.

2. Li Shuzeng, Sun Yujie, and Ren Jinjian, *Philosophy in China*, 1441–43.

3. Song Yingxing, *Works of Heaven*, chap. 3, no. 16, 13a.

4. Ibid., no. 11, 52a–63b.

5. Song Yingxing, *Talks about Heaven*, chap. *ri shuo* 3, 105–6.

6. Cang Xiuliang, "Zhu Xi zizhi tongjian gangmu" [Zhu Xi and the Comprehensive Mirror for Aid in Government], *Anhui shixue*, no. 1: 19.

7. Song Yingxing, *Works of Heaven*, preface, 1b; the story Song was referring to can be found in Confucius, *Kongzi jiayu* [The School Sayings of Confucius], ed. Robert Kramers and Su Wang, chap. *zhisi bian*, 12.

8. Song Yingxing, *Works of Heaven*, Preface, 1b.

9. Zhang Hua, *Bowuzhi xiaozheng* [An Annotated Correction of the Report of Extensive Matters], ed. Fan Ning and trans. Roger Greatrex, *The Bowu Zhi: An Annotated Translation*, chap. 3, 50. See also the translation of the Ming dynastic recension of Zhang Hua's work 52–58.

10. Brook, *The Confusions of Pleasure*, 168–69.

EPILOGUE

1. Song Yingxing, *Works of Heaven*, chap. 2, no. 13, 70b–71b.

2. Ibid., preface, 1a–1b.

3. Song Yingxing, *An Oppositionist's Deliberations*, chap. *shiqi yi*, 12.

4. Song Yingxing, *Works of Heaven*, preface, 3b.

5. Scarlett Jang, "Form, Content, and Audience: A Common Theme in Painting and Woodblock-Printed Books of the Ming Dynasty," *Ars Orientalis* 27: 21.

6. Shmuel N. Eisenstadt and Luis Roniger, *Patrons, Clients, and Friends: Interpersonal Relations and the Structure of Trust in Society*, 173–84.

7. Song Yingxing, *Works of Heaven*, preface 2b.

8. The full biography of Tu Shaokui can be found in Yang Zhouxian and Zhao Yuemian, *Xinjian Xianzhi* [Local Gazetteer of Xinjian County] (1680), chap. 25, 34. Tu is also mentioned in his father's official biography in Zhang, et al., *Ming History*, vol. 20, chap. 233, 6082. Tu's pupil/acquaintance Xiong Wenju wrote a poem about his master that was published in the anthology *Xue Tang xiansheng wenxuan* [A Selection of Master Xue Tangs Writings] of the year 1655, chap. 15, 1–4a. Xiong Wenju, another Nanchang honorable, praised Tu Shaokui for his social concern. See also Song Yingsheng, "Cha Tu Taiyin mufu renwen" ["Fengxin yaji zangban," 1638], partly collated in *Collection of the Fangyu Halls*, vol. 4, chap. 7, 10–12. Bai Huang and Cha Zhenxing (Zha Shenxing), *Xijiang zhi* [Gazetteer of the Western River] (1720), chap. 70, 5a–b.

9. Song Yingxing, *Works of Heaven*, preface 2b.

10. Pan Jixing, *Biography of Song Yingxing*, 249–50.

11. Benjamin A. Elman, "Collecting and Classifying: Ming Dynasty Compendia and Encyclopedias (Leishu)," *Extrême orient, Extrême occident, hors série*, 132.

12. Of Yingxing's knowledge on mining see Donald Wagner, *Iron and Steel in Ancient China,* and "Chinese Blast Furnaces from the 10th to 14th Century," *Journal of Historical Metallurgy* 37, no. 1, and Peter Golas, *Mining,* pt. 13 of *Chemistry and Chemical Technology in China.*

13. Chen Hongxu, *Collected Works of Master Chen Shiye,* chap. 4, 16–17.

14. Chen Hongxu, *You yang shanfang zangshu ji* [Account about the Book Collection of the Youyang Mountain Refuge] (1622). Chen Hongxu published another bibliography in 1637 under the title *Xushu muji* [Continued Report about Books] in 6 chapters. The library drowned when Chen Hongxu took flight in 1645 from the invading Manchu. Chap. *Hongxu zhizhou yueyuan shu* [Chen Hongxu's Book of the Amusement Park] in the "Shizhuang ji " [Collections of the Stone Cavern], in Chen Hongxu, *Collected Works of Master Chen Shiye,* appendix.

15. Song Yingxing, *Talks about Heaven,* chap. *rishuo* 1, 103, chap. *rishuo* 3, 105. Song Yingxing, *Works of Heaven,* chap. 3, no. 15, 29b.

16. Chen Hongxu, *Collected Works of Master Chen Shiye,* chap. 3b, 14a, 23b, chap. 9, 18a. See Pan Jixing, *Biography of Song Yingxing,* 216–19 for a full overview of Chen's guests.

17. Most bibliophiles carefully regulated access to prevent stealing. Joseph Peter McDermott, *Social History of the Chinese Book: Books and Literati Culture in Late Imperial China,* 138.

18. Chen Hongxu, *Collected Works of Master Chen Shiye,* chap. 1, 6, chap. 6, 104.

19. McDermott, *Social History of the Chinese Book,* 155–62.

20. Chen Hongxu, *Collected Works of Master Chen Shiye,* chap. 4, 16–17, chap. 2, 35–36.

21. Chen Hongxu, *Hanye lu* [Record of a Winter's Night] (1637; published 1650–80).

22. Xu Shipu, *Yu Dun ji* [Collection of Yu Dun] (*Kangxi yuankeben* 1691), chap. *wenshang,* 7.

23. Liu Tongsheng, *Jinlin shiji* [Anthology of Poetry by Jin Lin], vol. 5, chap. 13, 13; Zhang et al., *Ming History,* vol. 19, chap. 216, 5710.

24. R. A. Donkin, *Beyond Price—Pearls and Pearl-fishing: Origins to the Age of Discoveries,* 165–68.

25. Song Yingxing, *An Oppositionist's Deliberations,* preface, 3a/b.

26. Ge Quan, *Liming yu zhongcheng yi shiren zhengzhi jingshen de dianxing fenxi* [Order and Loyalty—Analysis of the Characteristics of the Spirit of Scholarly Politics], chap. 2, 34–72. For *ye* see 7–9.

27. Song Yingxing, *An Oppositionist's Deliberations,* chap. *shiqi yi,* 12–13.

28. Ma Tai-loi, "The Local Education Officials of Ming China, 1368–1644," *Oriens Extremus* 22, no. 1: 13.

29. Song Yingxing, *An Oppositionist's Deliberations,* chap. *xuezheng yi,* 31–34.

30. Song Yingxing, *Works of Heaven,* preface, 3b.

31. *Book of Changes,* chap. 1 *dachu,* 25–26.

32. Bol, *Intellectual Transitions in T'ang and Sung China,* 301.

33. Song Yingxing, *Works of Heaven,* chap. 3, no. 17, 48a.

34. Poo Mu-chou, "The Use and Abuse of Wine in Ancient China," *Journal of the*

Economic and Social History of the Orient 42, no. 2: 123–51; Kandice Hauf, "The Community Covenant in Sixteenth Century Ji'an Prefecture, Jiangxi," *Late Imperial China* 17, no. 2: 11.

35. Chan Hok-lam, *Legitimation in Imperial China: Discussions under the Jurchen-Chin Dynasty (1115–1234)*, 48.

36. In the case of the great ritual controversy, for example, during the reign of Emperor Shizong the question whether the object of loyalty was the emperor himself or the institution of the dynasty provoked a court crisis. Chu Hung-lam, "Review of *The Chosen One: Succession and Adoption in the Court of Ming Shizong*, by Carney T. Fisher," *Harvard Journal of Asian Studies* 54, no. 1: 276. Vast amounts of such discussions can be found since the Song dynasty. In the same way as scholars had to qualify for their advancement, the emperor had to validate his authority in terms of educational effort as well as the performance of Confucian morals. Dieter Kuhn, "Family Rituals," *Monumenta Serica* 40: 377, and Ho Ping-ti, *The Ladder of Success in Imperial China: Aspects of Social Mobility, 1368–1911*, chap. 2, 5, 6.

37. Song Yingsheng revised and compiled the *Enping xianzhi* [Gazetteer of Enping County] between the years 1637 and 1639. The preface is dated 1638. The history of the Ming publication is documented in Feng Shiyuan and Shi Tai, *Enping xianzhi* [Gazetteer of Enping County], preface, 5b. Song Yingsheng furthermore compiled his private jottings *Collection of the Fangyu Halls*, in 49 chapters (*juan*) with prefaces 1638 by Wu Ruijiang and Zhao Shijin and the author himself. The original print is now deposited in the Provincial Library of Hu'nan (Hu'nan shiyan tushuguan). A photocopy of these documents is stored in the Song Yingxing Museum. It was listed as a banned book. Sun Dianqi, *Qingdai jinshu zhijianlu* [Record of Known Banned Books of the Qing Dynasty], 24. All five sons of Yingsheng refused to serve the Qing dynasty (1644–1912) claiming loyalty to the Ming court. Song Liquan and Song Yude, *Family Annals of the House of Song*, chap. 22, 12; Lü Maoxian and Shuai Fangwei, *Local Gazetter of Fengxin*, chap. *renwuzhi*, 7b.

38. Yang Zhouxian and Zhao Yuemian, *Local Gazetteer of Xinjian County*] (1680, Kangxi keben), chap. 25, 34. See also the edition of the *daoguang* period of the year 1824 by Cui Deng'ao, *Xinjian xianzhi* [Local Gazetteers of Xinjian], Zhongguo fangzhi congshu, Huazhong difang 884, chap. 25, 34, chap. 40, 36–37.

39. Song Yingxing, *Yearnings*, chap. *lianyu shi*, poem 17, 133.

40. Hegel, *Reading Illustrated Fiction in Late Imperial China*, 98–103.

41. Chen Hongxu, *Collected Works of Master Chen Shiye*, preface, 3a. For the general situation, see McDermott, *Social History of the Chinese Book*, 115–41.

42. Lynn A. Struve, *The Ming-Qing Conflict, 1619–1683: A Historiography and Source Guide*, chap. 2, 25–27.

43. See originals in Bibliothéque Nationale France and Beijing National library. Pan Jixing, *Investigation of the Tiangong kaiwu*, 140–47.

44. Lucille Chia, *Printing for Profit: The Commercial Publishers of Jianyang, Fujian (11th–17th Centuries)*, 252.

45. Both covers of the second edition are also copied in Pan Jixing, *Investigation of the Tiangong kaiwu*, 143.

46. Wang Yeqiu, *Liulichang shihua* [Historical Chats about Liulichang], 12–18. For the distribution networks of smaller publishing houses in a regional frame see also Cyn-

thia J. Brokaw, "Commercial Publishing in Late Imperial China: The Zou and Ma Family Businesses of Sibao, Fujian," *Late Imperial China* 17, no. 1: 76–78. According to this source, the branches were often financially independent of the mother company. However, they distributed their books in one network of marketing, transport, and shops.

47. I am grateful to Lucille Chia, who brought these issues to my attention. She answered my questions with aptitude and much kindness. *Chunqiu Zuo Zhuan gangmu dingzhu* [Annotations and Remarks to the Spring and Autumn Commentaries of the Master Zuo Traditions systematically rearranged] supposedly compiled by Li Tingji (1542–1616), Naikaku Bunko, ed., *Kaitei Naikaku bunko kanseki bunrui mokuroku* [A Classified Catalogue of Chinese Literature from the Diet Library], 274 han, 158 hao. Zhao Wanli, *Guoli Beiping tushuguan shanben shumu* [Collection of Rare Books by the Beiping Library], jingbu 2576. Li Tingji was a native of Jinjiang in southern Fujian and held the position of grand secretary during the early seventeenth century. More than 25 Jianyang imprints—especially examination literature—were attributed to him. He became famous because he placed first in the metropolitan examination 1583. See also "Li T'ing-chi," in L. Carrington Goodrich, *Dictionary of Ming Biography, 1368–1644*, 1: 329.

48. Zhang Xiumin, *Zhang Xiumin yinshua shi lunwen ji* [Collected Papers on the History of Printing in China]. K. T. Wu, "Ming Printing and Printers," *Harvard Journal of Asiatic Studies* 13: 213.

49. Brokaw, "Commercial Publishing in Late Imperial China," 72, 76–77, indicates the small dimensions which a distribution network could have. See also Bao Fasheng, "Sibao diaoban yinshuaye de qingkuang diaocha" [Investigation of the Sibao Publishing Industry], *Liancheng wenshi ziliao* 18: 73.

50. The *Works of Heaven* arrived in France around 1724. The exact date is not known. A reprint was published in Japan in 1771. For further details see appendix.

51. Benjamin A. Elman, "The Story of a Chapter: Changing Views of the 'Artificer's Record' ('*Kaogong ji*') and the *Zhouli*," in *Statecraft and Classical Learning: The Rituals of Zhou in East Asian History.*

52. Fang Chaoying, "O-er t'ai," in Arthur Hummel, ed., *Eminent Chinese of the Ch'ing Period*, 601–3.

BIBLIOGRAPHY

Allsen, Thomas T. "The Circulation of Military Technology in the Mongolian Empire." In *Warfare in Inner Asian History (500–800)*. Edited by Nicola Di Cosmo. Handbook of Oriental Studies 6. Leiden: Brill, 2002.

———. *Commodity and Exchange in the Mongol Empire: A Cultural History of Islamic Textiles*. Cambridge Studies in Islamic Civilization. Cambridge: Cambridge University Press, 1997.

An Yanming. "Liang Shuming and Henri Bergson on Intuition: Cultural Context and the Evolution of Terms." *Philosophy East and West* 47, no. 3 (1997): 337–62.

Asim, Ina. "The Merchant Wang Zhen, 1424–1495." In *The Human Tradition in Premodern China*. Edited by Kenneth J. Hammond. The Human Tradition around the World 4. Wilmington: Scholarly Resources, 2002.

Bai Huang 白潢, and Zha Shenxing 查慎行. *Xijiang zhi* 西江志 [Gazetteer of the Western River]. 1720. Zhongguo fangzhi congshu, Huazhong difang 783. Reprint, Taibei: Chengwen chubanshe, 1989.

Ban Gu 班固. *Han shu* 漢書 [History of the Han Dynasty]. Annotated by Yan Shigu 顏師古. Xianggang: Xianggang Zhonghua shuju, 1970. References are to the 1970 edition.

———. *Baihu tong de lun* 白虎通德論 [Discussions of Virtue in the White Tiger Hall]. *Baizi quanji* 百字全集 [Complete Anthology of the Hundred Masters]. Shanghai: Zhejiang renmin chubanshe, 1984.

Bao Fasheng 包發生. "Sibao diaoban yinshuaye de qingkuang diaocha" 四堡雕版印刷業情況調查 [Investigation of the Sibao Publishing Industry]. *Liancheng wenshi ziliao* 18 (1993): 70–82.

Black, Alison Harley. *Man and Nature in the Philosophical Thought of Wang Fu-chih*. Seattle: University of Washington Press, 1989.

Bloom, Irene. *Knowledge Painfully Acquired: The K'un-chih chi by Lo Ch'in-shun*. New York: Columbia University Press, 1987.

Bloor, David. *Knowledge and Social Imagery*. 1976. Reprint, Chicago: University of Chicago Press, 1991. References are to the 1991 edition.

Bol, Peter K. *"This Culture of Ours": Intellectual Transitions in T'ang and Sung China*. Stanford, CA: Stanford University Press, 1992.

Bray, Francesca. "Agricultural Illustrations: Blueprint or Icon." In *Graphics and Text in the Production of Technical Knowledge in China: The Warp and the Weft.* Edited by Francesca Bray, Vera Dorofeeva-Lichtmann, and Georges Métailié, 521–67. Sinica Leidensia 79. Leiden: Brill, 2007.

———. *Agriculture.* Pt. 2 of *Biology and Biological Technology,* vol. 6 in Science and Civilisation in China, edited by Joseph Needham. Cambridge: Cambridge University Press, 1984.

———. *Technology and Gender: Fabrics of Power in Late Imperial China.* Berkeley: University of California Press, 1997.

Bray, Francesca, Vera Dorofeeva-Lichtmann, Georges Métailié, eds. *Graphics and Text in the Production of Technical Knowledge in China: The Warp and the Weft.* Sinica Leidensia 79. Leiden: Brill, 2007.

Brockey, Liam Matthew. *Journey to the East: The Jesuit Mission to China, 1579–1724.* Cambridge: The Belknap Press of Harvard University Press, 2007.

Brokaw, Cynthia J. "Commercial Publishing in Late Imperial China: The Zou and Ma Family Businesses of Sibao, Fujian." *Late Imperial China* 17, no. 1 (1996): 49–92.

———. *The Ledgers of Merit and Demerit: Social Change and Moral Order in Late Imperial China.* Princeton: Princeton University Press, 1991.

Brook, Timothy. *The Confusion of Pleasure: Commerce and Culture in Ming China.* Berkeley: University of California Press, 1998.

———. *Praying for Power: Buddhism and the Formation of Gentry Society in Late-Ming China.* Harvard-Yenching Institute Monograph Series 38. Cambridge: Harvard University and the Harvard-Yenching Institute, 1993.

Cai Wenluan 蔡文鸞 (*juren* 1663), Lin Yulan 林育蘭 (17th–18th cent.), *Fenyi xian zhi : Jiangxi sheng* 分宜縣志： 江西省 [Gazetteer of Fenyi County: Jiangxi Province]. 1683. Zhongguo fangzhi congshu. Huazhong difang 752. Reprint, Taibei: Chengwen chubanshe, 1989.

Cabezón, José Ignacio. "Buddhism and Science: On the Nature of the Dialogue." In *Buddhism and Science: Breaking New Ground.* Edited by Alan B. Wallace. New York: Columbia University Press, 2003.

Campany, Robert Ford. *Strange Writing: Anomaly Accounts in Early Medieval China.* SUNY Series in Chinese Philosophy and Culture. Albany: State University of New York Press, 1996.

Cang Xiuliang 倉修良. "Zhu Xi zizhi tongjian gangmu 朱熹資治通鑑綱目 [Zhu Xi and the Comprehensive Mirror for Aid in Government]." *Anhui shixue,* no. 1 (2007): 18–24.

Chai Jiguang 柴繼光. "Guanyu Song Yingxing Tiangong kaiwu zhong 'chiyan' bufen yixie wenti de bianshi 關於宋應星天工開物中'池鹽'部分一些問題的辨識 [A Discrimination of Some Questions in the Part of 'Salt lake' in the Tiangong kaiwu]." *Yanye shi yanjiu* 1 (1994): 30–32.

Chan Hok-lam 陳學霖. *Legitimation in Imperial China: Discussions under the Jurchen-Chin Dynasty (1115–1234).* Seattle: University of Washington Press, 1984.

Chan, Wing-tsit. *A Source Book in Chinese Philosophy.* Princeton: Princeton University Press, 1969.

Chang, Chun-Shu, and Shelley Hsueh-lun Chang. *Crisis and Transformation in Seventeenth-century China: Society, Culture, and Modernity in Li Yü's World.* Ann Arbor: University of Michigan Press, 1990.

Chang Fuyuan 常福元. *Li Zicheng Shanbei shishi yanjiu* 李自成陝北史事研究 [Research on Li Zicheng and the Historical Affairs in Northern Shanxi]. Lanzhou: Gansu renmin chubanshe, 2006.

Chang, Kang-I Sun. *The Late-Ming Poet Ch'en Tzu-lung: Crises of Love and Loyalism.* New Haven: Yale University Press, 1991.

Chen Cheng-Yih. "A Re-visit of the Work of Zhu Zaiyu in Acoustics." In *Current Perspectives in the History of Science in East Asia.* Edited by Kim Yung Sik and Francesca Bray. Seoul: Seoul National University Press, 1999.

Chen Guying 陳鼓應. *Bencao gangmu tongshi* 本草綱目通釋 [General Explanation of the Systematic Materia Medica] by Li Shizhen 李時珍 (1518–93). 1596. Reprint, Beijing: Xuefan chubanshe, 1992. References are to the 1992 edition.

Chen Guying 陳鼓應, Xin Guanjie 辛冠洁, and Ge Rongjin 葛榮晉, eds. *Ming Qing shixue sichao shi* 明清實學思潮史 [The History of the Trend of Practical Learning during the Ming and Qing Dynasty]. 2 vols. Ji'nan: Qilu shushe, 1989.

Chen Hongxu 陳宏緒. *Chen Shiye xiansheng ji* 陳士業先生集 [Collected Works of Master Chen Shiye]. 1687. Hsinchu: T'sing Hua University Microfilms.

———. *Hanye lu* 寒夜錄 [Record of a Winter's Night]. 1637. Published between 1650–80. Baibu congshu jicheng 24. Xuehai leibian 1399. Reprint, Taibei: Yiwen yinshuguan, 1965–70.

———. *Youyang shanfang cangshu ji* 酉陽山房藏書紀 [Record of the Book Collection of the Youyang Mountain Refuge]. 1622. Hsinchu. Tsing Hua University Microfilms.

Chen Jiujin 陳久金, and Yang Yi 楊怡. *Zhongguo gudai de tianwen yu lifa* 中國古代的天文與歷法 [Ancient Chinese Astrology and Astronomy]. Taibei: Taiwan shangwu yinshuguan, 1993.

Chen Juanjuan 陳娟娟. *Sichou shihua* 絲綢史話 [Silk Manufacture and Trade]. Beijing: Zhonghua shuju, Xinhua shudian Beijing faxing suo faxing, 1980.

Chen Shou 陳壽. *Sanguo zhi* 三國志 [Record of the Three States]. Reprint, Xianggang: Zhonghua shuju, 1971.

Chia, Lucille. *Printing for Profit: The Commercial Publishers of Jianyang, Fujian (11th–17th Centuries).* Harvard-Yenching Institute Monograph Series 56. Cambridge: Harvard University Asia Center for the Harvard-Yenching Institute, 2002.

Ching, Julia. *The Religious Thought of Chu Hsi.* Oxford: University Press, 2000.

Chiu Pengsheng. "The Discourse on Insolvency and Negligence in Eighteenth-Century China." In *Writing and Law in Late Imperial China, Crime, Conflict, and Judgment.* Edited by Robert E. Hegel and Katherine Carlitz. Asian Law Series 18. Seattle: University of Washington Press, 2007.

Chow Kai-wing. *The Rise of Confucian Ritualism in Late Imperial China: Ethics, Classics, and Lineage Discourse.* Stanford, CA: Stanford University Press, 1994.

———. "Writing for Success: Printing, Examinations, and Intellectual Change in Late Ming China." *Late Imperial China* 17, no. 1 (1996): 120–57.

Chu Hsi [Zhu Xi] and Lü Tsu-ch'ien [Lü Zuqian], eds. *Reflections on Things at Hand: The*

Neo-Confucian Anthology. Translated, with notes, by Wing-Tsit Chan. Records of Civilization 75. New York: Columbia University Press, 1967.

Chu Hung-lam 朱鴻林. "Ch'iu Chün's Ta-hsüeh yen-i-pu and its Influence in the Sixteenth and Seventeenth Centuries." *Ming Studies* 22 (1986): 1–32.

———. "Review of *The Chosen One: Succession and Adoption in the Court of Ming Shizong,* by Carney T. Fisher." *Harvard Journal of Asiatic Studies* 54, no. 1 (June 1994): 266–77.

Clunas, Craig. *Fruitful Sites: Garden Culture in Ming Dynasty China.* London: Reaktion Books, 1996.

———. *Superfluous Things: Culture and Social Status in Early Modern China.* Cambridge: Polity Press, 1991.

Crisciani, Chiara. "History, Novelty, and Progress in Scholastic Medicine." *Osiris,* 2nd ser., 6 (1990): 118–39.

Cua, Antonio S. *The Unity of Knowledge and Action: A Study in Wang Yang-ming's Moral Psychology.* Honolulu: University of Hawai'i Press, 1982.

Cui Deng'ao 崔登鰲. *Xinjian xianzhi* 新建縣志 [Local Gazetteers of Xinjian]. 1849. National Library of China. Call number 250.15/36.29.

Cui Shu 催述, "Kexue de gushijia 科学的古史家" [Cui Shu as a Scientific Historian of Antiquity]. In *Cui Dongbi yi shu* 催东壁遗书 [An Epitaph of Cui Shi]. Edited by Gu Jiegang. 1936. Reprint, Shanghai: Shanghai guji chubanshe, 1989.

Cullen, Christopher. *Astronomy and Mathematics in Ancient China: The Zhoubi suanjing.* Needham Research Institute studies 1. Cambridge: Cambridge University Press, 1996.

———. "The Science/Technology Interface in Seventeenth-Century China: Song Yingxing 宋應星 on Qi 氣 and Wuxing 五行." *Bulletin of the School of Oriental and African Studies* 53 no. 2 (1990): 295–318.

Cutter, Robert Joe. "Cao Zhi's (192–232) Symposium Poems." *Chinese Literature: Essays, Articles, Reviews* (CLEAR) 6, no. 1/2 (1984): 1–32.

Dai Nianzu 戴念祖 et al. *Zhongguo wuli xueshi da xi* 中國物理學史大系; *Guangxue shi* 光學史; *Lixue shi* 力學史; *Zhongwai wuli jiaoliu shi* 中外物理交流史; *Shengxue shi* 聲學史 [A History of Ancient Physics; A History of Optics; A History of Mechanics; A History of the Exchange on Physics Between China and Abroad; A History of Acoustics]. *Zhongguo wulishi da xi* 中國物理史大系 [A Series of Books on History of Physics in China]. 4 vols., to be continued. Changsha: Hu'nan jiaoyu chubanshe, 2000–2004.

Dardess, John W. *Blood and History in China: The Donglin Faction and its Repression, 1620–1627.* Berkeley: University of California Press, 1996.

———. *A Ming Society: T'ai-ho County, Kiangsi, Fourteenth to Seventeenth Centuries.* Berkeley: University of California Press, 1996.

Daston, Lorraine. "The Nature of Nature in Early Modern Europe." *Configurations* 6, no. 2 (1998): 149–72.

———. "Objectivity versus Truth." In *Wissenschaft als kulturelle Praxis, 1750–1900.* Edited by Hans Erich Bödeker, Peter Hanns Reill, and Jürgen Schlumbohm. Göttingen: Vandenhoek und Ruprecht, 1999.

Daston, Lorraine, and Katharine Park. *Wonders and the Order of Nature, 1150–1750.* New York: Zone Books, 1998.

Dear, Peter. "Cultural History of Science: An Overview with Reflections." *Science Technology Human Values* 20, no. 2 (1995): 150–70.

———. "Totius in Verba: Rhetoric and Authority in the Early Royal Society." *Isis* 76, no. 2 (1985): 145–61.

Deng Guangming 鄧廣銘. *Yue Fei zhuan* 岳飛傳 [Biography of Yue Fei]. Beijing: Sanlian shudian, 1955.

Deng Hongbo 鄧洪波 et al. *Zhongguo shuyuan zhangcheng* 中國書院章程 [Regulations of Academies in China]. Changsha: Hu'nan daxue chubanshe, 2000.

Dennis, Joseph. "Between Lineage and State: Extended Family and Gazetteer Compilation in Xinchang County." *Ming Studies* 45–46 (2001): 69–113.

DeWoskin, Kenneth J. "Picturing Performance: The Suite of Evidence for Music Culture in Warring States China." In *La pluridisciplinarité en Archéologie Musicale: IVe rencontres internationales du Groupe d'études sur l'Archéologie Musicale de l'ICTM (8–12 octobre 1990)*. Edited by Catherine Homo-Lechner and Annie Bélis. Paris: Maison des Sciences de l'Homme, 1994.

———. *A Song for One or Two: Music and the Concept of Art in Early China*. Michigan Monographs in Chinese Studies 42. Ann Arbor: University of Michigan Press, 1982.

Dolby, William. "The Origins of Chinese Puppetry." *Bulletin of the School of Oriental and African Studies* 41, no. 1 (1978): 97–120.

Donkin, R. A. *Beyond Price—Pearls and Pearl-fishing: Origins to the Age of Discoveries*. Philadelphia: American Philosophical Society, 1998.

Du Wan 杜綰 (fl. 1126). *Yunlin shipu* 雲林石譜 [Stone Catalogue of Cloudy Forest]. Edited by Zhu Jiuding 諸九鼎 and Gao Zhao 高兆. Congshu jicheng chubian 1720. Reprint, Beijing: Zhonghua shuju, 1935, 1985. References are to the 1985 edition.

Eisenstadt, Shmuel N., and Luis Roniger. *Patrons, Clients, and Friends: Interpersonal Relations and the Structure of Trust in Society*. Cambridge: Cambridge University Press, 1984.

Elman, Benjamin A. *A Cultural History of Civil Examinations in Late Imperial China*. Taipei: SMC Publishing Inc., 2000.

———. "Collecting and Classifying: Ming Dynasty Compendia and Encyclopedias (Leishu)." *Extrême-Orient, Extrême-Occident, hors série* (2007): 131–53.

———. "The Story of a Chapter: Changing Views of the 'Artificer's Record' ('*Kaogong ji*') and the *Zhouli*." In *Statecraft and Classical Learning: The Rituals of Zhou in East Asian History*. Edited by Benjamin Elman and Martin Kern. Leiden: Brill, 2010.

———. "Imperial Politics and Confucian Societies in Late Imperial China: The Hanlin and Donglin Academies." *Modern China* 15, no. 4 (1989): 379–418.

———. *On Their Own Terms: Science in China, 1550–1900*. Cambridge: Harvard University Press, 2005.

Emerson, John. "Yang Chu's Discovery of the Body." *Philosophy East and West* 46, no. 4 (1996): 533–66.

Falkenhausen, Lothar von. *Suspended Music: Chime-Bells in the Culture of Bronze Age China*. Berkeley: University of California Press, 1993.

Fang Yizhi 方以智. *Fang Yizhi quanshu* 方以智全書 [Complete Works of Fang Yizhi]. 2 vols. Reprint, Shanghai: Shanghai guji chubanshe, 1988. References are to the 1988 edition.

———. *Wuli xiao zhi* 物理小識 [Notes on the Principle of Things]. 12 vols. 1785. Reprint, Taibei: Taiwan shangwu yinshuguan, 1981. References are to the 1981 edition.

Fan Jinmin 范金民, ed. *Ming Qing Jiangnan shangye de fazhan* 明清江南商業的發展 [The development of the economy in Jiangnan during the Ming and Qing Dynasties]. Nanjing: Nanjing daxue chubanshe, 1998.

Fan Ye 范曄. *Hou Han shu* 後漢書 [History of the Former Han Dynasty]. Annotated by Li Xian 李賢 et al. 1965. Reprint, Beijing: Zhonghua shuju, 1973. References are to the 1973 reprint.

Fei Siyen. "We Must be Taxed: A Case of Populist Urban Fiscal Reform in Ming Nanjing (1368–1644)." *Late Imperial China* 28, no. 2 (2007): 1–40.

Feng Shiyuan 馮師元, and Shi Tai 石臺. *Enping xianzhi* 恩平縣志 [Gazetteer of Enping County]. 18 vols. 1825. Reprint, Taibei: Chengwen chubanshe, 1966.

Fogel, Joshua A. *The Cultural Dimensions of Sino-Japanese Relations: Essays on the Nineteenth and Twentieth Centuries.* Armonk, NY: M. E. Sharpe, 1995.

Foucault, Michel. *The Order of Things: An Archaeology of the Human Sciences.* New York: Vintage Books, 1973.

Fu Daiwie 傅大為. "A Contextual and Taxonomic Study of the 'Divine Marvels' and 'Strange Occurrences' in the Mengxi bitan." *Chinese Science* 11 (1993–94): 3–35.

———. "When Shen Kuo Encountered the 'Natural World.'" Unpublished paper, quoted with permission of the author.

——— and Lei Xianglin 雷祥麟. "Mengxi shen de yuyan yu xiangsixing- dui Mengxi bitan zhong renmingyun zhi yuzhi ji shenqi, bishi ermen de yanjiu 夢溪神的語言與相似性－對夢溪筆談中人民運之預知及神奇畢事二門之研究 [Language and Similarity in the Dream Brook: A Study of Prognostication, Divine Oddities, and Strange Events in "Mengxi bitan"]." *Ts'ing-hua Hsueh-bao* 3 (1994): 31–60.

Fung Yu-lan [Feng Youlan]. *A Short History of Chinese Philosophy.* 2 vols. Edited by Derk Bodde. 1948. Reprint, New York: Macmillan, 1959. References are to the 1959 reprint.

Fu she dang'an 復社檔案 [Archives of the Restoration Society]. Collection of Manuscripts, Beijing Library.

Gardner, Daniel K. "Chu Hsi [Zhu Xi] and the Transformation of the Confucian Tradition." In *Learning to Be a Sage: Selections from the Conversations of Master Chu, Arranged Topically,* by Chu Hsi [Zhu Xi], translated with commentary by Daniel Gardner. Berkeley: University of California Press, 1990.

Gauvin, Jean Francois. "Artisans, Machines, and Descartes's Organon." *History of Science* 44 (2006): 187–216.

Ge Quan 葛荃. *Liming yu zhongcheng: Shiren zhengzhi jingshen de dianxing fenxi* 立命與忠誠：士人政治精神的典型分析 [Order and Loyalty: Analysis of the Characteristics of the Spirit of Scholarly Politics]. Taibei: Xingding shiwenhua, 2002.

Ge Rongjin 葛荣晉, ed. *Zhongguo shixue sixiangshi* 中國實學思想史 [A History of the Ideas of Practical Learning in China]. Beijing: Shoudu shifan daxue chubanshe, 1994.

Gilbert, William. *Tractus, sive, physiologica nova de magnete, magneticisque corporibus et de magneete tellure: sex libris comprehensus*. Compiled 1600. Third edition, Stettin, 1633.

Girardot, Norman J. *Myth and Meaning in Early Taoism: The Theme of Chaos (hun-tun).* Berkeley: University of California Press, 1983.

Glahn, Richard von. *Fountain of Fortune: Money and Monetary Policy in China, 1000–1700.* Berkeley: University of California Press, 1996.

Golas, Peter. *Mining.* Pt. 13 of *Chemistry and Chemical Technology in China,* vol. 5 in Science and Civilisation in China, edited by Joseph Needham. Cambridge: Cambridge University Press, 1999.

———. "Like Obtaining a Great Treasure: The Illustrations in Song Yingxing's The Exploitation of the Works of Nature." In *Graphics and Text in the Production of Technical Knowledge in China: The Warp and the Weft.* Edited by Francesca Bray, Vera Dorofeeva-Lichtmann, and Georges Métailié, 569–614. Sinica Leidensia 79. Leiden: Brill, 2007.

Gombrich, E. H. "Standards of Truth: The Arrested Image and the Moving Eye." *Critical Inquiry* 7, no. 2 (1980): 237–73.

Goodrich, L. Carrington. *Dictionary of Ming Biography, 1368–1644.* 2 vols. New York: Columbia University Press, 1976.

Greatrex, Roger. *The Bowu Zhi: An Annotated Translation.* Stockholm: Skrifter utgivna av Föreningen för Orientaliska Studier, 1987.

Gulik, Robert H. van. "The Lore of the Chinese Lute: An Essay in Ch'in Ideology (Continued)." *Monumenta Nipponica* 2, no. 1 (1939): 75–99.

Guo Jinhai 郭金海. "Mingdai Nanjing chengqiang zhuan mingwen luelun 明代南京城墻磚銘文略論 [A Concise Introduction of the Inscriptions on the Bricks of City Wall of Nanjing in the Ming Dynasty]." *Dongnan wenhua,* no. 1 (2001): 75–78.

Guo Shuchun 郭書春. "Jia Xian 'Huang Di jiu zhang suan jing xicao' chutan" 贾宪《黄帝九章算經细草》初探 [An Initial Inquiry Into Jia Xian's "Detailed Solutions of the Problems in the Nine Mathematics Chapters of Huang Di"]. *Ziran kexue shi yanjiu* 7, no. 4 (1988): 328–34.

Guo Shusen 郭樹森. "Shi lun Song Yingxing dui yuanqi benti lun de fengfu he fazhan 試論宋應星對元氣本體論的豐富和發展 [A Discourse about the Richness and Development of Song's Concept of yuanqi]." *Jiangxi shehui kexue* 5 (1984): 118–23.

Guo Yu 國語 [Discourses from the States]. Attributed to Zuo Qiuming 左邱明. 1900. Reprint, Shanghai: Guji chubanshe, 1978.

Hackin, John, Clément Huart, Raymonde Linossier et al. *Asiatic Mythology: A Detailed Description and Explanation of the Mythologies of All the Great Nations of Asia.* New York: Thomas Y. Crowell, 1963.

Han Kuo-Huang, and Judith Gray. "The Modern Chinese Orchestra." *Asian Music* 11, no. 1 (1979): 1–43.

Handlin, Joanna F. *Action in Late Ming Thought: The Reorientation of Lü K'un and Other Scholar-Officials.* Berkeley: University of California Press, 1983.

Harrell, Stevan. "On the Holes in Chinese Genealogies." *Late Imperial China* 8, no. 2 (1987): 53–77.

Hartman, Charles. *Han Yü and the T'ang Search for Unity.* Princeton: Princeton University Press, 1986.

———. "The Reluctant Historian: San Ti, Chu Hsi, and the Fall of the Northern Sung." *T'oung Pao* 89, no. 2 (2003): 100–148.

Hauf, Kandice. "The Community Covenant in Sixteenth Century Ji'an Prefecture, Jiangxi." *Late Imperial China* 17, no. 2 (1996): 1–50.

Hazelton, Keith. "Patrilines and the Development of Localized Lineages: The Wu of Hsiu-ning City, Hui-chou, to 1528." In *Kinship Organization in Late Imperial China, 1400–1900.* Edited by Patricia B. Ebrey and James L. Watson. Berkeley: University of California, 1986.

He Zhaowu 何兆武. "Lun Song Yingxing de sixiang" 論宋應星的思想 [A Discussion of Song Yingxing's Thought]. *Zhongguo shi yanjiu* 2 (1978): 149–60.

Hegel, Robert E. *Reading Illustrated Fiction in Late Imperial China.* Stanford, CA: Stanford University Press, 1998.

Ho Peng Yoke [He Bingyu]. *Li, Qi and Shu: An Introduction to Science and Civilization in China.* Hong Kong: Hong Kong University Press, 1985.

Ho Ping-ti [He Bingdi] 何炳棣. *The Ladder of Success in Imperial China: Aspects of Social Mobility, 1368–1911.* New York: Columbia University Press, 1962.

Hu Daojing 胡道靜. *Zhongguo gudai de leishu* 中國古代的類書 [Encyclopedias of Ancient China]. Beijing: Zhonghua shuju, 1982.

Hu Jichuang 胡寄窗. *Zhongguo jingji sixiangshi jianbian* 中國經濟思想史簡編 [A Concise History of Chinese Economic Thought]. Shanghai: Lixin huiji chubanshe, 1997.

Hu Xingmin 户幸民. "Shilun Song Yingxing de ziran zhexue sixiang 試論宋應星的自然哲學思想 [A Discourse about Song Yingxing's Philosophy of Nature]." *Jiangxi shehui kexue* 5 (1982): 69–76.

Huang, H. T. *Fermentation and Food Science.* Pt. 5 of *Biology and Biological Technology,* vol. 6 in Science and Civilisation in China, edited by Joseph Needham. Cambridge: Cambridge University Press, 2000.

Huang Mingtong 黄明同. "Cong Lun Qi kan Song Yingxing de ziran guan 從論氣看宋應星的自然觀 [Song Yingxing's Notion of Nature according to his Treatise *On Qi*]." *Huanan shifan xuebao shehui kexueban* 4 (1982): 24–28, 34.

Huang, Ray. *1587, A Year of No Significance: The Ming Dynasty in Decline.* New Haven: Yale University Press, 1981.

Huang Xiangpeng 黄翔鵬. *Zhongguoren de yinyue he yinyuexue* 中國人的音樂和音樂學 [Music and Musicology of the Chinese]. Ji'nan: Shandong wenyi chubanshe, 1997.

Hummel, Arthur, ed. *Eminent Chinese of the Ch'ing Period (1644–1912).* 1943. Reprint, Taibei: SMC, 1991.

Hunt, Frederick Vinton. *Origins in Acoustics: The Science of Sound from Antiquity to the Age of Newton.* With a foreword by Robert Edmund Apfel. Ann Arbor, MI: UMI, 1978.

Ikoma Shō 生駒晶. "Minsho kakyo gôkaku sha no shusshin ni kan suru ichi kōsatsu 明初科舉合格者の出身に關する一考察 [A Study of Birthplaces of Successful Examination Candidates during the Early Ming]." In *Yamane Yukio kyōju taikyū kinen Mindaishi ronsō* 山根幸夫教授退休紀念明代史論叢 [Essays on Ming History in Honor of Professor Yamane Yukio's Retirement]. Edited by Yamane Yukio and Mindaishi Kenkyūkai. Tōkyō: Kyūko shōin, 1990.

Ivanhoe, Philip J. *Ethics in the Confucian Tradition: The Thought of Mencius and Wang Yang-ming.* American Academy of Religion Academy Series 70. Atlanta: Scholars Press, 1990.

Jang, Scarlett. "Form, Content, and Audience: A Common Theme in Painting and Woodblock-Printed Books of the Ming Dynasty." *Ars Orientalis* 27 (1997): 1–26.

Janousch, Andreas. "Salt Production Methods and Salt Cults at Xiechi Salt Lake in Southern Shanxi." Conference paper presented at the Conference of ISHEASTM, Baltimore, 2008.

Jia Sixie 賈思勰 (6th cent.). *Qimin yaoshu* 齊民要術 [Essential Techniques for the Peasantry]. Annotated and edited by Miao Qiyu 繆啟愉. Chengdu: Bashu shushe, 1988.

Jiang Xiaoyuan 江曉源. *Tianxue waishi* 天學外史 [Unofficial History of Astronomy]. Shanghai: Shanghai renmin chubanshe, 1999.

Kasoff, Ira E. *The Thought of Chang Tsai (1020–1077).* New York: Cambridge University Press, 1984.

Kaufmann, Walter. *Musical References in the Chinese Classics.* Detroit Monographs in Musicology 5. Detroit: University Press, 1976.

Kerr, Rose, and Nigel Wood, with Ts'ai Mei-fen and Zhang Fukang. *Ceramic Technology.* Pt. 12 of *Chemistry and Chemical Technology,* vol. 5 in Science and Civilisation in China, edited by Joseph Needham. Cambridge: Cambridge University Press, 2004.

Kim Youngmin. "Luo Qinshun (1465–1547) and His Intellectual Context." *T'oung Pao* 89, no. 4 (2003): 367–441.

Kramers, Robert Paul. *K'ung Tzu Chia yü [Kong Zi jiayu]: The School Sayings of Confucius. Introduction. Translation of Sections 1–10.* Sinica Leidensia 7. Leiden: Brill, 1949.

Kuhn, Dieter. "Family Rituals." *Monumenta Serica* 40 (1992): 369–85.

Kuhn, Thomas S. *The Structure of Scientific Revolutions.* 1963. Reprint, Chicago: University of Chicago Press, 1970. References are to the 1970 edition.

Lackner, Michael, Friedrich Reiman, and Michael Friedrich, eds. *Chang Tsai [Zhang Zai]: Rechtes Auflichten: Cheng-meng.* Übersetzt aus dem Chinesischen mit Einleitung und Kommentar versehen. Philosophische Bibliothek 419. Hamburg: Felix Meiner Verlag, 1996.

Lam, Joseph S. C. "Huizong's Ritual and Musical Insignia." *Journal of Ritual Studies* 19, no. 1 (2005): 1–18.

———. *State Sacrifices and Music in Ming China: Orthodoxy, Creativity, and Expressiveness.* SUNY Series in Chinese Local Studies. Albany: State University of New York Press, 1998.

Lanling Xiaoxiao Sheng 藍領笑笑生. *Zhang Zhupo piping Jinping mei* 張竹坡批評金瓶梅 [Blossom in a Golden Vase, annotated by Zhang Zhupo 張竹坡 (1670–98)]. Edited by Wang Rumei 王汝梅. 2 vols. Ji'nan: Qi Lu shushe, 1991.

Lao Zi 老子. *Laozi xinbian jiaoshi* 老子新編校識 [New Text Laozi, with Explanatory Commentary]. Edited by Wang Xing. Shenyang: Liaoshen shushe, 1990.

Lau, D. C. 劉殿爵,Chen Fangzheng 陳方正, and He Zhihua 何志華, eds. Cao Zhi 曹植. *Cao Zhi ji zhu zi suoyin* 曹植集逐字索引 [A Concordance to the Works of Cao Zhi], ICS Concordances to Works of Wei-Jin and the Northern and Southern Dynasties. Collected Works No. 9. Xianggang: Zhongwen daxue chubanshe, 2001.

Le Blanc, Charles. "From Cosmology to Ontology through Resonance: A Chinese Interpretation of Reality." In *Beyond Textuality: Asceticism and Violence in Anthropological Interpretation.* Edited by Gilles Bibeau and Ellen Corin. Berlin: Mouton de Gruyter, 1995.

———. "Résonance: Une Interprétation Chinoise de la Réalité." In *Mythe et Philosophie à l'Aube de la Chine Impériale: Etudes sur le Huainan Zi.* Edited by Charles Le Blanc and Rémi Mathieu. Montreal: Presses de l'Université de Montréal, 1992.

Ledderose, Lothar. *Ten Thousand Things: Module and Mass Production in Chinese Art.* Princeton: Princeton University Press, 2000.

Lee, Jig-chuen. "Wang Yang-ming, Chu Hsi, and the Investigation of Things." *Philosophy East and West* 37, no. 1 (1987): 24–35.

Legge, James, ed. and trans. *The Chinese Classics: With a Translation, Critical and Exegetical Notes, Prolegomena, and Copious Indexes.* 5 vols. Hong Kong: Hong Kong University Press, 1960. References are to the 1960 edition.

Leibold, Michael. *Die handhabbare Welt: Der pragmatische Konfuzianismus Wang Tingxiangs (1474–1544).* Heidelberg: edition forum, 2001.

Levenson, Joseph R. "The Amateur Ideal in Ming and Early Ch'ing Society: Evidence from Painting." In *Chinese Thought and Institutions.* Edited by John K. Fairbank. Chicago: University of Chicago Press, 1957.

Liang Fang-chung 梁方仲. *The Single-Whip Method (I-t'iao-pien fa 一條鞭法) of Taxation in China.* Translated by Wang Yü-ch'uan 王毓銓. Cambridge: Harvard University Press, 1956.

Li Shaoqiang 李紹強, and Xu Jianqing 徐建青. *Zhongguo shougongye jingji tongshi: Ming Qing juan* 中國手工業經濟通史：明清卷 [A Comprehensive History of Chinese Handicraft Economy: Ming and Qing]. Fuzhou: Fujian renmin chubanshe, 2004.

Li Shuzeng 李書增, Sun Yujie 孫玉杰, and Ren Jinjian 任金鑒, eds. *Zhongguo mingdai zhexue* 中國明代哲學 [Philosophy in China during the Ming Dynasty]. Zhengzhou: He'nan renmin chubanshe, 2002.

Li Tingji 李廷機 (1542–1616). *Chunqiu Zuo Zhuan gangmu dingzhu* 春秋左傳綱目定訂 [Annotations and Remarks to the Spring and Autumn Commentaries of Master Zuo Traditions]. Naikaku Bunko, ed., *Kaitei Naikaku bunko kanseki bunrui mokuroku* [A Classified Catalogue of Chinese Literature from the Diet Library], 274 han, 158 hao.

Liu An 劉安. *Huainan zi* 淮南子[The Master of Huainan]. (120 BC) Reprint, Changchun: Jilin renmin chubanshe, 1999.

Liu Baonan 劉寶楠 (1791–1855), Liu Gongmian 劉恭冕 (1824–83), and Song Xiangfeng 宋翔鳳 (1776–1860), eds. *Lunyu zhengyi* 論語正義 [Annotated Commentary to the Analects]. 2 vols. Beijing: Zhonghua shuju, 1990.

Liu Jun 劉軍, Mo Fushan 莫福山, and Wu Yazhi 吴雅芝. *Zhongguo gudai de jiu yu yinjiu* 中國古代的酒與飲酒 [China's Ancient Wine and Wine-Drinking]. Zhongguo gudai shehui shenghuo congshu 11. Taibei: Taiwan shangwu yinshuguan, 1998.

Liu Tongsheng 劉同升. *Jinlin shiji* 錦鱗詩集 [Anthology of Poetry by Jin Lin]. 18 vols. 1640. Reprinted and reedited, Nanchang, 1937. References are to the 1937 edition.

Liu Yeqiu 劉葉秋. *Lidai biji gaishu* 歷代筆記概述 [Survey of the Private Jottings of Past Dynasties]. Beijing: Zhonghua shuju, 1980.

Liu Zepu 劉澤溥, and Gao Bojiu 高博九, eds. *Bozhou zhi* 亳州志 [Local Gazetteer of Bozhou]. 1656. Beijing National Library Rara Collection.

Li Yinqing 李寅清 (19th cent.), Xia Congding 夏琮鼎, Yan Shengwei 嚴升偉. *Tongzhi Fenyi xianzhi* 同治分宜縣志 [Gazetteer of Fenyi county]. Jiangxi, 1871. See also the reprint. Zhongguo difangzhi jicheng. Jiangxi fuxianzhiji 35 中國地方志集成. 江西府縣志輯.Nanjing: Jiangsu guji chubanshe, 1996.

Liu Yuxi 劉禹錫 (772–842). *Liu Binke jiahua lu* 劉賓客嘉話錄 [Record about Liu Binke's Auspicious Words]. Edited by Wei Xuan 韋絢 (9th cent.), Zhao Lin 趙璘 (fl. 836–46), and Su E 蘇鶚 (jin shi 886). Taibei: Taiwan shangwu yinshuguan, 1966.

Lloyd, Geoffrey, and Nathan Sivin. *The Way and the World: Science and Medicine in Early China and Greece.* New Haven: Yale University Press, 2003.

Loewe, Michael. "The Oracles of the Clouds and the Winds." *Bulletin of the School of Oriental and African Studies* 51, no. 3 (1988): 500–520.

Lu Jiansan 陸鑒三. "Chuchu dou zhu zu yiyi bu neng qu: Yuan Ming Qing Hangzhou de lüyou" 處處逗駐足依依不能去：元明清杭州的旅游 [Tourism in Hangzhou Thirteenth to Nineteenth Century]. In *Yuan, Ming, Qing mingcheng Hangzhou* 元明清名城杭州 [A Famous City in the Yuan, Ming and Qing Dynasties: Hangzhou]. Edited by Zhou Feng 周峰. Hangzhou: Zhejiang renmin chubanshe, 1997.

Lü Maoxian 呂懋先 and Shuai Fangwei 帥方蔚. *Fengxin xianzhi* 奉新縣志 [Local Gazetteer of Fengxin]. 16 vols. Nanchang: Jiangxi Provincial Library, 1871.

Lü Weiqi 呂維祺. *Mingde xiansheng wenji* 明德先生文集 [The Collected Works of Lü Weiqi]. Tainan xian Liuying xiang: Zhuang yan chubanshe, 1997.

Lufrano, Richard John. *Honorable Merchants: Commerce and Self-Cultivation in Late Imperial China.* Studies of the East Asian Institute. Honolulu: University of Hawai'i Press, 1997.

Lu Gwei-Djen, and Joseph Needham. *Celestial Lancets: A History and Rationale of Acupuncture and Moxa.* Cambridge: Cambridge University Press, 1980.

Lu Jiying 盧繼影, ed. *Quanbu da Yan Song; you ming, Kaishan fu: Gailiang jingxiben* 全部打嚴嵩；又名開山府： 改良京戲本 [A Complete Version of Beating Yan Song; or, Kaishan Prefecture: Corrected Beijing Opera Libretto]. Shanghai: Luohan chubanshe, 1935.

Lunyu [Analects]. Reprint. Beijing: Zhongguo shehui kexue chubanshe, 2003.

Luo Qi 羅頎. *Wu yuan* 物原 [Origin of Things]. Congshu jicheng jianbian 64. Taibei: Taiwan shangwu yinshuguan, 1966.

Luo Qinshun 羅欽順. *Kunzhi ji* 困知記 [Record on Knowledge Painfully Acquired]. 1528. Taibei: Taiwan shangwu yinshuguan, 1983.

Lu Shiyi 陸世儀. "Ming ji Fushe jilüe" 明季復社紀略[A Report about the Restoration Society]. In *Zhongguo yeshi jicui* 中国野史集粹 [Anthology of Superior Unofficial Histories of China]. Edited by Chen Li 陳力, vol. 1. Chengdu: Bashi shushe, 2000.

Ma Zuchang 馬祖常 (1279–1338). *Shi tian wen ji* 石田文集 [Collected Works of Ma Zucheng]. Taibei: Shangwu yinshuguan, 1976.

Ma Tai-loi. "The Local Education Officials of Ming China, 1368–1644." *Oriens Extremus* 22, no. 1 (1975): 11–28.

Major, John S. *Heaven and Earth in Early Han Thought: Chapters Three, Four, and Five of the Huainanzi.* Appendix by Christopher Cullen. SUNY Series in Chinese Philosophy and Culture. Albany: State University of New York Press, 1993.

———, and Jenny F. So. "Music in Late Bronze Age." In *Music in the Age of Confucius.* Edited by Jenny F. So. Washington: Freer Gallery of Art and Arthur M. Sackler Gallery (Smithonian Institution), 2000.

Mao Deqi 毛德琦 (18th cent.). *Bailu shuyuan zhi* 白鹿書院志 [On the White Deer Academy]. 1718. Reprint, 1975, Taibei: Chenwen chubanshe, 1989.

Marmé, Michael. *Suzhou: Where the Goods of All the Provinces Converge.* Stanford, CA: Stanford University Press, 2005.

McDermott, Joseph P. *A Social History of the Chinese Book: Books and Literati Culture in Late Imperial China*. Hong Kong: Hong Kong University Press, 2006.

Mencius [Mengzi] 孟子. *Mengzi jizhu* 孟子記住 [An Annotated Collection of Mengzi]. Edition of *Sishu ji zhu zhang ju* [The Four Books and the Five Classics with commentaries by Zhu Xi]. Tianjin: Tianjin guji shudian yingyin, 1988.

Millinger, James Ferguson. "Ch'i Chi-kuang, Chinese Military Official: A Study of Civil-Military Roles and Relations in the Career of a Sixteenth Century Warrior." PhD diss., Yale University, 1968.

Milne, William Charles. *A Retrospect of the First Ten Years of the Protestant Mission to China.* Malacca: Anglo-Chinese Press, 1820.

Mote, Frederick W. *Imperial China 900–1800.* Cambridge: Harvard University Press, 1999.

———, and Denis Twitchett, eds. *The Cambridge History of China.* Vols. 7 and 8, *The Ming Dynasty, 1368–1644.* Cambridge: Cambridge University Press, 1988.

Naikaku Bunko 内閣文庫, ed. *Kaitei Naikaku bunko kanseki bunrui mokuroku* 改訂内閣文庫漢籍分類目錄 [A Classified Catalogue of Chinese Literature from the Diet Library]. Tokyo: Naikaku bunko, 1971.

Needham, Joseph. *Physics.* Pt. 1 of *Physics and Physical Technology,* vol. 4 in Science and Civilisation in China. Cambridge: Cambridge University Press, 1962.

———. *Military Technology: The Gunpowder Epic.* Pt. 7 of *Chemistry and Chemical Technology,* vol. 5 in Science and Civilisation in China. Cambridge: Cambridge University Press, 1986.

Ogawa Shogo 小川省吾. *Kinsei shokusengaku kōyō* 近世色染学綱要 [Systematic Studies About Dyeing in the Early Modern Period]. Tokyo: Kogyo Tosho, 1936.

Okanobori Teiji 岡登貞治. *Senshoku seigi* 染色精義 [Correct Meanings in Dyeing]. Osaka: Tōyō Tōsho, 1950.

Owen, Stephen. *Readings in Chinese Literary Thought.* Cambridge, MA: Council on East Asian Studies, 1992.

Pan Jixing 潘吉星. *Mingdai kexuejia Song Yingxing* 明代科学家宋應星 [The Scientist Song Yingxing of the Ming Dynasty]. Beijing: Kexue chubanshe, 1981.

———. *Song Yingxing pingzhuan* 宋應星評傳 [A Critical Biography of Song Yingxing]. Nanjing: Nanjing daxue chubanshe, 1990.

———. *Tiangong kaiwu jiaozhu yu yanjiu* 天工開物校注及研究 [Critical Edition of the Tiangong kaiwu, with a study]. Chengdu: Bashu shushe, 1989.

Peterson, Willard J. *Bitter Gourd: Fang I-chih and the Impetus for Intellectual Change.* New Haven: Yale University Press, 1979.

Poo Mu-chou [Pu Muzhou] 蒲慕州. "The Use and Abuse of Wine in Ancient China." *Journal of the Economic and Social History of the Orient* 42, no. 2 (1999): 123–51.

Potter, Donald. "[Biography of] Wen T'i-jen." In *Dictionary of Ming Biography, 1368–1644.* Ming Biographical History Project of the Association for Asian Studies, edited by L. C. Goodrich and Fang Chaoying. Vol. 2. New York: Columbia University Press, 1976.

Powers, Martin J. *Pattern and Person: Ornament, Society, and Self in Classical China.* Harvard East Asian Monographs 262. Cambridge: Harvard University Asia Center, 2006.

Pregadio, Fabrizio. *Great Clarity: Daoism and Alchemy in Early Medieval China.* Asian Religions and Cultures. Stanford, CA: Stanford University Press, 2005.

———, ed. *The Encyclopedia of Daoism.* New York: Routledge, 2008.

Puett, Michael J. *The Ambivalence of Creation: Debates Concerning Innovation and Artifice in Early China.* Stanford, CA: Stanford University Press, 2001.

Qian Chaochen 錢超塵. *Wang Qingren yanjiu jicheng* 王清任研究集成 [Anthology of Studies of Wang Qingren]. Beijing: Zhongyi guji chubanshe, 2002.

Qiu Jun 邱濬 (1421–95). *Daxue yanyi bu* 大學衍義補 [Extended Meaning of the Great Learning]. 165 *juan.* Edited by Zhen Dexiu 眞德秀 (1178–1235). Edition of Wenyuan ge Siku quanshu [Wenyuan Pavillion Siku quanshu edition]. Taiwan: Taiwan shang wu yinshuguan, 1983–86.

Quirin, Michael. "Scholarship, Value, Method and Hermeneutics in Kaozheng: Some Reflections on Cui Shu (1740–1816) and the Confucian Classics." *History and Theory* 35, no. 4 (1996): 34–53.

Rawson, Jessica. *Chinese Jades: From the Neolithic to the Qing.* London: British Museum Press, 2002.

———. "The Many Meanings of the Past in China." In *Perception of Antiquity in Chinese Civilization.* Edited by Dieter Kuhn and Helga Stahl. Heidelberg: edition forum, 2001.

Rheinberger, Hans-Jörg. *Toward a History of Epistemic Things: Synthesizing Proteins in the Test Tube.* Stanford, CA: Stanford University Press, 1997.

Roberts, Lissa, Simon Schaffer, and Peter Dear, eds. *The Mindful Hand: Inquiry and Invention from the Late Renaissance to Early Industrialisation.* History of Science and Scholarship in the Netherlands 9. Amsterdam: KNAW 2007.

Robinson, David M. *Bandits, Eunuchs, and the Son of Heaven: Rebellion and the Economy of Violence in Mid-Ming China.* Honolulu: University of Hawai'i Press, 2001.

Rostoker, William, Bennet Bronson, and James Dvorak. "The Cast-Iron Bells of China." *Technology and Culture* 25, no. 4 (1984): 750–67.

Rowe, William T. *Saving the World: Chen Hongmou and Elite Consciousness in Eighteenth-Century China.* Stanford, CA: Stanford University Press, 2001.

———. "Success Stories: Lineage and Elite Status in Hanyang County, Hubei, c. 1368–1949." In *Chinese Local Elites and Patterns of Dominance.* Edited by Joseph W. Esherick and Mary Backus Rankin. Berkeley: University of California Press, 1990.

Rowley, George A. "A Chinese Scroll of the Ming Dynasty: Ming Huang and Yang Kuei-fei Listening to Music." *Attribus Asiae* 31, no. 1 (1969): 5–31.

Ruan Yuan 阮元 (1764–1849), Chen Changqi 陳昌齊 (1743–1820), et al. *Guangdong tongzhi* 廣東通志 [General gazetteer of Guangdong]. 1864. Reprint, Shanghai: Shanghai Guji chubanshe, 1934. References are to the 1934 edition.

Sakai Tadao. "Confucianism and Popular Educational Works." In *Self and Society in Ming Thought,* edited William Theodore de Bary et al., 331–66. New York: Columbia University Press, 1970.

Schafer, Edward H. "The Pearl Fisheries of Ho-p'u." *Journal of the American Oriental Society* 72, no. 4 (1952): 155–68.

———. *Tu Wan's Stone Catalogue of Cloudy Forest.* Berkeley: University of California Press, 1961.

Schäfer, Dagmar. *Des Kaisers seidene Kleider: Staatliche Seidenmanufakturen in der Ming-Zeit (1368–1644).* Heidelberg: edition forum, 1998.

Shapin, Steven. *A Social History of Truth: Civility and Science in Seventeenth-Century England.* Chicago: University of Chicago Press, 1994.

———, and Simon Schaffer. *Leviathan and the Air Pump: Hobbes, Boyle, and the Experimental Life.* Princeton: Princeton University Press, 1985.

Shapiro, Barbara J. *A Culture of Fact: England 1550–1720.* Ithaca: Cornell University Press, 2000.

Scheid, Hildegard. "Die Entwicklung der Staatlichen Seidenweberei in der Ming-Dynastie (1368–1644)." Master's thesis, University of Würzburg, 1994.

Shen Gua [Shen Kuo] 沈括 (1031–95). "Hun yi yi" 渾儀議 [On the Armillary Sphere]. In *Lidai tianwen lü lideng zhi huibian* 歷代天文律曆等志彙編 [A Collection of Historical Documents on Astronomy]. Edited by Zhonghua shuju bianji bu 中華書局編輯部. Beijing: Zhonghua shu ju, 1975.

———. *Mengxi [Mengqi] bitan jiaozheng* 夢溪筆談校證 [Brush talks from the Dream Brook]. Edited by Hu Daojing 胡道靜. Shanghai: Shanghai guji chubanshe, 1958, 1960, 1987.

Shen Shixing 申時行 (1535–1614) and Li Dongyang 李東陽 (1447–1516). *Da Ming huidian* 大明會典 [Collected Statutes of the Great Ming]. 1511, 1587. Reprint, Taibei: Xin Wenfeng chuban gongsi, 1976.

Shujing 書經 [Book of Documents]. Edition of *Sishu wujing* [The Four Books and the Five Classics]. Tianjin: Tianjing guji shudian yingyin, 1988.

Shuoyuan jiaozheng 說苑校證 [Collection of Speeches with Corrections and Annotations] *Shuo yuan* 說苑 by Liu Xiang 劉向 77?–6? BC. Edited by Xiang Zonglu 向宗魯 (1895–1941). Beijing: Zhonghua shuju, 1987.

Siebert, Martina. "Making Technology History." Conference paper presented at Max Planck Institute for the History of Science, From Invention to Innovation, July 2006. Forthcoming.

———. *Pulu: Abhandlungen und Auflistungen zu materieller Kultur und Naturkunde im traditionellen China.* Opera sinologica 17. Wiesbaden: Harrasowitz, 2006.

Sima Qian 司馬遷. *Shiji* 史記 [Records of the Grand Historian]. 10 vols. Reprint, Xianggang: Zhonghua shuju, 1969. References are to the 1969 edition.

Sivin, Nathan. *Chinese Alchemy: Preliminary Studies.* Harvard Monographs in the History of Science. Cambridge: Harvard University Press, 1968.

———. *Granting the Seasons: The Chinese Astronomical Reform of 1280, With a Study of its Many Dimensions and a Translation of its Records* 老子新編校識. New York: Springer, 2009.

———. "Shen Kua." In *Dictionary of Scientific Biography.* Edited by Charles C. Gillespie. New York: Charles Scribner's Sons, 1970–80.

———. "Wang Hsi-Shan." In *Science in Ancient China: Researches and Reflections.* Variorum Collected Studies Series CS 506, V. Aldershot, Hampshire: Variorum, 1995.

———. "Why the Scientific Revolution Did Not Take Place in China—Or Didn't It?" *Chinese Science* 5 (1982): 45–66.

———. *Traditional Medicine in Contemporary China: A Partial Translation of Revised Outline of Chinese Medicine (1972) with an Introductory Study on Change in Present-day and Early Medicine*. Ann Arbor: Center for Chinese Studies, University of Michigan, 1987.

Smith, Kidder. "Sima Tan and the Invention of Daoism, 'Legalism,' 'et cetera.'" *Journal of Asian Studies* 62, no. 1 (2003): 129–56.

Smith, Pamela H. *The Body of the Artisan: Art and Experience in the Scientific Revolution*. Chicago: University of Chicago Press, 2004.

Song Liquan 宋立權, and Song Yude 宋育德. *Baxiu Xinwu Ya qi Song shi zongpu* 八修新吳雅溪宋氏宗譜 [Eighth Supplement to the Family Annals of the House of Song from Xinwu]. Copy of Fengxin Qinlu tang cangban: Song Yingxing bowuguan, 1934.

Song Yingsheng 宋應昇. *Fangyu tang ji* 方玉堂集 [Collection of the Fangyu Hall]. 1638. Hsinchu: Tsing Hua University Microfilms. Original stored at Fengxin, Yaxi. See also the facsimile edition, Siku jinhui shu congkan 四庫禁毀書叢刊, vol. 165, 153–464. Beijing: Beijing chubanshe, 2000.

Song Yingxing. "Huayin guizheng" 畫音歸正 [A Return to Orthodoxy] (lost), "Silian Shi" 思憐詩 [Yearnings], "Ye Yi" 野議 [An Oppositionist's Deliberations], "Tan Tian" 談天 [Talks about Heaven], and "Lun Qi" 論氣 [On Qi]. In *Ye Yi; Lun Qi; Tan Tian; Silian Shi* 野議，談天，論氣，思憐詩 [An Oppositionist's Deliberations; On Qi; Talks about Heaven; Yearnings]. 1637. Reprint, Shanghai: Shanghai renmin chubanshe, 1976. References are to the 1976 edition.

———. *Tiangong kaiwu* 天工開物 [The Works of Heaven and the Inception of Things]. 1637. Reprint. Guangzhou: Guangdong renmin chubanshe, 1976. References are to the 1976 edition.

———. *Tiangong kaiwu* 天工開物 [The Works of Heaven and the Inception of Things]. Annotated by Zhong Guangyan 鍾廣言. Xianggang: Zhonghua shuju Xianggang fenju, 1988.

Sterckx, Roel. "Transforming the Beasts: Animals and Music in Early China." *T'oung Pao* 136 (2000): 1–46.

Struve, Lynn A. *The Ming-Qing Conflict, 1619–1683: A Historiography and Source Guide*. Monograph and Occasional Paper Series 56. Ann Arbor, MI: Association for Asian Studies, 1998.

Su Dongpo 蘇東坡, *Dongpo zhilin* 東坡志林 [The Forest of Records by Su Shi]. Beijing: Zhonghua shuju, 1981.

Sun Dianqi 孫殿起. *Qingdai jinshu zhijianlu* 清代禁書知見錄 [Record of Known Banned Books of the Qing Dynasty]. Shanghai: Shangwu yinshuguan, 1957.

Sun Laichen. "Military Technology Transfers from Ming China and the Emergence of Northern Mainland Southeast Asia (c. 1390–1527)." *Journal of Southeast Asian Studies* 34, no. 3 (2003): 495–517.

Sun Xiaochun and Jacob Kistemaker. *The Chinese Sky during the Han: Constellating Stars and Society*. Sinica Leidensia 38. Leiden: Brill, 1997.

Sung Ying-hsing [Song Yingxing]. *Chinese Technology in the Seventeenth Century: T'ien-kung k'ai-wu*. Translated from the Chinese and annotated by E-Tu Zen Sun and Sun

Shiou-Chuan. Mineola, NY: Dover Publications, 1997. First published 1966, University Park: Pennsylvania State University. References are to the 1997 edition.

Tan Qian 談遷 (1594–1657), Zhang Zongxiang 張宗祥. *Guo que* 國榷 [Deliberations about the State]. Reprint, Beijing: Guji chubanshe, 1958.

Tan Qiao 譚峭 (10th cent.). *Hua shu* 化書 [A Book about Transformation]. Edited by Ding Zhenyan 丁禎彥 and Li Sizhen 李似珍. Reprint, Beijing: Zhonghua shuju, 1996.

Tang Ying 唐英. *Tang Ying ji* 唐英集 [Collected Writings of Tang Ying]. Edited by Zhang Faying 張發穎 and Diao Yunzhan 刁雲展. Shenyang: Liaoshen shushe, 1991.

Tong, James W. *Disorder under Heaven: Collective Violence in the Ming Dynasty.* Stanford, CA: Stanford University Press, 1991.

Tong Shinan 佟世男. *Enping xianzhi* 恩平縣志 [Local Gazetteer of Enping County]. 1478, revised by Song Yingsheng 宋應昇 ca. 1640. Reprint, Tianjin: Tianjin tushuguan gujibu Fuzhi, 1984. References are to the 1984 edition.

Tsai, Shih-shan Henry. *The Eunuchs in the Ming Dynasty.* Albany: State University of New York Press, 1996.

Tu Wei-ming [Weiming]. *Centrality and Commonality: An Essay on Chung-yung.* Monographs of the Society of Asian and Comparative Philosophy 3. Honolulu: University Press of Hawai'i, 1976.

———. "The Continuity of Being: Chinese Visions of Nature." In *On Nature.* Edited by Leroy S. Rouner. Notre Dame, IN: University of Notre Dame Press, 1984.

Tuo Tuo 脫脫. *Song shi* 宋史 [Song History]. Beijing: Zhonghua shuju, 2000.

Volkmar, Barbara. *Die Fallgeschichten des Arztes Wan Quan (1500–1585?): Medizinisches Denken und Handeln in der Ming Zeit.* München: Elsevier Urban & Fischer, 2007.

Wagner, Donald. "Chinese Blast Furnaces from the 10th to the 14th Century." *Historical Metallurgy* 37, no. 1 (2003): 25–37.

———. *Iron and Steel in Ancient China.* Leiden: Brill, 1993.

———. "Song Yingxing's Illustrations of Iron Production." In *Graphics and Text in the Production of Technical Knowledge in China: The Warp and the Weft.* Edited by Francesca Bray, Vera Dorofeeva-Lichtmann, and Georges Métailié, 615–32. Sinica Leidensia 79. Leiden: Brill, 2007.

Wakeman, Frederic E., Jr., "Boundaries of the Public Sphere in Ming and Qing China." *Daedalus* 127, no. 3 (1998): 167–89.

———. "China and the Seventeenth-Century Crisis." *Late Imperial China* 7, no. 1 (1986): 1–26.

———. *The Fall of Imperial China.* New York: Free Press, 1975.

———. "Localism and Loyalism during the Qing Conquest of Jiangnan: The Tragedy of Jiangyin." In *Telling Chinese History: A Selection of Essays.* Edited by Lea H. Wakeman. Berkeley: University of California Press, 2009.

Waley-Cohen, Joanna. *The Culture of War in China: Empire and the Military under the Qing Dynasty.* New York: I. B. Tauris and Co., 2006.

Waltner, Ann. "Review Essay: Building on the Ladder of Success: The Ladder of Success in Imperial China and Recent Work on Social Mobility." *Ming Studies* 17 (1983): 30–36.

Wang Anshi 王安石 (1021–86). *Wang Wengong wenji* 王文公文集 [Collected Writings of Wang Anshi]. Shanghai: Shanghai renmin chubanshe, 1974.

Wang Chong 王充 (27–97?). "Lun Heng" 論衡 [Discourse Weighed in Balance]. In *Lun Heng xigu* 論衡析詁 [Analysis of the "Discourse Weighed in Balance" with analytic notes]. Edited by Zheng Wen 鄭文. Chengdu: Bashu shuji, 1999.

Wang Dang 王讜 (fl. 1101–10). *Tang Yulin: Fujiao kanji* 唐語林：附校勘記 [A Forest of Sayings of the Tang Dynasty, Collated]. 1935; Beijing: Zhonghua shuju, 1985.

Wang Fuzhi 王夫之 (1619–99). *Chuanshan quanji* 船山全集 [Complete Writings of Master Quanshan]. Taizhong: Taiyuan wenhua fuwushe, 1965. Reprint of the 1933 edition with the title Chuanshan yishu 船山遺書. Shanghai: Taipingyang shudian, 1933. References are to the 1965 edition.

——— and Huang Zongxi 黃宗羲 (1610–95). *Lizhou Chuanshu wushu* 梨州船山五書 [Five Books by (Huang) Lizhou and (Wang) Chuanshan]. Taibei: Shijie shuju, 1974.

——— and Zhang Zai 張載 (1020–77). *Zhang Zi Zheng Meng zhu* 張子正蒙注 [Commentary of the Discourse for Beginners of Master Zhang]. Beijing: Zhonghua shuju, 1956.

Wang Jingxian 王靖憲, ed. *Zhongguo lidai huihua: Gugong bowuyuan canghua ji* 中國歷代繪畫: 故宮博物院藏畫集 [Chinese Paintings from Successive Dynasties of the Beijing Palace Museum Collection]. Vol. 1. Beijing: Renmin meishu chubanshe, 1978.

Wang Shimao 王世懋 (1536–88). *Minbu shu* 閩部疏 [Memorials about Fujian Province]. Xuxiu Siku quanshu 734. Shanghai: Shanghai guji chubanshe, 2002.

Wang Tingxiang 王廷相. *Wang Shi jia cang ji* 王氏家藏集 [Anthology of Writings Stored in the House of the Wang Clan]. 1536, edited in 1655. Taibei: Weiwen tushu chubanshe, 1976.

Wang Xichan 王錫闡 (1628–82). *Xiao an xin fa* 曉菴新法 [New Methods of Xiao'an]. Congshu jicheng jianbian 425. Taibei: Taiwan shangwu yinshuguan, 1965.

Wang Yang-ming [Yangming]. *Instructions for Practical Living and Other Neo-Confucian Writings.* Translated by Wing-tsit Chan. New York: Columbia University Press, 1963.

Wang Yeqiu 王冶秋. *Liulichang shihua* 琉璃廠史話 [Historical Chats about Liulichang]. Beijing: Xinzhi sanlian shudian, 1963.

Wang Zhen 王禎. *Nongshu* 農書 [Book of Agriculture]. Edited by Wang Yuhu 王毓瑚. 1313. Reprint, Beijing: Nongye chubanshe, 1981.

Wang Zichen 王咨臣, and Xiong Fei 熊飛. *Song Yingxing xueshu zhuzuo sizhong* 宋應星學術著作四種 [Four Treatises on Song Yingxing's Tenets]. Nanchang: Jiangxi renmin chubanshe, 1988.

Ward, Julian. *Xu Xiake (1587–1641): The Art of Travel Writing.* Richmond: Curzon, 2001.

Weatherford, Jack. *Genghis Khan and the Making of the Modern World.* New York: Crown Publishers, 2004.

Wei Qingyuan 韋慶遠. *Zhang Juzheng he Mingdai zhonghou qi zhengju* 張居正和明代中后期政局 [Zhang Juzheng and the Policy Circumstances of the Mid- and Late Ming Period]. Guangzhou: Guangdong gaodeng jiaoyu chubanshe, 1999.

Wen Tiren 溫體仁, Zhang Zhifa 張至發 (jinshi 1601), et al., eds. "Huaizong shilu 懷宗實錄 [Veritable Records of the Reign of Emperor of the Emperor Huaizong]." In *Ming shilu* 明實錄 [Veritable Records of the Ming Dynasty]. 1577. Reprint, Taibei: Zhongyang yanjiuyuan lishi yuyuan yanjiusuo, 1962–66.

Wills, John E. Jr. *Mountain of Fame: Portraits in Chinese History.* Princeton: Princeton University Press, 1994.

Wu Guodong 伍國棟. *Zhongguo yinyue* 中國音樂 [China's Music]. Shanghai: Shanghai waiyu jiaoyu chubanshe, 1999.

Wu, K. T. 吳光清. "Ming Printing and Printers." *Harvard Journal of Asiatic Studies* 7, no. 3 (1943): 203–60.

Xia Minghua 夏明華. "Jingzhou gucheng leming zhuan yu wule gongming" 荊州古城勒名磚與物勒工名 [The Craftsmen Mark System and the Engraved Bricks of Jingzhou City Walls]. *Jianghan kaogu* 87, no. 2 (2003): 66–72.

Xi Zezong. "Chinese Studies in the History of Astronomy, 1949–1979." *Isis* 72, no. 3 (1981): 456–70.

Xu Zi 徐鼒. *Xiaotian jizhuan* 小腆紀傳 [Biographies of an Era of Little Prosperity]. 2 vols. Beijing: Zhonghua shuju, 1958.

Xu Guangqi 徐光啓 (1562–1633). *Nongzheng quanshu* 農政全書 [Comprehensive Treatise on Agricultural Administration]. 1639. Edition of Wenyuan ge Siku quanshu. Taiwan: Taiwan shang wu yinshuguan, 1983–86.

Xu Hongzu 徐弘祖 (1586–1641). *Xu Xiake youji* 徐霞客遊記 [Travelogues of Xu Xiake]. Edited by Chu Shaotang 褚紹唐 and Wu Yingshou 吳應壽. 2nd ed. Shanghai: Shanghai guji chubanshe, 1987.

Xu Shipu 徐世溥, *Yu Dun ji* 榆墩集 [Collection of Yu Dun]. 1691. Nanchang: Jiangxi sheng tushuguan, Ref. Nr. 3455/23.

Xiong Wenju 熊文舉. *Xue Tang xiansheng wenxuan* 雪堂先生文選 [A Selection of Master Xue Tang's Writings]. Original, 1655. Nanchang. Jiangxi Provincial Library.

Yang Chung-Chieh 楊中傑. "Fojing linxu chen—zuizhong jiben lizi, zhenkong ji liangzi zhi yuan? 佛經鄰虛塵—最終基本粒子,真空及量子之源 [Buddhistic Linshitron: The Ultimate Fundamental Particle, Vaccum and the Origin of Quantum?]." *Fojiao yu kexue* 7, no. 1 (2006): 34–45.

Yang Weizeng 楊維增. *Song Yingxing sixiang yanjiu shi ji shiwen zhuyi* 宋應星思想研究試及詩文註譯 [Research about the Ideas of Song Yingxing and an Annotated Translation of his Poetry]. Guangzhou: Zhongshan daxue chubanshe, 1987.

Yang Zhouxian 楊周憲, and Zhao Yuemian 趙曰冕. *Xinjian Xianzhi* 新建縣志 [Local Gazetteer of Xinjian County]. 1680. Zhongguo fangzhi congshu. Huazhong difang 884. Reprint, Taibei: Chengwen chubanshe, 1989.

Yijing 易經 [Book of Changes]. (2nd cent. BC) Edition of *Sishu wujing* [The Four Books and the Five Classics]. Tianjin: Tianjin guji shudian yingyin, 1988.

Yong Huang. "A Neo-Confucian Conception of Wisdom: Wang Yangming on the Innate Moral Knowledge (*liangzhi*)." *Journal of Chinese Philosophy* 33, no. 3 (2006): 393–408.

Yoshida Tora. *Salt Production Techniques in Ancient China: The Aobo tu*. Translated and revised by Hans Ulrich Vogel. Sinica Leidensia 27. Leiden: Brill, 1993.

"Yueji" 樂記 [Record on Music]. In *Liji* 禮記 [Book of Rites]. Edition of *Sishu wujing* [The Four Books and the Five Classics]. Tianjin: Tianjing guji shudian yingyin, 1988.

Yung, Bell, ed. *Celestial Airs of Antiquity: Music of the Seven-String Zither of China*. Recent Researchers in the Oral Tradition of Music 5. Madison, WI: A-R Editions, 1997.

Zhang Hua 張華. *Bowuzhi jiaozheng* 博物志校證 [An Annotated Correction of the Report of Extensive Matters]. Edited by Fan Ning 范寧. Beijing: Zhonghua shuju, 1984.

Zhang Hui. *Songdai biji yanjiu* 宋代筆記研究 [Research on Song Dynastic Private Jottings]. Wuchang: Huazhong shifan daxue, 1993.

Zhang Juzheng 張居正 (1525–82) et al. "Shizong shilu" 世宗實錄 [Veritable Records of the Reign of the Emperor Shizong]. 1577. In *Ming shi lu* 明史錄 [Ming Veritable Records]. Edited by Huang Zhangjian 黃彰健.Reprint, Taibei: Zhongyang yanjiuyuan, Lishi yuyuan yanjiusuo, 1966. References are to the 1966 edition.

———. *Sishu jizhu* 張居正著四書集註 [The Four Books, with annotations]. Edition of Wenyuan ge Siku quanshu. Taiwan: Taiwan shangwu yinshuguan, 1983–86.

Zhang Liwen 張 立文. *Song Ming lixue luoji jiegou de yanhua* 宋明理學邏輯結構的演化 [The Transformation of the Logic in Structural Argumentations of the School of Principle during the Song and Ming Dynasties]. Taibei: Sanmin shuju, 1994.

Zhang Lüxiang 張履祥. *Bu* [*Shenshi*] *Nongshu* 補[深氏]農書[Supplement to (Shen's) Agricultural Book]. 1658. Reprint, Beijing: Zhonghua shuju, 1956. References are to the 1956 edition.

Zhang Qizhi 張豈之. *Ruxue, lixue, shixue, xinxue* 儒學・理學・實學・新學 [New Studies about the *Ru*-School of Confucianism, the *Li*-School of Principle, and the *Shi*-School of Practical Learning]. Xi'an: Shaanxi renmin jiaoyu chubanshe, 1994.

Zhang Tingyu 張廷玉 (1672–1755) et al. *Ming shi* 明史 [Ming History]. 1736. Reprint, Beijing: Zhonghua shuju, 1991. References are to the 1991 edition.

Zhang Wei-hua. "Music in Ming Daily Life, as Portrayed in the Narrative 'Jin Ping Mei.'" *Asian Music* 23, no. 2 (Spring–Summer, 1992): 105–34.

Zhang Xiumin 張秀民. *Zhang Xiumin yinshua shi lunwen ji* 張秀民印刷史論文集 [Collected Papers on the History of Printing in China]. Beijing: Yinshua gongye chubanshe, 1988.

Zhang Yunming. "Ancient Chinese Sulphur Manufacturing Processes." *Isis* 77, no. 3 (1986): 487–97.

Zhang Zai 張載 (1020–77). "Zheng meng 正蒙 [Correct Discipline for Beginners]." In *Zhang Zai ji* 張載集 [Collected Writings of Zhang Zai]. Edited by Zhang Xishen 章錫琛. Beijing: Zhonghua shuju, 1978.

Zhang Zhuo 張鷟 (ca. 660–740). *Chao ye qian cai* 朝野僉載 [Records of Official and Unofficial Matters]. Taibei: Taiwan shangwu yinshuguan, 1966.

Zhao Jie. "Ties That Bind: The Craft of Political Networking in Late Ming Chiang-nan." *T'oung Pao* 86, nos. 1–3 (2000): 136–64.

Zhao Wanli 趙萬里. *Guoli Beiping tushuguan shanben shumu* 國立北平圖書館善本書目 [Collection of Rare Books in the Beiping Library]. [Beiping: Guoli Beiping tushuguan], 1933.

Zhao Zhongwei. "Chinese Genealogies as a Source for Demographic Research: A Further Assessment of Their Reliability and Biases." *Population Studies* 55, no. 2 (2001): 181–93.

Zheng Da 鄭達 (17th cent.), ed. *Yeshi wuwen* 野史無文 [Unofficial History Without Documents]. Reprint, Beijing: Zhonghua shuju, 1962.

Zheng Qiao 鄭樵. *Tongzhi yiwen lue* 通志藝文略 [Essentials on the Arts of the *Tongzhi*]. 6 vols. 1161. Collated by Wang Qishu 汪啟淑, 1749. University of Hong Kong Library.

———. "Ji Fang libu shu [. . .]." In *Tongzhi ershi lüe* 通志二十略 [Selections of Twenty

Essential Treatises of the *tongzhi*]. Edited by Wang Shumin 王樹民. 2 vols. Beijing: Zhonghua shuju, 1995.

Zheng Yun 鄭澐 (*jinshi* 1727). *Hangzhou Fuzhi* 杭州府治 [Hangzhou Gazetteer]. 1784. Xuxiu Siku quanshu 701–3. Reprint, Shanghai: Shanghai guji chubanshe, 1995.

Zhong Tai 鍾泰, ed. *Bozhou zhi* 亳州志 [Gazetteer of Bozhou]. 1894. Reprint, Taibei: Chengwen chubanshe, 1985. References are to the 1985 edition.

Zhou Guidian 周桂鈿. *Zhongguo guren luntian* 中國古人論天 [Ancient Chinese Views of the Heavens]. Beijing: Xinhua chubanshe, 1993.

Zhu Yan 朱琰. *Tao Shuo* 陶說 [Description of Ceramics] from 1774. Edited by Bianzuan wuyuan hui. Xuxiu siku quanshu. Shanghai: Shanghai guji chubanshe, 1995–2000.

Zhu Xi. *Yupi zizhi tongjian gangmu* 御批資治通鑑綱目 [Comprehensive Mirror for Aid in Government]. Edition of Wenyuan ge Siku quanshu [Wenyuan Pavillion Siku quanshu edition]. Taiwan: Taiwan shangwu yinshuguan, 1983–86. Complement in *Siku quanshu zhenben*-edition.

———. *Hui an xiansheng Zhu Wen gong wenji* 晦庵先生朱文公文集 [Zhu Xi's Literary Writings]. Taibei: Taiwan shangwu yinshuguan, 1980.

———. *Zhuzi yulei* 朱子語類 [Classified Conversations of Master Zhu Xi]. Edited by Li Jingde 黎靖德 (13th cent.), 1270. Lixue congshu. Reprint, Beijing: Zhonghua shuju, 1986.

Zuo Qiuming 左丘明. *Zuo Zhuan* 左傳 [Commentaries of Master Zuo Traditions]. Edition of *Sishu wujing*. Tianjin: Tianjin guji shudian yingyin, 1988.

INDEX

This index includes a glossary of significant terms. English entries of these terms refer to the Chinese transcriptions (pinyin) and characters, as well as the page numbers where they are discussed.